Lake Verevi, Estonia – A Highly Stratified Hypertrophic Lake

Developments in Hydrobiology 182

Series editor

K. Martens

Lake Verevi, Estonia – A Highly Stratified Hypertrophic Lake

Edited by
Ingmar Ott & Toomas Kõiv

Estonian Agricultural University, Estonia

Springer

Library of Congress Cataloging-in-Publication Data

A C.I.P. Catalogue record for this book is available from the Library of Congress.

ISBN-13 978-90-481-7014-2 e-ISBN-13 978-1-4020-4363-5

Published by Springer,
P.O. Box 17, 3300 AA Dordrecht, The Netherlands

Cover illustration: Southern part of Lake Verevi in May 2005. Photo I. Ott

Printed on acid-free paper

TABLE OF CONTENTS

Hydrobiologia (2005) 547:vii
I. Ott & T. Kõiv (eds), Lake Verevi, Estonia – A Highly Stratified Hypertrophic Lake
DOI 10.1007/s10750-005-4137-y

Preface

Ingmar Ott

Estonia is the fourth country in Europe by proportional area of lakes (4.8%). The main part of this area belongs to the lakes of Peipsi and Võrtsjärv. The numerous others are small and shallow, devoid of any special economic value. Lakes in the urban areas are used mostly for recreation. They are attractive also to limnologists. Lake Verevi (surface 12.6 ha, maximum depth 11.0 m) is located in the small town of Elva (6400 inhabitants). Tartu, the second city in Estonia, lies at a distance of 25 km. The relatively recent settlement (ca 115 years) is suitable for leisure with a hilly pine forest and several small lakes. Small wooden private houses and summer cottages without special industry are dominant. Elva has been attractive for tourists and holidaymakers during its whole existence. The first swimming pool and a beach hall, one of the best at that time in Estonia, were built already in 1929.

Prof. H. Riikoja, the founder of Estonian limnology, performed a survey of Estonian lakes in the first half of the 20th century, including a study of Lake Verevi in 1929. This time the lake was in its natural state with good ecological quality. The same was noticed in the 1950s during the complex investigation led by the next grand Estonian limnologist, N. Mikelsaar. Alarming appearances of deterioration of the lake had been noticed since the 1970s. All phenomena connected with rapid eutrophication have been revealed, among these one of the highest values of phytoplankton biomass ever recorded in Estonia – 724 g m^{-3} wet weight. Since 1984 the lake was investigated yearly except 1987 and 1992. In the 1980s, phenomena connected with hypertrophic conditions prevailed in each year. After that period, the lake became really unstable, with alternating communities and ecological status in different years. The water column attained a mosaic character with narrow spatial microhabitats. These findings led to an idea to study the functioning of the whole ecosystem in more detail. The most profound seasonal studies took place in 2000 and 2001, with 25 researchers participating; a partly meromictic status of the lake was discovered. The most results are included in the present volume, while some autecological studies including those on diurnal migration are still awaiting publication. All the members are ready to cooperate with Estonian and foreign colleagues in more detailed investigations and experiments offered by the environment of this peculiar lake ecosystem.

The publication was made possible by several financial supporters – Estonian Science Foundation (G No. 3579, 4835, 5738, 3689, 4080, 1804, 4483), Estonian Ministry of Education and Science (core grant Nos. 0370208s98, 0362482s03 and 0362480s03), M. & T. Nessling Foundation (G No. 99084).

Võrtsjärv Limnological Station, 27 July 2004

Hydrobiologia (2005) 547:1–20
I. Ott & T. Kõiv (eds), Lake Verevi, Estonia – A Highly Stratified Hypertrophic Lake
DOI 10.1007/s10750-005-4138-x

General description of partly meromictic hypertrophic Lake Verevi, its ecological status, changes during the past eight decades, and restoration problems

Ingmar Ott[1,*], Toomas Kõiv[1], Peeter Nõges[1], Anu Kisand[1], Ain Järvalt[1] & Enno Kirt[2]
[1]*Estonian Agricultural University, Institute of Zoology and Botany, Võrtsjärv Limnological Station, 61101 Rannu, Tartu County, Estonia*
[2]*OÜ Enno Projektid Ltd, Toome Str. 82, 10913 Tallinn, Estonia*
*(*Author for correspondence: E-mail: iott@zbi.ee)*

Key words: long-term ecosystem changes, ecological status, vertical distribution of substances and biota, lake restoration

Abstract

The present study describes generally the ecosystem of Lake Verevi while more detailed approaches are presented in the same issue. The main task of the article is to estimate long-term changes and find the best method for the restoration of good ecological status. Lake Verevi (surface 12.6 ha, mean depth 3.6 m, maximum depth 11 m, drainage area 1.1 km^2, water exchange 0.63-times per year) is a hypertrophic hard-water lake located in town Elva (6400 inhabitants). Long-term complex limnological investigations have taken place since 1929. The lake has been contaminated by irregular discharge of urban wastewaters from oxidation ponds since 1978, flood from streets, and infiltrated waters from the surrounding farms. The so-called spring meromixis occurred due to extremely warm springs in recent years. The index value of buffer capacity of Lake Verevi calculated from natural conditions is on the medium level. Water properties were analysed according to the requirements of the EU Water Framework Directive. According to the classification, water quality as a long-term average of surface layers is moderate-good, but the water quality of bottom layers is bad. Values in deeper layers usually exceed 20–30 times the calculated reference values by Vighi and Chiaudani's model. Naturally, at the beginning of the 20th century the limnological type of the lake was moderately eutrophic. During the 1980s and 1990s the ecosystem was out of balance by abiotic characteristics as well as by plankton indicators. Rapid fluctuations of species composition and abundance can be found in recent years. Seasonal variations are considerable and species composition differs remarkably also in the water column. The dominating macrophyte species vary from year to year. Since the annual amount of precipitation from the atmosphere onto the lake surface is several times higher, the impact of swimmers could be considered irrelevant. Some restoration methods were discussed. The first step, stopping external pollution, was completed by damming the inlet. Drainage (siphoning) of the hypolimnetic water is discussed. Secondary pollution occurs because Fe:P values are below the threshold. The authors propose to use phosphorus precipitation and hypolimnetic aeration instead of siphoning.

Introduction

Physically and chemically stratified lakes have a special ecosystem structure. Vertical gradients of environmental parameters became a limnological issue as early as in the study by Hutchinson (1938). The investigation of vertical distribution of biota developed from the descriptive stage through the investigation of ecological processes into a stage dealing with the ecological holistic approach to lake management (Ripl, 1976; Faafeng & Nilssen, 1981; Wolter, 1994; Lindenschmidt & Chorus, 1997). Refining and prediction of functioning of an ecosystem and the ecological status of stratified lakes is complicated for many reasons. One of the most important factors is the occurrence of many microhabitats for biota and their irregular interactions. Sometimes complexity of investigation seems unrealistic and smaller compartments need to be used (Pipp & Rott, 1995). Long-term data sets (since 1929) and the availability of complex data with references to the other articles about the same issue (hydrochemical and hydrophysical, sediments, bacterioplankton, phytoplankton, protozoa, metazooplankton, epiphyton, meio- and macrozoobenthos, macrophytes, fishes) serve as the basis of this study. The influence of wind and also of inflow in recent years is minimal, although usually core factors for the functioning of the ecosystem. To some extent these conditions make it easier to predict the functioning of the whole ecosystem. The present study describes generally the ecosystem of Lake Verevi, while more detailed approaches are presented in the same issue. The main task of the article is to estimate long-term changes and to find the best method for restoration.

Site description

Lake Verevi is located in a small town of Elva (6400 inhabitants). Tartu, the second city in Estonia, is at a distance of 25 km. The relatively young town (ca. 115 years-old) is suitable for leisure with a landscape of a hilly pine forest and several small lakes. Small wooden private houses and summer cottages without special industry dominate. Elva has been attracting for tourists and holidaymakers. One of the best swimming pools and beach halls in Estonia was built here in 1929.

The lake has an elongated shape in the north–south direction with the deepest and widest part near the southern end (Fig. 1). By origin Lake Verevi is a kettle lake formed by the melting of a buried ice block from the decaying glacier (Mäemets & Ennok, 1991). The drainage basin represents a hydrologically complex landscape – from the south and the south-east the lake is surrounded by sandy hills and dunes covered with pine forests. The densely populated eastern shore slopes steeply towards the lake. The area to the west is wetland and covered by quagmires and swamps.

Lake Verevi is a small and relatively deep lake (Table 1) with low water exchange (Loopmann, 1984). The high value of relative depth (the ratio of maximum depth as a percentage of the mean diameter of the lake on the surface; Wetzel, 1983) supports the idea that the water column does not mix easily. The lake is thermally sharply stratified and strong gradients of chemical substances occur during summer. Usually, the lake is dimictic, water mixing in spring has been incomplete in recent years even at homothermal conditions, which adds some temporal meromictic features to the lake (Nõges & Kangro, 2005; Ott et al., 2005). The metalimnion is progressively eroded during summer and autumn and a complete mixing usually takes place in November. The lake has up to 10 small inflows, but only three of them (Fig. 1; inflows 1, 4, and 5) are nearly permanent. Inflows 4 and 5 start from two spring-fed lakelets, Linajärv and Jaanijärv located in the northern part of the watershed. The main part of the annual inflow comes irregularly from inlet 10, which has been closed totally since 2002. Lake Verevi receives also a significant part of water as hardly measurable sub-surface run-off (Mäemets & Ennok, 1991). Small ditches and bottom springs in the narrow northern part form the bulk of the inflowing water. The lake has been contaminated by irregular discharge of urban wastewaters from oxidation ponds since 1978, flood from streets, and infiltrated waters from the surrounding farms. The outflow of the lake is located on the western shore (Fig. 1, N 7), and it flows into the Kavilda river valley. In dry years, the outflow becomes discontinuous. The ice-free period lasts mainly between April and November.

Materials and methods

Lake Verevi has been studied extensively over a long period, between 1929 and 2001 (in 1929, 1957, 1984, 1985, 1986, 1988, 1989, 1991 and each year between 1993 and 2001). The first data are available in the literature (Riikoja, 1930, 1940; Eesti Järved, 1968). Plankton and hydrochemical samples have been gathered mainly from the deepest

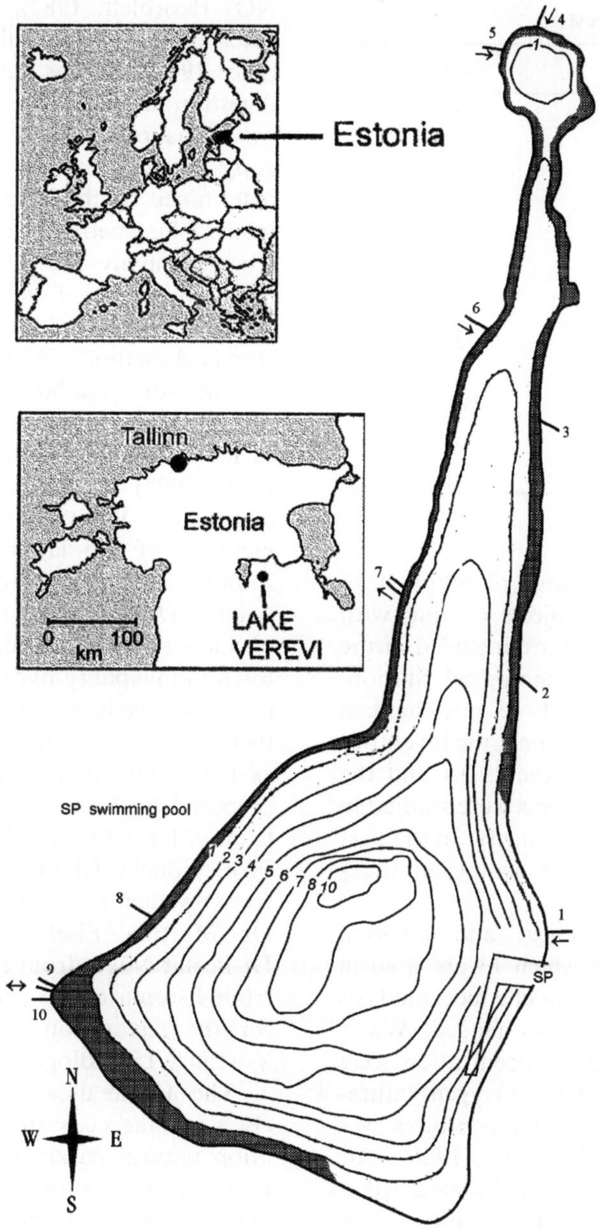

Figure 1. Location and map of Lake Verevi. Numbers: 1–6, 8–10 inlets, 7 outlet.

point of the lake from 3 to 4 layers. The Ruttner (volume 1 or 2 l) or the Van Dorn sampler (2 l) were used. In 2000 and 2001 phytoplankton and water samples were gathered from eight layers using a special vacuum probe. A Masterflex pump (model N 7533–60) with an easy-load pump head (model 7518–12) was used for pumping water to the surface. A hose with an inner Ø 8 mm was placed vertically into the water. The lower tip of the vertically placed hose was closed and the water was sucked through a horizontal tube in order to obtain water from the horizontal layers. The capacity of the complex device is approximately 2 l min^{-1}. Seasonal observations were carried out in 1988, 1991, 1993, 2000, and 2001; the other years in summer or during the vegetation period.

Table 1. General data of Lake Verevi

Parameter (unit)	Value
Length (m)	950
Maximum width (m)	320
Length of shoreline (m)	2125
Surface (ha)	12.6
Maximum depth (m)	11.0
Mean depth (m)	3.6
Relative depth %	2.7
Drainage area (km^2)	1.1
Duration of ice cover (months)	5
Times of water exchange per year	0.63
Lake type	hypertrophic
Water volume (10^6 m^{-3})	453.6

Since the 1980s the same group of people have been involved in different projects dealing with lake investigation. All the data are included in the database of the Võrtsjärv Limnological Station. The main goal of the investigations has changed during the decades. In the beginning, the inventory of the lake gained priority in the 1920s and the 1950s, then water quality problems were studied in the 1980–1990s gained the first priority. In the past years holistic investigations on ecological functioning have added.

Altogether 18 hydrochemical and physical parameters were studied, of which 13 are used in this article. The main methods of chemical analysis used since 1984 have not been changed. Water temperature and oxygen (O_2) concentration were measured by a thermo-oximeter. Oxygen saturation (O_2%) for different water temperatures was calculated according to Hellat et al., (1986). The pH of water was measured by a pH-meter. Alkalinity (HCO_3^-) was determined by titration using HCl (Unifitsirovannye ..., 1977). Chloride ion (Cl^-) was quantified mercurimetrically (Unifitsirovannye ..., 1977). Conductivity (EC) was measured by a JENWAY Model 4150 Conductivity Meter (Fall, 1996). Total phosphorus (TP) was determined after persulphate oxidation as sulphates (PO_4^{3-}). The content of PO_4^{3-} was determined by the molybdene blue method (Reports ..., 1977). NO_3^- was reduced to nitrit (NO_2^-). Sulphanil-amide and N-(1-naphthyl)-ethylenediamine dihydrochloride was used for the determination of NO_2^- (Koroleff, 1982). Total nitrogen (TN) was determined after persulphate oxidation as NO_2^-. Since 1995, TN was determined after persulphate digestion as NO_3^-. The content of nitrates (NO_3^-) was measured by second-derivative UV spectroscopy (Crumpton et al., 1992). Silicon (Si) was determined by the indophenol blue and silicomolybdic blue method (Hansen & Koroleff, 1999). Chemical analyses of different fractions of organic matter by means of dichromate and permanganate oxidation were performed titrimetrically using the standard methods (Alekin, 1959).

The other methods are described in different articles of this publication (Järvalt et al.; Kangro et al.; Kisand; Mäemets & Freiberg; Timm & Möls, 2005) as well as in an earlier publication about Lake Verevi (Timm, 1991). Hydrochemical methods of the laboratory of the Institute of Zoology and Botany are described also in a special issue of *Hydrobiologia* (Möls et al., 1996).

Lake restoration has been planned by the Elva town municipality over the past 15 years. Several projects have been carried out. The article discuss two of them. The project "Improvement of water exchange and pollution load of Lake Verevi" is prepared by Enno Project Ltd. An international team of limnologists (Prof. S. Björk from Sweden, Prof. Wilhelm Ripl from Germany, Dr Bo Verner from Sweden, Dr Gertrud Cronberg from Sweden, Dr Martina Eiseltova from Czech Republic, Dr Peeter Nõges from Estonia, Dr Arvo Tuvikene from Estonia et al.) proposed an ecological plan for the restoration during a workshop at the Võrtsjärv Limnological Station in 1993.

The article uses the index of buffer capacity (BC) of lake ecosystems. It takes into consideration natural parameters as important preconditions of the ecological status. A larger surface area is connected with better aeration. It means also the larger water volume, which grants stability to the ecosystem. Intensive water exchange besides aeration assures the inload of mineral and organic matter. Total alkalinity raises carbonate buffer capacity, and organic substances (humic compounds) can adsorb phosphates. Humic compounds are also the main factors of forming light conditions in the water column of Estonian lakes. The index characterises the ability of the ecosystem to tolerate eutrophication. An equation follows:

$$BC = \ln S * SWE * HCO_3; * COD_{Cr}/1000$$

where

$\ln S$ – natural logarithm from lake surface (ha)

SWE – times of water exchange per year on the subjective scale

<0.5 times – 1
0.5–2 – 2
2–4 – 4
4–10 – 4
>10 – 5

HCO_3; – total alkalinity (mg l^{-1})

COD_{Cr} – chemical oxygen demand (dichromate oxidizability, mg l^{-1})

The range of BC values is between 1 and 100 according to the calculated values of ~700 Estonian lakes. The soft-water lake types have the lowest values, and hard-water lake types such as mixotrophic and eutrophic have the highest.

Ecological status is estimated by the TSI-4 index:

$$TSI - 4 = 15.4\ pH - 7.4\ NS_{CLAD} - 20.9\ SD\ 36.8\ PA_{phyto} + 7\ NS_{Eugleno}$$

where

pH – pH in surface water

NS_{CLAD} – number of cladoceran species

SD – Secchi disc transparency (m)

PA_{phyto} – partial phytoplankton abundance (sum of relative species abundances of species/number of species). Estimations of relative abundances of species: 5 – highly abundant; 4 – abundant; 3 – moderate; 2 – few; 1 – rare; 0 – absent

$NS_{Eugleno}$ – number of euglenophyte species.

The values <70 are considered to be low, 70–130 medium and >130 high.

Water quality was estimated by the recommendations of the Water Framework Directive of the European Community. Estonia has a classification (Ott, 2001) on the basis of abiotic parameters. Four groups of lakes were distinguished. Lake Verevi is a the light-coloured hard-water lake. The quality of the ecosystems is divided into 5 classes (Table 2).

Total phosphorus concentrations of natural theoretical conditions can be calculated by the equation proposed by Vighi & Chiaudani (1985):

$$\text{Log TP} = 1.48 + 0.33(\pm 0.09)\ \text{Log MEI alk}$$

where

MEI alk – morphoedaphic index (mean depth m/alkalinity mg-eq. l^{-1})

Vollenweider (1975) proposed a possibility to estimate the phosphorus loading tolerance of lakes by the empirical model on the basis of the mean depth, water retention, and actual P loading (Fig. 2).

Results

Water properties and buffer capacity

The chemical and physical properties of water are different in the upper and lower parts of the water column (Table 3). Data in Table 3 are calculated for the summer stagnation period taking into account the whole database. According to the classification, water quality as a long-term average of surface layers is moderate-good (Table 2),

Table 2. The classification of water quality of light coloured hard-water lakes in Estonia (Ott, 2001)

Characteristic	I Class excellent	II Class good	III Class moderate	IV Class poor	V Class bad
Water transparency (m)	>3	2–3	1–2	<1	<1
pH at surface	7–8	8–8.3	8.3–8.8	8.8–9; 6–7	9>; <6
Tot-P (mg m^{-3})	<30	30–60	60–80	80–100	>100
Tot-N (mg m^{-3})	<500	500–700	700–1000	1000–1300	>1300
Organic matter content (Chemical oxygen demand by dichromate oxidizability) (mg l^{-1})	<15	15–30	30–40	40–50	>50
Range of metalimnion in summer (in stratified lakes) (m)	>6	5–6	3–4	2–3	<2
Chlorophyll *a* (mg m^{-3})	<10	10–20	20–40	40–50	>50
Sulphate concentration (except halotrophic lakes) (mg l^{-1})	<10	10–50	10–50	10–50	>50

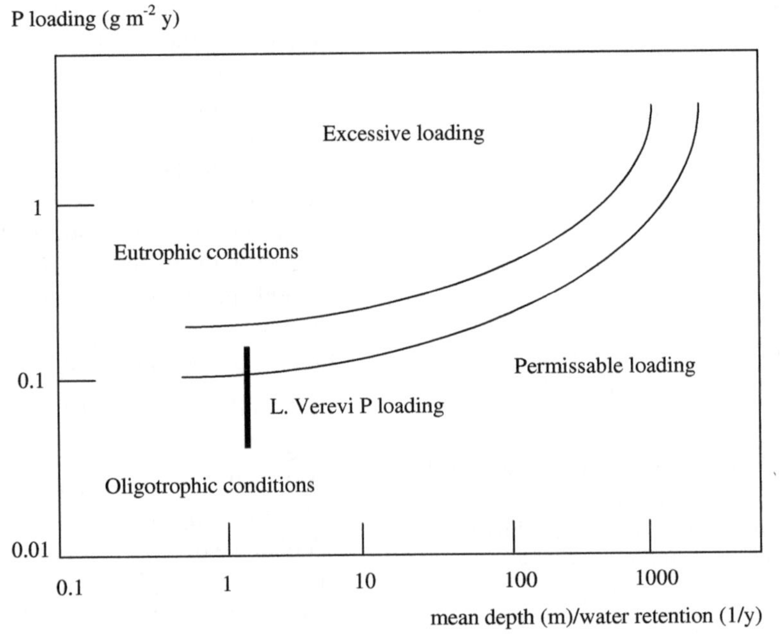

Figure 2. Phosphorus loading tolerance of Lake Verevi by Vollenweider's (1975) model.

Table 3. Average values of water property parameters in Lake Verevi in summer

Parameter	Period	Surface layers (<2.5 m)	Bottom layers (>5 m)
Secchi disc transparency (SD; m)	1929–2001	2.1 (n = 17)	
Ph	1929–2001	8.31 (34)	6.9 (n = 25)
Chemical oxygen demand(permanganate oxidizability; COD_{Mn}; mg l^{-1})	1929–2001	11 (21)	14.6 (23)
Chemical oxygen demand (dichromate oxidizability; COD_{Cr}; mg l^{-1})	1957–2001	31.2 (22)	38.7 (23)
Total alkalinity (mg HCO_3 l^{-1})	1957–2001	213.8 (31)	376.4 (30)
Total nitrogen (TN; mg m^{-3})	1984–2001	980 (28)	6322 (26)
Total phosphorus (TP; mg m^{-3})	1984–2001	55 (22)	830 (24)
Sulphate concentration (mg SO_4 l^{-1})	1929–2001	24.6 (12)	31 (16)

but water quality in the bottom layers is bad. Also the range of metalimnion, mainly 2.5 m in summer, reflects bad quality.

The value of the BC of Lake Verevi calculated from natural conditions is on the medium level (51). The ecosystem is characterized by a very high value of total alkalinity combined with weak water exchange. The value of the trophic state index (TSI-4) is high (139) and corresponds to the unstable ecosystem. Calculations on the basis Vighi & Chiaudani's equation (1985) yield total phosphorus values for the pristine state of the lake 31 and 35 mg m^{-3} in the upper and deeper layers, respectively. In some years, during the summer stagnation period, values in the epilimnion have been close to the pristine level, but values in deeper layers exceed usually 20–30 times the reference values.

P. Nõges (the present issue) calculated phosphorus loading between April and October in 1991 and 1993. P loading was in the range 6–21 kg depending on different proportions of the surface inflow. Annual loading could be considered 0.05–0.17 g P m^{-2}, presuming that between April and October the lake receives 20–80% of the annual total. P loading from the inflow does not exceed the tolerance limit of Lake Verevi (Fig. 2).

Long-term changes

Changes in the surrounding landscape and lake management

Lake Verevi is located in a town and therefore the surroundings have been redesigned repeatedly. The first swimming pool was built in 1929. Since then several smaller or bigger structures have been erected on the eastern shore for holidaymakers. Almost every year the lake shore and the swimming pool have been covered by new sand. Between the early 1970s and 1997 there were café and sports facility building situated just above the water on the eastern shore. In the summer of 1998, the water table was lowered 0.7 m in order to clean the bottom of the swimming pool and to prevent the growth of macrovegetation (Photo 1). The bottom was covered by geotextile approximately in the range of 30 m from the shore. The carpet was covered by gravel. Two photos taken in 1925 and 2001 show the general view from the southern shore to the lake (Photos 2, 3). More or less open landscape in 1925 has changed now into closed one. Brushwood as well as emergent plants surrounded the lake in 2001.

The treatment of municipal wastewaters has been crucial to the ecological status of the lake. In 1978 oxidation ponds were constructed just to the west of the lake, right beside the inflow. The treatment had a modest effect, because activated sludge treatment was unavailable and the sedimentation ponds were too small. This place is wetland and during the flood period in spring the sewage water flows into the lake a number of times. A new wastewater treatment plant was built in 1986, but in fact it began to work in April 1988 (Järvet, 1989). The old oxidation ponds were completely isolated by dams in 2002. It means that irregular contamination during floods could have occurred during the last 24 years. In 2000 and 2001

Photo 1. Renovation of the swimming pool in 1998 (water level was lowered 0.7 m, bottom was covered by geotextile). Photo I. Ott.

Photo 2. General view from the southern shore to the lake in 1925. Photo K. Leius.

the outflow ditch from the lake was dredged and cleaned.

In 2002 several restructurings were made on the western side of the lake concerning water exchange. Besides isolation of the old oxidation ponds, the inflow of the lake was stopped. The inlet and outlet were connected by digging a straight ditch. It means that the lake may be temporarily closed. It had happened already earlier in dry seasons but not by direct human activity. At the same time, the wetland and the forest near the inlet were reconstructed and a bridge was built over the inlet with the purpose of opening the view to the lake and allowing holidaymakers to visit the site as an attractive natural park and wetland. In spring 2002, the optimal level of the lake was established and the outflow was dammed at the corresponding height.

Hydrochemical and physical data

Long-term changes were estimated on the basis of summer data. Unfortunately, there are no data about nutrients before 1984. Naturally, at the beginning of the 20th century the limnological type of the lake was moderately eutrophic. It could be concluded on the basis of community description as well as on chemical data. Figure 3 shows the range of the aerobic layer. The dynamics of the main hydrochemical parameters were calculated separately for the surface (<2.5 m) and bottom layers (>5.5 m). Several characteristics (COD_{Mn}, alkalinity, sulphates; Figs. 4–6) in the surface have similar dynamics: the values were the lowest in 1929, 1957 and in the 2000s. In the bottom layers, the decrease of the same parameters over the past studied years is almost similar except COD_{Mn} in 2000 – this time there was an increase in content. The values are different for 1998 when the water table was lowered. The low values for some parameters in 1996 are remarkable. The dynamics of the nutrient content show considerable fluctuation (Figs. 7, 8). In the surface layers, very high values were measured again in the summer of 1998. In the bottom layers, the concentration of total phosphorus has increased since 1985 except

Photo 3. The same in 2001. Photo I. Ott.

for the last year of the investigations. Besides very high values in the bottom layers the fluctuation of nitrogen is remarkable while phosphorus is more stable. In fact, the content of substances in the deepest metre varies to a great extent.

Trophic state index

The trophic state index (TSI-4), calculated on the basis of both abiotic and biotic characteristics (Fig. 9), has a tendency to increase during the study period and is mostly on a high level. Two years differ from the general trend – 1996 and 2001. In both years the clear water phase of the ecosystem lasted longer than usual, and water transparency reached more than 4 m.

Fishes

According to the literature (Eesti Järved, 1968), Lake Verevi is considered to be a fish-productive lake. The predominant species included roach, perch, bream, and pike. Rudd, crucian carp, and tench occurs too. Presumably ruffe, eel, gudgeon, weatherfish, and sunbleak have been caught. In 1984, 1989 (Timm, 1991), 1994, 2001 and 2002 bream was not caught anymore, being earlier in the third place by abundance.

In comparison with the other 30 similar Estonian lakes, the catches in Lake Verevi are essentially lower. Despite the fact that accurate comparable data are absent, it is obvious that the species composition has become impoverished. Biomass and abundance of the dominating species have decreased. The rudd is the only species that has increased its abundance. The obvious prevalent fish is roach. Estimated catches of the fishes have decreased during the investigation period: 20 kg ha^{-1} in 1956, 18 in 1989, and 14 in 2001–2002 (for more details see Järvalt et al., 2005).

Phytoplankton

Phytoplankton showed three stages of changes between 1928 and 2001. The species composition was stable until 1957. During this period the

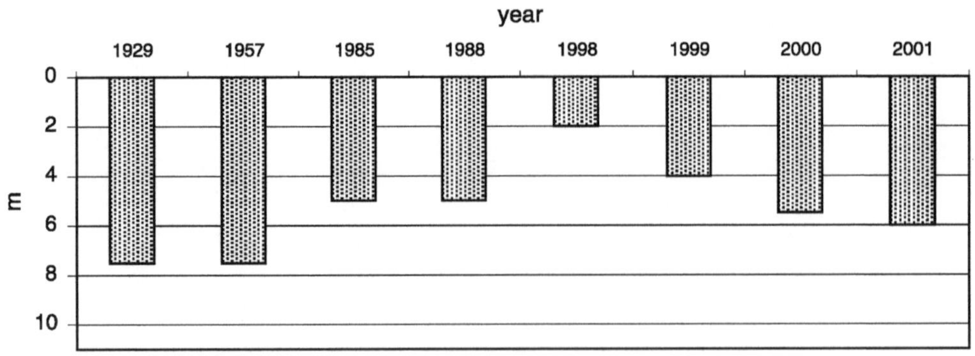

Figure 3. Long-term changes in the aerobic layer of Lake Verevi.

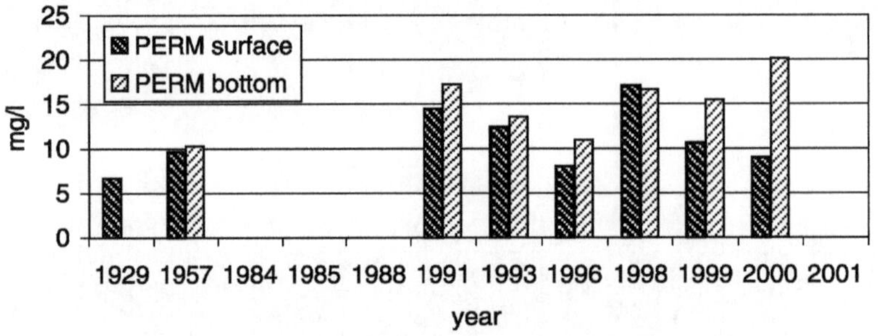

Figure 4. Permanganate oxidizability (PERM) in the surface and bottom layers.

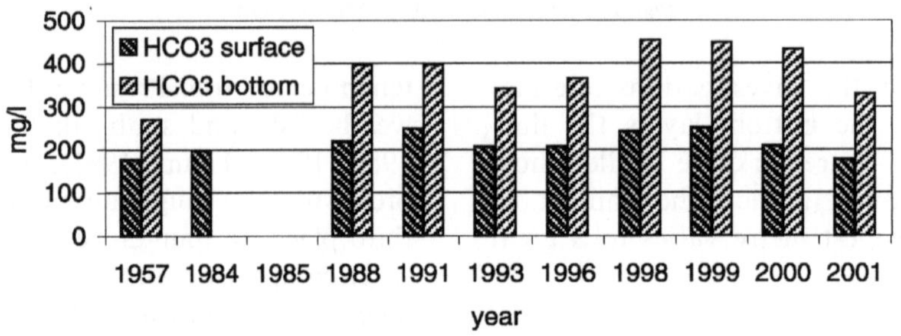

Figure 5. Alkalinity (HCO_3) in the surface and bottom layers.

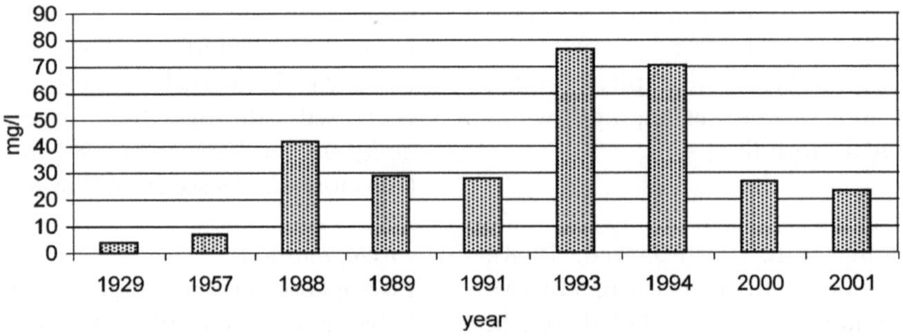

Figure 6. Sulphate content (SO_4) in the surface layers in summer.

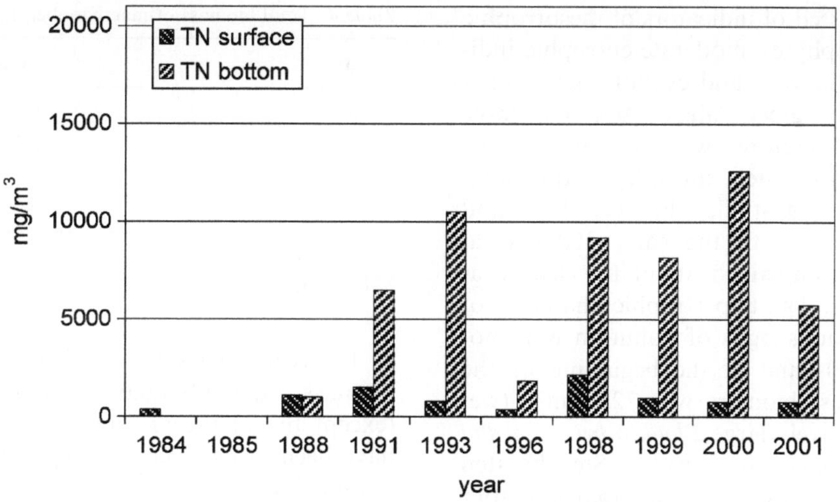

Figure 7. Total nitrogen (TN) in surface and bottom layers.

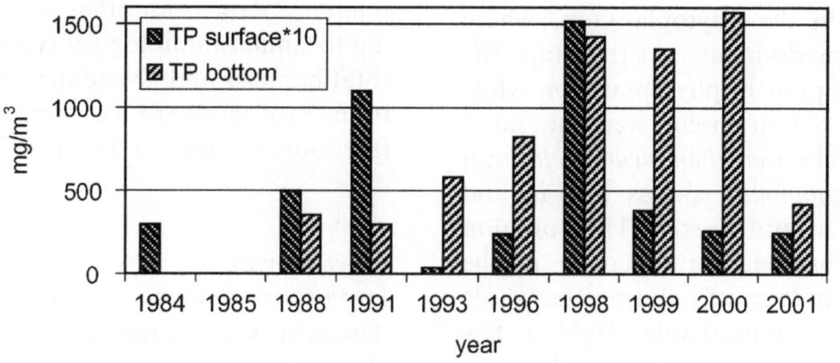

Figure 8. Total phosphorus (TP) in surface and bottom layers.

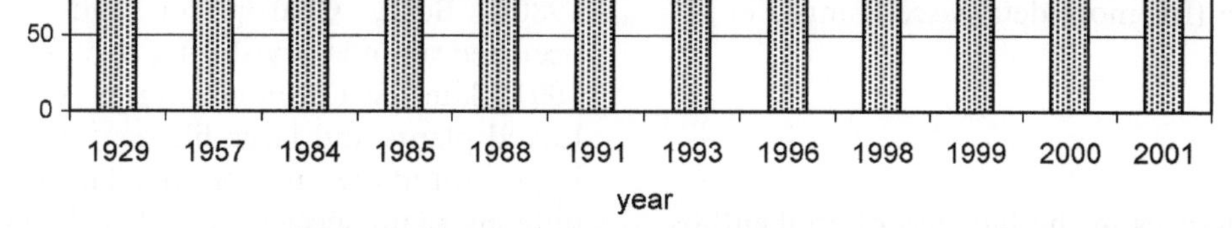

Figure 9. Dynamics of the trophic state index (TSI-4).

community consisted of indicators of mesotrophic status like chrysophytes, moderate eutrophic indicator species of diatoms and cyanobacteria, relatively large-sized green algae and eurytopic dinoflagellates. Therefore, we can consider the trophic state as moderately eutrophic at this time.

The lake was not studied between 1958 and 1983. Nevertheless, until the early 1980s local people had not complained about the deterioration of the status. The hypertrophic character of the lake and obvious signs of pollution were noticed in the 1980s and at the beginning of the 1990s. The highest biomass was 724 g m^{-3} (wet weight). The relatively large *Planktothrix agardhii* with its form *aequicrassa* prevailed. Step by step the amount of species requiring organic matter increased, like euglenophytes, green algae with flagella, and cryptophytes.

Recent years have witnessed the third stage of long-term changes in the phytoplankton, where various species are predominate. At this stage the epilimnion has been poor in phytoplankton while high biomasses of different species were measured in the deeper parts like *Ceratium hirundinella* at a depth of 5 m, *Cryptomonas* species at 6 m and euglenophytes at 7 m and deeper. The common feature of pond ecosystems, an increase in the number of euglenophyte species during the whole investigated period is remarkable. Only a few species of euglenophytes occurred during the early investigation period, but 14–23 different species occurred during the growth season in the counting sample in 2000 and 2001. Modified Nygaard's (1949) phytoplankton compound quotient (Ott & Laugaste, 1996), reflecting the trophic state of the lake and calculated on the basis of summer data, increased in the whole water column during the 1980s. Since 1986, the quotient has exceeded the hypertrophic limit (7) and has remained on a very high level throughout study period with an exception of 1994. As a rule, in summer epilimnetic values were lower in comparison with the whole water column (for more details see Kangro et al., 2005).

Zoobenthos

Long-term changes in the biomass of zoobenthos have reflected the impoverishment of the community at least since 1988 (Table 4).

Table 4. Total biomass of macrozoobenthos in Lake Verevi

Year	Total biomass (g m^{-2})
1957	6.37*
1984	8.27*
1988	2.78*
1998–2001	2.02**

*Timm (1991).
**Timm & Möls (2005).

In recent years the open water habitats, the biomass, and abundance of macrozoobenthos (except the phantom midge *Chaoborus flavicans*) have been rather constant beginning with the epilimnion up to the upper hypolimnion (depth 2–4 m). However, they have been low in the lower hypolimnion (depth 6 m), which was inhabited mainly by *Chaoborus*. Comparison with long-term reference data from other Estonian lakes, belonging to similar limnological types, indicated that the total biomass and abundance (without *Chaoborus*) in the profundal of Lake Verevi was extremely low (for more details see Timm & Möls, 2005).

Macrophytes

Long-term changes are mostly related to abundance; species composition has been more stable. However, three recently important species, *Lemna trisulca*, *Potamogeton friesii* and *P. pectinatus*, had not be mentioned before the 1980s (*Potamogeton* spp. were not identified as species in 1929). *Myriophyllum spicatum*, being lush in 1929 (Riikoja, 1940), is absent today, as well as *Stratiotes aloides*. The decline of *Schoenoplectus lacustris* and *Equisetum fluviatile* was marked already in 1957, as well as the increasing abundance of *Potamogeton natans*. All these processes continued during the following four decades. The most remarkable change is the replacement of the dominating charophytes by *Ceratophyllum demersum* in the 1980s. Both, Charophyta and *Ceratophyllum* occurred abundantly during 4–5 years (Mäemets, 1991); later on *Ceratophyllum* prevailed. Floating-leaved plants and large filamentous algae became important in the same period. The growth areas of different plant species have also changed. Only the charophytes have continuously preferred the northern part that is rich in springs. The depth limit of the submerged belt was slightly decreasing before 1998, being 3.8 m in 1957 and 3.0 m in 1988. At the time of water lowering in order to restore the swimming pool in the summer of 1998, the shallow northern end of the lake as well as the whole transition zone between the emergent vegetation and the submerged vegetation were denuded. After the restoration of the water level in summer 1999, the submerged plants reached a depth of 2.0 m; in 2000 2.5 m. In 2001–2002, the maximum depth of submerged plants was 4.0 m, but mostly it was up to 3.0 m. In the following years after the water table recovery the main phenomenon was the alternation of the prevailing species while the species composition and abundance became more stable step by step (for more details see Mäemets & Freiberg, 2005).

Discussion

Hydrochemical and –physical data

Long-term changes in the hydrochemical properties are usually explained by changes in diffuse or point pollution derived from agriculture, wastewater from settlements, melioration or landscape design, climate changes, etc. The main reason for changes in Lake Verevi could be irregular discharge of the wastewater from the oxidation ponds during the flood seasons. It was noticed several times during fieldwork in spring, but since water overflowed from the ponds and infiltrated through the wetland into the lake, no one measured the influx. Neither the regular measurements of the inflows between 1991 and 1993 nor the used method of analogy calculated from the daily runoff of the River Elva (in 1991,1993 and 2000, 2001; Nõges, 2005) showed heavy pollution. Generally, Lake Verevi can tolerate the calculated phosphorus loading from the inflows (Fig. 2). At the same time sediments should be taken into account.

The dynamics of the sulphate content of the lakes is related to human activity. Oil shale mines, fertilisers, and traffic are the main resources of the sulphur inload to the lakes in Estonia (Eesti Loodus, 1995). The increase in the sulphate concentration over the past half-century is the most reliable in comparison with the other ions in the natural inland waterbodies (Mäemets et al., 1994). Since the mid-1990s sulphate emission has decreased essentially due to depletion of the oil-shale industry and nature conservation measures (Kört et al., 2001). In recent years, depending on the weather conditions, the spring overturn of the water was absent. This affected the content and distribution of nutrients, as well as the ecological state. The decrease in nutrients, sulphates and alkalinity was noticeable in 2001 compared to the previous year (Figs. 5–8). It was noticeable during the whole vegetation period in 2001. The winter 2000/2001 had little snow because the warm period lasted 19 days at the end of November and in December 2000 (data of the Estonian Meteorological Institute, Tõravere Weather Station). Precipitation in winter 1999/2000 was 182 mm and in 2000/2001 it amounted to 160 mm. On 12 February 2001 no oxygen depletion was discovered in the water column. These conditions favoured mineralisation of the organic matter. Because of rapid stratification after the ice break-up, dissolved organic and nutrient substances were not mixed up from the deeper layers. Therefore, the concentration of substances stayed on a low level. The low values for several hydrochemical characteristics in the summer of 1996 were probably due to the same reasons. Unfortunately, in the spring of 1996, stratification parameters were not measured. Comparison of precipitation in the winter 1994/1995 (323 mm) and 1995/1996 (173 mm) yielded the same results as in 1999/2000 and 2000/2001.

Phytoplankton

The ecological status of the lake was close to natural in the 1920s and the 1950s. Species composition of the first stage of the long-term study period corresponded to the conception of natural reference conditions of hard-water light-coloured stratified lakes. Species richness is a feature that characterises natural conditions of the phytoplankton community. This is a well-known phenomenon in hard-water lakes (Gollerbach, 1977; Heinonen, 1980; Ott & Kõiv, 1999, Moss et al., 2003), and it is one of the main features of climax communities (Lampert & Sommer, 1997). The next distinct characteristic is high evenness

(Willen, 2000). The balanced status is also reflected by noticeable amount of small-sized cyanobacteria like species from the genus *Merismopedia* (Rosenström & Lepistö, 1996; Nixdorf et al., 2001) and the same amount of large-sized green algae species like *Pediastrum*, *Staurastrum*, *Botryococcus* (Ott et al., 1997; Willen, 2000), which occurred in Lake Verevi during the early studies.

During the 1980s and the 1990s the ecosystem was unbalanced and indicators of hypertrophic conditions prevailed. Recent years witnessed rapid fluctuations in species composition and abundance. Seasonal variations are considerable and species composition differs to a remarkable degree also in the water column. Phytoplankton of stratified hypertrophic lakes is characterised by uneven vertical distribution.

The situation is different from other lakes in Estonia, which generally follow the dynamics of agricultural diffuse and point pollution and also climatic variability. Lake Verevi has contaminated sediments and internal loading is the main reason for high nutrient concentrations. There are communities characteristic of hypertrophic conditions and high phytoplankton biomasses. The changes between the 1950s and 1980s caused by human impact were also favoured by the small lake volume and weak water exchange, which made the lake vulnerable to pollution. The seasonal dynamics of phytoplankton is rather similar from year to year in lakes with a moderate trophic state (Hutchinson, 1967). However, increased eutrophication makes the biomass of cyanobacteria and seasonal patterns less predictable. Lake Verevi witnessed the rise of cyanobacterial importance in the community in the 1980s. The presence or absence of *P. agardhii* was also an important factor determining the community structure.

The sharp gradients of light and nutrients, brought about already in early spring in recent years by strong thermal stratification, caused the occurrence of a large number of vertically narrow niches in the water column. This leads to biota specialisation and stratification, which helps to avoid competition. During this short time of changes highly different species could take advantage, but this early formed community persisted for the whole vegetation period. Next year another starting community could be the winner.

Fishes

During the eutrophication cyprinid species domination in the lakes is well-known. The tendency of the increase in small-sized non-valuable fishes (Premazzi & Chiaudani, 1992; Pihu, 1998) while the total catch could even increase. The decrease in the catches could be caused by the widening of the hypoxic water volume in Lake Verevi.

Macrophytes

Obviously due to steep stratification and a higher trophic state, the long-term changes in Lake Verevi are different from the lakes with the same morphometry. In recent years regulation of the water level and partly meromixis are the main factors affecting the growth of macrophytes. One could suggest the following reasons for the long- and last year short-term changes could be concluded (Mäemets & Freiberg, 2005).

The replacement of rooted submerged plants by unrooted plants may be caused by a decrease in sediment density, resulting in worse rooting conditions and lower fertility. *Ceratophyllum demersum* is relatively shadow-tolerant, and expansion of *Ceratophyllum* in Lake Verevi took place in the period of low summer transparency of the water. A fine foliage structure favours its growth in poor light conditions, while unrooted plants are able to rise into the upper water layer where illumination is better. Moreover, they can use reserves stored near the bottom in spring. Utilisation of nutrients released in the course of the decomposition of other plant species is possible. Replacement of charophytes by *C. demersum* in the 1980s was probably also caused by changes in the sediment structure. Changes after the water lowering in 1998 demonstrated a strong impact of the aeration, mineralisation, and thickening of bottom sediments, resulting in the mass development of *Ranunculus circinatus* and recovery of charophytes to some extent. Temporary water lowerings stimulate the germination of the propagules of *R. circinatus*. Obviously, a provisional increase in the abundance and biomass of *Potamogeton friesii* and *P. pectinatus* as well as the frequent occurrence of other nutrient-demanding species – *Elodea canadensis* and *Potamogeton crispus* – are related to sediment changes. Even the extraordinarily intensive development of *Cladophora* and other filamentous algae in 2000 may have been favoured by better availability of nutrients in the water column. Possibly, the subsequent decline in *Ranunculus* and some *Potamogeton* species was caused by the exhaustion of sediments or by competition.

Rapid changes in Lake Verevi may be related to the disturbed hypertrophic ecosystem, temporary changes in sediment fertility, and weather conditions. The responses of different species to the temporary water lowering have proceeded at different rates. The research findings suggest the existence of significant differences in matter circulation, which depend on the dominating species.

Influence of swimmers

It is estimated that on a hot summer-day 2000–3000 holidaymakers visit the beach of Lake Verevi. Schulz (1981), Ott & Lokk (1996) determined the amount of substances during swimming (Table 5).

Mäemets & Ennok (1991) calculated the nutrient loading from swimmers into Lake Verevi assuming that 20,000 swimmers visit the lake per year and that 10% of them behave improperly in the water in some sense. The swimming season lasts approximately 100 days per year. The lake received 205 g P and 4.65 kg N, which makes 1.6 mg P m^{-2} (appr. 1.5% from the annual loading through inlets, for more details see Nõges, 2005) and 3.7 mg N m^{-2} (0.6%). Ott & Lokk (1996) made similar calculations for the closed Lake Viitna Pikkjärv that has a low buffer capacity and softwater. They concluded that since annually the atmosphere precipitated 6-times more phosphorus and 25-times more nitrogen onto the lake surface, the impact of swimmers could be considered negligible. Taking into account the buffer capacity of Lake Verevi and the amount of nutrient resources in the sediments (Kisand, 2005), this impact could be considered even less insignificant.

Table 5. Amount of substances excreted from swimmers

Substance	Ott & Lokk (1996)	Schulz (1981)
Total phosphorus (mg from person)	1.4 ± 0.56	1.077
Phosphates (mg)	0.63 ± 0.33	
Total nitrogen (mg inimeselt)	140 ± 55	115
Nitrates (mg)	0.3 ± 0.21	
Phosphorus from urine (mg)		93
Nitrogen from urine (mg)		1400
Sodium (10^3 swimmers g d^{-1})		7780
Potassium (10^3 swimmers g d^{-1})		7350
Chlorine (10^3 swimmers g d^{-1})		13300
Calcium (10^3 swimmers g d^{-1})		380

Restoration projects

The best results could be achieved taking into account technological solutions as well as the ecological features of waterbodies. It includes the character of matter circulation, trophic level, species composition, their relations, etc., not neglecting biogeochemical processes in the water–sediment interface. In fact, restoration needs a holistic ecological approach of whole the landscape (Eiseltova, 1994).

Several preconditions should be met before real restoration starts. It is important to stop external loading. Also, one has to set goals for the restoration and the final design of the ecosystem (Björk, 1994a). In the case of Lake Verevi the first step was completed in 2002 by stopping external loading. Goals for the future include the lowering of the trophic level down to natural. The final design of the ecosystem as well as surroundings has not been decided as yet.

The ecosystem morphometry and limnological lake type induces possible methods of restoration (Björk, 1988, 1994a). The decisive features for the selection of the proper method for Lake Verevi include stratification, polluted sediments, and weak water exchange. The number of cases where the combination of different methods is used at the same time increases step by step (Jeppesen et al., 1990a; Berge, 1993). One of the reasons is economy of the finances, but the combination of methods too, can give better results and cover more aspects of the functioning of the ecosystem.

In recent years, the ecological status in the epilimnion of Lake Verevi has been good partly due to meromictic conditions. This situation is not stable. If a strong spring overturn occurs, it will bring a lot of nutrients into the water column and one could predict heavy water blooms with the accompanying phenomena. Sediment removal (Björk, 1988, 1994b) or sediment treatment (Ripl, 1976, 1994; Wolter, 1994) can be used for rapid restoration. Biomanipulation (Richter, 1986; Faafeng & Brabrand, 1990; Van Donk & Gulati, 1991; Gulati, 1995b) does not lower the trophic level that much but raises the efficiency of the matter circulation and top-down control of the food chain. The long-term stability after short-term biomanipulation is questionable and can be achieved mainly in shallow lakes (Richter, 1986; Jeppesen et al., 1990b; Gulati, 1995a; Kasprzak et al., 2002). Biomanipulation in stratified lakes has more obstacles (Dawidowicz et al., 2002) and good results can be achieved after long-term efforts (Krienitz et al., 1996). Also, biomanipulation is rather expensive (Kasprzak, 1995).

Drainage (siphoning) of the hypolimnetic water is one of the simplest ways to reduce the nutrient content of the lake. Nõges (2005) calculated the water and mass balance for different conditions of Lake Verevi. A decrease in TP content can be taken into account only if the balance is based on the in and outflow as well as the TP content in the lake (Table 6). Theoretically, during 1 year of siphoning, TP concentration decreases from 165.5 mg m^{-3} to 63.6–92 mg m^{-3}, that is by a factor 1.8–2.6. Presumably, the natural level of P content can be achieved in 2 years. Due to the exclusion of other P resources, as the groundwater supply, diffuse loading, precipitation, and especially diffusion from the sediments, the noticed eventual P value during siphoning should be somewhat higher.

The phosphorus content in the sediments is not very high in the Lake Verevi in comparison with other lakes (Table 7). Taking into account the prevailing anoxic conditions in the hypolimnion, low pH, and redox potential, a great part of the sedimented P could be released into the water.

The ratio of Fe–P is crucial for binding P in the sediments. Fe:P mass ratio >15 can control the release of reactive P if pH is <8 in the surface of the bottom in oxic conditions of shallow lakes (Jensen et al., 1992; Søndergaard et al., 1993). The minimum limit is considered to be 10 (Wolter, 1994). We measured these values in the thin 5–10 cm layer (9.6) and deeper than 30 cm in the sediments of Lake Verevi. Calculated as residual from this ratio, the released P amount is theoretically 27 kg P, which seems to be an underestimated value. Scheffer (1998) explains that in the sediments with a high content of organic matter, iron can become unavailable for phosphorus immobilisation due to precipitation with sulphide as insoluble FeS. Steenberger et al. (1993) provided the sequence of molecules utilised as electron acceptors during microbial processes. Iron is used before sulphate. In Lake Verevi, according to our measurements, the content of sulphides in the hypolimnion is remarkable – up to possible saturation limit. Iron does not bind P due to high sulphide content in the sediments, which sometimes are oxygenated. In the hypolimnetic area of bottom, reduced conditions prevail, and Fe cannot bind P anyway. Generally, the amount of Fe in the Lake Verevi seems to be low – average value was 17.2 in the upper 20 cm layer. Wolter (1994) described the successful restoration of stratified Lake Gross-Glienecker, where the Fe content in sediments was 20 g kg^{-1} dry matter, and this was considered too low.

Chemical fractions of phosphorus in the sediments give also information about the possible P release into the water. We measured weakly bound P, aluminium and iron bound P, apatite P and residual P in 1994 and 2001 (for details see Kisand, 2005). Three first fractions can be mobile in the water–sediment interface in different conditions. The calculated mean P content in 10 cm thick layer of hypolimnetic bottom area (40% from the bottom), was 551 kg, and potentially released P 301 kg (Kisand, 2005).

When selecting the restoration method, one of the most important conditions is the accuracy of results. During siphoning of the hypolimnetic water, geochemical processes in the interface of water–sediment will be in a great extent unpredictable. Presumably oxidation conditions will change and secondary pollution can even increase. One reason could be the breaking of stratification. It is difficult to calculate the intensity of phosphorus removal/inactivation in sediments in time. Cooke et al. (1993) considered disadvantages during siphoning the pollution of downstream waterbodies and the bad smell of the outflowing water. Uncertainties of results will decrease if one uses phosphorus precipitation and hypolimnetic aeration instead of siphoning. However, it will add to the expenses.

Table 6. Annual theoretical (predicted) changes in phosphorus content (total phosphorus, TP) during siphoning of hypolimnetic water in Lake Verevi

Characteristic	Value
Maximum TP discharge from inflow (kg)	26.25
Maximum inflow (m^3)	328048
Lake volume (m^3)	465270
Average TP concentration in the lake (mg m^{-3})	165.5
Average amount of TP in the lake (kg)	77
Average volume of the epilimnion, (m^{-3})	314670
Average volume of the meta- and hypolimnion (m^{-3})	150601
Average TP concentration in the meta- and hypolimnion (mg m^{-3})	285.5
Average amount of TP in the meta- and hypolimnion (kg)	43
Average TP concentration in the epilimnion (mg m^{-3})	108
Average amount of TP in the epilimnion (kg)	34
Total amount of P in the lake including discharge (kg)	103
Proportion of the volume of epilimnion from the total (%)	68
Proportion of the volume of meta- and hypolimnion from the total (%)	32
Maximum annual discharge of TP to the epilimnion (kg)	17.75
Amount of TP in the epilimnion including discharge (kg)	51.75
TP concentration in the epilimnion including discharge (mg m^{-3})	164.5
Maximum annual discharge of TP to the meta- and hypolimnion (kg)	8.5
Amount of TP in the meta- and hypolimnion including discharge (kg)	51.5
TP concentration in the meta- and hypolimnion including discharge (mg m^{-3})	341.9
Outflow, maximum–minimum (m^3)	283944–203302
Amount of TP, outflowing from the meta- and hypolimnion (kg)	51.75
Water volume outflowing annually from the epilimnion, maximum–minimum (m^3)	133343–52701
Amount of TP, outflowing from the epilimnion, maximum–minimum, (kg)	21.9–9
TP outflowing from the lake, maximum–minimum (kg)	73.7 – 60.4
TP retention, minimum–maximum (kg)	29.6–43
Retention TP concentration, minimum–maximum (mg m^{-3})s	63.6–92

Table 7. Average total phosphorus concentration in sediments of some Estonian lakes

Lake	Total phosphorus content in the sediment g kg^{-1} DW
Verevi (Rummi, 1991), upper 1 m layer	1.19
Verevi (Kisand, 2005), upper 40 cm	1.17
Arbi (Kisand, 2005), upper 35 cm	4.04
Neitsijärv (Prede et al., 1999), upper 40 cm	1.69
Võrtsjärv (Kisand, 2005), upper 50 cm	0.78

Acknowledgements

The study was supported by the core grants of the Ministry of Education Nos. 0370208s98, 0362482s03 and by grants of the Estonian Science Foundation Nos. 3579 and 4835. We would like thank the entire research team (T. Nõges, K. Kangro, P. Zingel, K. Kübar, E. Lill, H. Tammert, T. Timm, H. Timm, H. Mäemets, L. Freiberg, V. Kisand, H. Starast, T. Möls, A. Lindpere, I. Solovjova, M. Reunanen, T. Krause, A. Palm, R. Laugaste, K. Ott, J. Zirk, M. Tammert etc.) who participated in the project. We would also like to thank Dr Enn Veldi, for revising the English. Anne Jõeveer and Ingrid Niklus from the Tõravere Station of the Estonian Institute of Meteorology and Hydrology provided climatic data.

References

Alekin, O. A., 1959. Metody issledovaniya fizicheskih svojstv i himicheskovo svoistva vody In Pavlovski, E. N. & V. I. Zhadin, (eds), Zhizn' presnyh vod SSSR. 4, 2: 213–300 [Methods for investigation of physical and chemical properties of water. In Russian].

Berge, D., 1993. Restaureringsstrategi for eutrofierte innsjøer. NIVA – rapport. 73 pp. [Restoration strategies for eutrophied lakes. In Norwegian].

Björk, S., 1988. Redevelopment of lake ecosystems – a case-study. Approach -Ambio 17(2): 90–98.

Björk, S., 1994a. Planning and accomplishment of redevelopment and restoration projects. In Eiseltova', M. (ed.), Restoration of Lake Ecosystems, A Holistic Approach, 59–61 pp.

Björk, S., 1994b. Restoration methods and techniques. Sediment removal. In: Eiseltova', M. (ed.), Restoration of Lake Ecosystems, A Holistic Approach, 82–89 pp.

Cooke, G. D., E. B. Welch, S. A. Peterson & P. R. Newroth, 1993. Restoration and Management of Lakes and Reservoirs. Lewis Publishers, Boca Raton, 548 pp.

Crumpton, W. G., T. M. Isenhart & P. D. Mitchell, 1992. Nitrate and organic N analyses with second-derivative spectroscopy. Limnology and Oceanography 37(4): 907–913.

Dawidowicz, P., A. Perjs, A. Engelmayer, A. Martyniak, J. Kozlowski, L. Kufel & M. Paradowska, 2002. Hypolimnetic anoxia hampers top-down food-web manipulation in a eutrophic lake. Freshwater Biology 47: 2401–2409.

Eesti Loodus, 1995. Compiled by A. Raukas. Publishing Office "Valgus". Tallinn. 606 pp. [Estonia. Nature. In Estonian, Russian and English summary].

Eesti Järved, 1968. Tallinn, "Valgus". 532 pp. [Estonian lakes. In Estonian].

Eiseltova', M., (ed.). 1994. Restoration of Lake Ecosystems, A Holistic Approach. Preface, 7–8 pp.

Faafeng, B. A. & Å. Brabrand, 1990. Biomanipulation of a small, urban lake – removal of fish exclude bluegreen blooms. Verhandlungen Internationale Vereinigung für theoretische und angewandte Limnologie 24: 597–602.

Faafeng, B. A. & J. P. Nilssen, 1981. A twenty-year study of eutrophication in a deep Soft-water lake. Verhandlungen Internationale Vereinigung für theoretische und angewandte Limnologie. 21: 412–424.

Fall, M. J, 1996. JENWAY Model 4150 Conductivity Meter manual. Jenway Limited, Gransmore Green, Felsted, Dunmow, Essex, CM6 3LB, England, 19 pp.

Gollerbach, M. (eds), 1977. Zhizn rastenij. Tom Tretij. Vodorosli i lishainiki. Moskva, Prosveshtshenije p. 487 Life of plants. Third part. Algae and Lichens. In Russian.

Gulati, R. D., 1995a. Food-chain manipulation as a tool in management of small lakes in the Netherlands: The Lake Zwemlust example. In: Guidelines for Lake management. Vol. 7. de Bernardi R. & G. Giussani (eds), Biomanipulation in Lakes and Reservoirs management, 147–163 pp.

Gulati, R. D., 1995b. Manipulation of Fish Population for Lake Recovery from Eutrophication in the Temperate Region. In: Guidelines for Lake Management. Vol. 7. de Bernardi R. & G. Giussani (eds), Biomanipulation in Lakes and Reservoirs management, 53–81 pp.

Hansen, H. P. & F. Koroleff, 1999. Determination of nutrients. In Grasshoff, K., K. Kremling & M. Ehrhardt (eds), Methods of Seawater Analysis. WILEY-VCH, Weinheim. New York. Chichester. Brisbane. Singapore. Toronto, 600 p.

Heinonen, P., 1980. Quantity and composition of phytoplankton in Finnish Inland Waters. *Publication of Water Research Institute*, Helsinki, 37, 91.

Hellat, K., A. Malirin, L. Mei & T. Tenno, 1986. Metrologicheskoe obespechenie sredstv analiza kisloroda v vode. Acta et commentationes Universitatis Tartuensis 757: 184–193 Metrological garanteeing of devices for measuring oxygen in the water. In Russian.

Hutchinson, G. E., 1938. Chemical stratification and lake morphology. Proceedings of the National Academy of Sciences 4: 63–69.

Hutchinson, G. E., 1967. A Treatise on Limnology. II. Introduction to Lake Biology and the Limnoplankton. John Wiley and Sons, New York, 1115.

Järvalt, A., T. Krause & A. Palm, 2005. Diel migration and spatial distribution of fish in a small stratified lake. Hydrobiologia 547: 197–203.

Järvet, A., 1989. Veekogude kasutamise vastuolud Kavilda oja näitel In: Põllumajandus ja keskkonnakaitse. Tallinn – Elva. pp. 84–90 [Contradictions of management of water bodies on example of Kavilda stream. In Estonian].

Jensen, H. S., P. Kristensen, E. Jeppesen & A. Skytthe, 1992. Iron:phosphorus ratio in surface sediment as an indicator of phosphate release from aerobic sediments in shallow lakes. Hydrobiologia 235/236: 731–743.

Jeppesen, E., J. P. Jensen, P. Kristensen, M. Søndergaard, E. Mortensen, O. Sortkjær, K. & Olrik, 1990b. Fish manipulation as a lake restoration tool in shallow, eutrophic, temperate lakes 2: threshold levels, long-term stability and conlusions. – Hydrobiologia 200/201: 219–227. Gulati, R. D., Lammens, E. H. R. R., Meijer, M.-L., van Donk, E (eds), Biomanipulation – Tool for Water Management.

Jeppesen, E., M. Søndergaard, E. Mortensen, P. Kristensen, B. Riemann, H. J. Jensen, J. P. Müller, O. Sortkjær, J. P. Jensen, K. Christofferson, S. Bosselmann, E. Dall, & 1990a. Fish manipulation as a lake restoration tool in shallow, eutrophic, temperate lakes 1: cross– analysis of three Danish case studies. – Hydrobiologia 200/201: 205–218. Gulati, R. D., Lammens, E. H. R. R., Meijer, M.-L., van Donk, E., (eds), Biomanipulation – Tool for Water Management.

Kangro, K., R. Laugaste, P. Nõges & I. Ott, 2005. Long-term changes and seasonal development of phytoplankton in a strongly stratified hypertrophic lake. Hydrobiologia 547: 91–103.

Kasprzak, P., 1995. Objectives of biomanipulation. In: De Bernardi, R., Giussani, G. (eds), Guidelines of lake management. 7. de Bernardi, R. and Giussani, G. (eds), Biomanipulation in lakes and reservoirs management, 1–15 pp.

Kasprzak, P., J. Benndorf & T. R. Mehner Koschel, 2002. Biomanipulation of lake ecosystems: an introduction. Freshwater Biology 47: 2277–2281.

Kisand, A., 2005. Distribution of sediment phosphorus fractions in hypertrophic strongly stratified Lake Verevi. Hydrobiologia 547: 33–39.

Koroleff, F., 1982. Total and organic nitrogen. In Grasshoff, K. (ed.), Methods of Seawater Analysis. Verlag chemie, 162–168.

Kört, M., T. Truuts & K. Pajuste, 2002. Õhu saasteainete kaugkande seire. Eesti Keskkonnaseire 2001. 23–25. [Monitoring of long-distance air pollution. In Estonian].

Krienitz, L. & P. R. Kasprzak Koschkel, 1996. Long term study on the influence of eutrophication, restoration and biomanipulation on the structure and development of phytoplankton communities in Feldberger Haussee (Baltic Lake District, Germany). Hydrobiologia 330: 89–110.

Lampert, W. & U. Sommer, 1997. Limnoecology: The Ecology of Lakes and Streams. Oxford Univeristy Press, 382 pp.

Lindenschmidt, K.-E. & I. Chorus, 1997. The effect of aeration on stratification and phytoplankton populations in Lake Tegel. Berlin Archiv Für Hydrobiologie 139(3): 317–346.

Loopmann, A., 1984. Suuremate Eesti järvede morfomeetrilised andmed ja Veevahetus. Tallinn, 150 lk. [Morphometrical data and water exchange of larger Estonian lakes. In Estonian].

Mäemets, Aare & K. Ennok, 1991. Valgala iseloom, sissevoolude vee keemiline koostis ja järve resotsukoormus. In Timm, H. (ed.), State of Lake Verevi. Hydrobiological researches XVII. pp. 34–44. [Catchment features, chemical composition of water of inflows and pollution loading. In Estonian].

Mäemets, Aime, 1991. Suurtaimestik. In H. Timm (ed.), State of Lake Verevi (Hydrobiological Researches XVII), Tartu: 95–106 [Macrovegetation. In Estonian].

Mäemets, A., I. Ott & A. Mäemets, 1994. Eesti väikejärvede seisundi muutused ja kaitse. Kogumik: Eesti jõgede ja järvede seisund ning kaitse. Toim. A. Järvekülg. Teaduste Akadeemia kirjastus. Tallinn, lk. 32–47. [Changes an protection of Estonian small lakes). In Estonian, English summary].

Mäemets, H. & L. Freiberg, 2005. Long- and short-term changes of the macrophyte vegetation in strongly stratified hypertrophic Lake Verevi. Hydrobiologia 547: 175–184.

Möls, T., H. Starast, A. Milius & A. Lindpere, 1996. The hydrochemical state of Lake Peipsi–Pihkva. Hydrobiologia 338: 37–47.

Moss, B., D. Stephen, C. Alvarez, E. Becares, W. Van de Bund, S. E. Collings, E. Van Donk, E. De Eyto, T. Feldmann, C. Fernández-Aláez, M. Fernández-Aláez, R. J. M. Frankeng, F. García-Criado, E. Gross, M. Gyllström, L.-A. Hansson, K. Irvine, A. Järvalt, J.-P. Jenssen, E. Jeppesen, T. Kairesalo, R. Kornijów, T. Krause, H. Künnap, A. Laas, E. Lill, B. Lorens, H. Luup, M. R. Miracle, P. Nõges, T. Nõges, M. Nykänen, I. Ott, W. Peczula, E. T. H. M. Peeters, G. Phillips, S. Romo, V. Russell, J. Salujõe, M. Scheffer, K. Siewertsen, H. Smal, C. Tesch, H. Timm, L. Tuvikene, I. Tõnno, T. Virro & D. Wilson, 2003. The determination of ecological quality in shallow lakes – a tested system (ECOFRAME) for implementation of the European Water Framework Directive. Aquatic Conservation: Marine and Freshwater Ecosystems, 13: 507–549.

Nixdorf, B., J. Mischke, U. Hoffmann, A. Hemm, M. & E. Hoehn, 2001. Classification and assessment of lakes in Germany according to the bilogical indicator phytoplankton – first results. Classification of Ecological Statud of Lakes and Rivers. Saara Bäck & Krister Karttunen (eds). TemaNord 2001: 584, 24–27 pp.

Nõges, P., 2005. Water and nutrient mass balance of temperate partly meromictic Lake Verevi Hydrobiologia 547: 21–31.

Nõges, T. & K. Kangro, 2005. Primary production of phytoplankton in a strongly stratified temperate lake. Hydrobiologia 547: 105–122.

Nygaard, G., 1949. Hydrobiological Studies on some Danish Ponds and Lakes II: The quotient hypothesis and some little known or new phytoplankton organisms. Kunglige Danske Vidensk, Selskab. 7, 242 pp.

Ott, I., 2001. Typology and ecological classification of Estonian lakes. In Classification of ecological status of Lakes and Rivers. TemaNord 2001: 584, 62–64 pp.

Ott, I. & T. Kõiv, 1999. Eesti väikejärvede eripära ja muutused. Estonian Small Lakes: Special Features and Changes. Tallinn. 128 pp.

Ott, I. & R. Laugaste, 1996. Fütoplanktoni koondindeks Üldistus Eesti väikejärvede kohta Eesti vabariigi Keskkonnaministeeriumi Infoleht 3/96. lk. 7–8. [Phytoplankton compound quotient. Conclusion about Estonian small lakes. In Estonian].

Ott, I., R. Laugaste, S. Lokk & A. Mäemets, 1997. Plankton changes in Estonian small lakes in 1951–93. – Proceedings of Estonian Academy of Sciences Biology. Ecology. 46, ½, 58–79.

Ott, I. & S. Lokk, 1996. Viitna Pikkjärv ja puhkajad – Eesti Loodus.174–176. [Lake Viitna Pikkjärv and holidaymakers. In Estonian, English summary].

Ott, I., A. Rakko, D. Sarik, P. Nõges & K. Ott, 2005. Sedimentation rate of seston during the formation of temperature stratification after ice break-up in the partly meromictic Lake Verevi. Hydrobiologia 547: 51–61.

Pihu, E., 1998. Fishes and fisheries management in Lake Võrtsjärv. Limnologica 28(1): 91–94.

Pipp, E. & E. Rott, 1995. A phytoplankton compartment model for a small meromictic lake with special reference to species-specific niches and long-term changes. Ecological Modellling 78: 129–148.

Prede, M., I. Ott, A. Kisand, R. Laugaste, H. Mäemets, H. Timm, A. Järvalt, E. Kirt & T. Oja, 1999. A. Maastik (ed.), Lakescape of Otepää: Past, present and future. Tacis–Phare, 24 pp.

Premazzi, G. & G. Chiaudani, 1992. Current approaches to assess water quality in lakes. In ECSC-EEC-EAEC, Brussels, Luxembourg, 249–308.

Reports of the Baltic Intercalibration Workshop, 1977, Kiel: 27–28.

Richter, A. F., 1986. Biomanipulation and its feasibility for water quality management in shallow eutrophic management in shallow eutrophic water bodies in the Netherlands. Hydrobiological Bulletin 20(1/2): 165–172.

Riikoja, H., 1930. Zur Morphometrie eineiger Seen Eestis. Andmeid Eesti ala järvede uurimiseks 15: 116–201.

Riikoja, H., 1940. Zur Kenntnis eineiger Seen Ost-Eestis, insbesonderere ihrerWasserchemie. Loodusvarade Instituudi Limnoloogia Laboratooriumi Avaldised, 2.

Ripl, W., 1976. Biochemical oxidation of polluted lake sediment with nitrate. A new restoration method. Ambio 5: 132–135.

Ripl, W., 1994. Restoration methods and techniques. Sediment treatment. In Eiseltova', M. (ed.), Restoration of Lake Ecosystems, a holistic approach, 75–82 pp.

Rosenström, U. & L. Lepistö, 1996. Phytoplankton indicator species of different types of boreal lakes. Algological Studies 82 = Archiv Für Hydrobiologie Supplementum 116, 131–140 pp.

Rummi, P., T. Mägi, J. Ütsi, H. Mäemets, A. Lindpere & A. Mäemets, 1991. Põhjasetted. In Timm, H. (ed.), State of Lake Verevi. Hydrobiological Researches XVII. 22–33 pp. [Bottom sediments. In Estonian].

Scheffer, M., 1998. Ecology of shallow lakes. M.B. Busher (ed.), Population and Community Biology Series. Chapman & Hall Publisher, 357 pp.

Schulz, L., 1981. Nährstoffeintrag in Seen durch Badegäste. Zentralblatt Für Bakteriologie., 1B, 6.

Søndergaard, M., P. Kristensen & E. Jeppesen, 1993. Eight years of internal phosphorus loading and changes in the sediment phosphorus profile of Lake Søbygård Denmark. – Hydrobiologia, 253: 345–356.

Steenberger, C. L. M., J.-P. R. A. Sweerts & T. E. Cappenberg. 1993. Microbial biogeochemical processes in lakes: stratification and eutrophication, 69–99 pp.

Timm, H. (ed.), 1991. Verevi järve seisund. A monograph. Tartu. 139 pp. [State of Lake Verevi. In Estonian, English and Russian summary].

Timm, H. & T. Möls, 2005. Macrozoobenthos of lake Verevi. Hydrobiologia 547: 185–195.

Unifitsirovannye metody issledovaniya kachestva vod, 1977. I, Moskva: 831 pp. [Standardized methods for investigation of water quality. In Russian].

Van Donk, E., & R. D. Gulati, 1991. Ecological management of aquatic ecosystems: a complementary technique to reduce eutrophication-related preturbians. In Terrestrial and aquatic ecosystems: preturbation and recovery. Published by Ellis Horword Limited, 566–575 pp.

Vighi, M. & G. Chiaudani, 1985. A simple method to estimate lake phosphorus concentrations resulting from natural background loadings. Water Research 19: 987–991.

Vollenweider, R. A., 1975. Input – output models with special reference to the phosphorus loading concept in limnology. Scweizerische Zeitschrift für Hydrobiologie 37: 53–84.

Wetzel, R. G., 1983. Limnology. Saunders College Publishing. 767 pp.

Willen, E., 2000. Phytoplankton in water quality assessment – an indicator concept. In Heinonen, P., G. Ziglio, & A. Van der Beken (eds), Hydrological and limnological aspects of lake monitoring: 58–80. John Wiley & Sons Ltd.

Wolter, K.-D., 1994. Restoration methods and techniques. Phosphorus precipitation. In Eiseltova', M. (ed.), Restoration of Lake Ecosystems, a holistic approach, 63–69 pp.

Hydrobiologia (2005) 547:21–31
I. Ott & T. Kõiv (eds), Lake Verevi, Estonia – A Highly Stratified Hypertrophic Lake
DOI 10.1007/s10750-005-4140-3

Water and nutrient mass balance of the partly meromictic temperate Lake Verevi

Peeter Nõges[1,2]

[1]*Institute of Zoology and Botany, Estonian Agricultural University, Võrtsjärv Limnological Station, 61101, Rannu, Estonia*
[2]*EC Joint Research Centre, Institute for Environment and Sustainability, TP 290, I-21020, Ispra (VA), Italy*
E-mail: peeter.noges@jrc.it

Key words: water balance, nitrogen balance, phosphorus balance, stratification

Abstract

Mass balances of total nitrogen and total phosphorus were calculated for Lake Verevi (area 0.126 km^2, maximum depth 11 m, mean depth 3.6 m), a sharply stratified small lake located in South Estonia within the borders of the town Elva. The lake has up to 10 small inflows but only three of them are nearly permanent. Accidental overflows from near-by oxidation ponds during high floods have been the major source of the nutrient load of the lake in the past. L. Verevi receives a significant part of its inflow from groundwater, which is difficult to measure. In dry years the outflow is temporary. During summer the lake is sharply thermally and chemically stratified. The spring turnover is often incomplete even in homothermal conditions, thus giving the lake some meromictic features. The influx of nitrogen exceeded the outflux at any supposed proportion (20%, 50%, 80%) of surface runoff. The lake retained 45–90% of the nitrogen influx by sedimentation and/or by denitrification. The largest nitrogen losses with loss rates more than 10 kg N d^{-1} occurred in May and June. The calculated phosphorus retention rate became strongly negative during mixing periods. From June to November, phosphorus release from the sediment exceeded sedimentation by 205 kg in 1991 and by 79 kg in 1993. Earlier stagnation and absence of a full spring turnover in the 2000 has slowed down the recovery of the lake because less phosphorus is flushed out. However, the stronger stratification and significantly smaller phosphorus content in the epilimnion limits biological activity and as a result improves the water quality of the surface layer.

Introduction

Construction of a sound mass balance entails two main subjects: (a) evaluation of the water balance, and (b) evaluation of the balance of any substance in question. As far as possible, these evaluations should be based on measurements in the catchment area and in the lake. If direct measurements of some component, e.g., the groundwater inflow to lakes, are not available, indirect estimates can be made. However, as warned by Jørgensen & Vollenweider (1988), indirect estimates must be used and interpreted with caution. Nutrient balances have provided data on a large variety of lakes, from those acting as efficient sinks of phosphorus up to the lakes in which almost continuous phosphorus release occurs from anaerobic (Laugaste, 1994) or aerobic sediments (Löfgren, 1987). If combined with sediment investigations, nutrient balances become a useful tool even for measuring intimate processes like denitrification (Jensen et al., 1990, 1992; Dudel & Kohl, 1991; Ahlgren et al., 1994). The greatest value of mass balances is that they give at least a clue of the magnitude of the processes that is needed to understand the functioning of the ecosystem. A mass balance calculation for shallow L. Võrtsjärv, for example, showed that the internal phosphate load from the sediment during a one-day storm exceed the total annual load from the watershed (Nõges & Kisand, 1999).

Water balance and mass balances of total nitrogen and total phosphorus in a sharply stratified small lake are discussed in the present paper on a daily and vegetation period scale. Development of thermal stratification is analysed with respect to its effect on the external and internal nutrient balance.

Description of the study area

L. Verevi (area 0.126 km^2, maximum depth 11 m, mean depth 3.6 m) is located within the borders of the small town Elva in South Estonia. The lake has an elongated shape in the north-to-south direction with the deepest and widest part near the southern end of the lake (Fig. 1). By origin L. Verevi is a kettle lake formed by melting of a buried ice block of a decaying glacier (Mäemets, 1991). The 1.1 km^2 large drainage basin (including the lake area) represents a hydrologically complex landscape: in south and southeast the lake is surrounded by sandy hills and dunes covered with pinewoods. The densely populated east shore slopes strongly towards the lake. The area to the west is low and swampy. During high floods, overflow from oxidation ponds of the wastewater treatment plant was discharged to the lake via this swampy area and can be considered the major source of nutrients in the lake. In the year 2001, a soil barrier was built to exclude the ponds from the drainage area. The lake has up to 10 small inflows but only three of them (Fig. 1: 1, 4 and 5) are nearly permanent. Inflows 4 and 5 begin from two spring-fed lakelets, Linajärv and Jaani järv, which are located in the northern part of the watershed. L. Verevi gets a significant part of its water as hardly measurable subsurface runoff (Mäemets et al., 1991). The outflow of the lake is located on the west shore of the lake and flows into River Kavilda. In dry years the outflow becomes discontinuous.

The funnel-like bathymetry (Fig. 1) causes a sharp thermal and chemical stratification of the lake during summer (Nõges & Kangro; Ott et al., 2005a). The spring turnover is often incomplete even in homothermal conditions, thus giving the lake some meromictic features. The metalimnion is progressively eroded during summer and autumn and complete mixing takes usually place in November.

Material and methods

Water balance calculations for the years 1991 and 1993 were based on the general equation:

$$I + P - E - O \pm \Delta V = 0 \qquad (1)$$

in which

I – inflow (surface runoff + groundwater);
P – precipitation onto the lake surface;
E – evaporation from the water surface;
O – outflow;
ΔV – change in storage during the period in question.

All components of the water balance were expressed in m^3 d^{-1}. The inflow to the lake was calculated using the method of analogy. Calculations were based on daily runoff data of the nearby River Elva (gauging station Elva, watershed 239 km^2) measured by the Estonian Institute of Meteorology and Hydrology. Daily precipitation data were provided by the Tõravere meteorological station located at a distance of 5 km from the lake. To take into account the losses through evaporation, we used the long-term average evaporation measured at Tiirikoja (N-E Estonia), 460 mm y^{-1}, multiplied by 1.33, the coefficient for transition from land to open water areas (Pärn & Eipre, 1983). The seasonal distribution of evaporation is given in Table 1.

Outflow from the lake was modelled using the overflow weir equation (Eloranta, 1992) that relates the flow rate (Q, L s^{-1}) to the head on the weir (H, m) and width of the weir (B, m):

$$Q = 1.8 \cdot B \cdot H^{3/2} \qquad (2)$$

As an initial condition the maximum depth of the lake was set to be 10 m and supposed to characterise the winter low flow period. The width of the weir and the initial head were used as variables for model calibration. Calculations were made with a daily

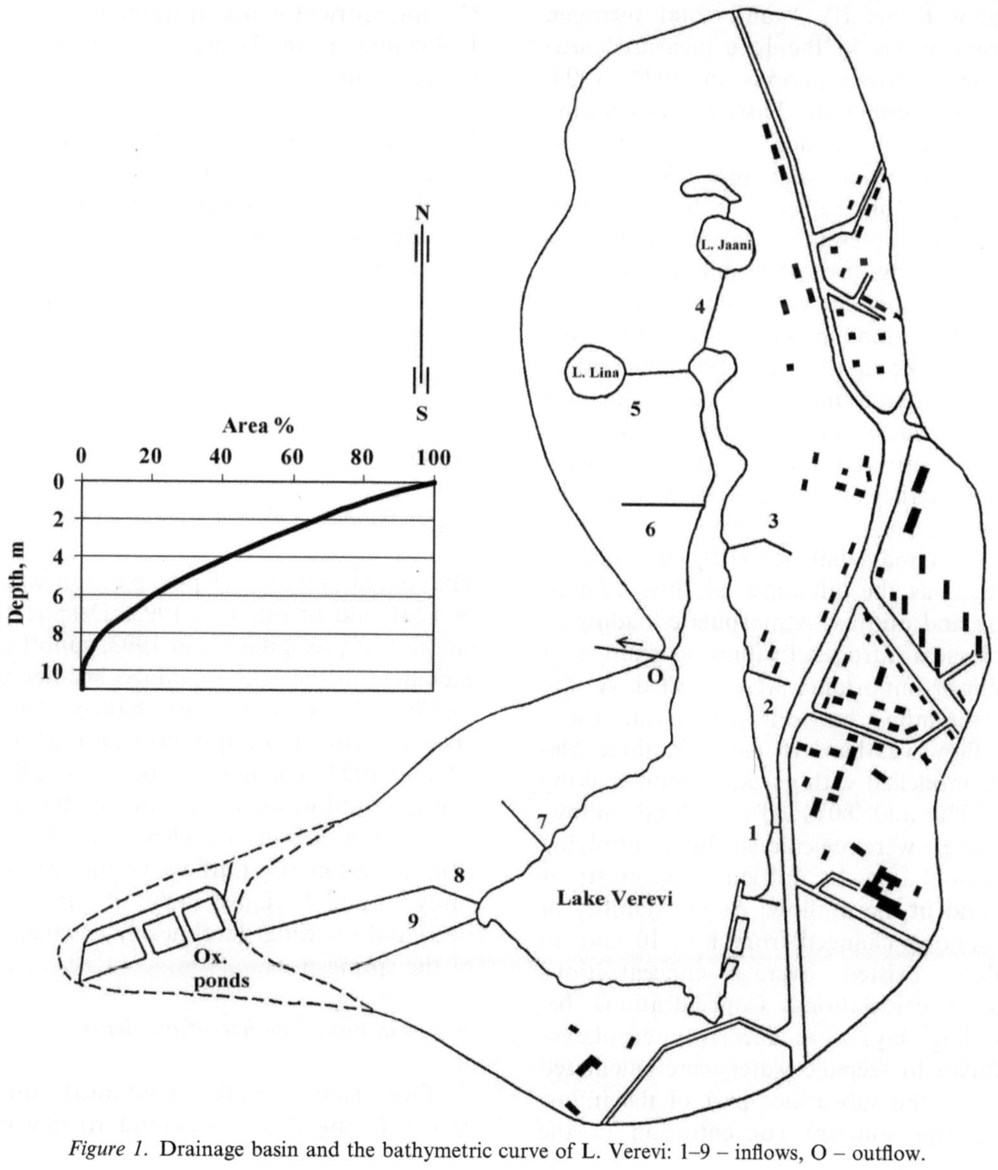

Figure 1. Drainage basin and the bathymetric curve of L. Verevi: 1–9 – inflows, O – outflow.

step, i.e. the lake volume at day 2 was calculated by adding the inflow and precipitation of day 1 to the initial volume and by subtracting from the sum the evaporation and outflow. The relationship between the lake volume and water level was used to calculate the new head and the corresponding outflow for day 2. The amplitude of water level fluctuation and the interruption of the outflow registered in summer 1993 were the parameters used for model verification.

Table 1. Seasonal distribution of evaporation at the Tiirikoja hydrometeorological station as percentage of the annual sum (after Pärn & Eipre, 1983)

Jan	Feb	Mar	Apr	May	Jun	Jul	Aug	Sep	Oct	Nov	Dec
0	2	5	11	17	21	22	11	6	4	1	0

Total phosphorus (P_{tot}) and total nitrogen (N_{tot}) concentrations in the lake measured seasonally during ice-free periods in 1991, 1993, 2000 and 2001 formed the basis for calculating the internal nutrient balance. Water samples at the deepest point of the lake were taken with a Ruttner or Van Dorn sampler or by a vacuum probe (Ott et al., present issue) from 3 to 8 horizons depending on the development of thermal stratification. Seasonal measurements of nutrient concentrations in the inflows and in the outflow of the lake in 1991 and 1993 enabled to calculate the external nutrient balance for ice-free periods of those years. Samples taken from four dug wells around the lake in 1991 were used to characterise the nutrient content of the groundwater.

The external mass balance (EB) of nutrients was calculated as the difference of their content in the influx and outflux. Atmospheric loading as well as biological nitrogen fixation, as sources of probably minor importance were omitted. As the division of the inflow between surface runoff and subsurface flow was hard to measure, three scenarios were modelled with surface runoff making up 20%, 50% and 80% of the total inflow. Nutrient fluxes were calculated by multiplying daily water discharges by nutrient concentrations in inflows and in the outflow. As the number of detectable inflows changed from 1 to 10 and no major inflows existed, average concentrations were used in calculations. Concentrations between sampling days were linearly interpolated. Nutrient fluxes in seepage water were calculated as products of the subsurface part of the inflow and the average nutrient concentration in the wells.

Total nutrient contents in the epilimnion, in the non-mixed layer, and in the whole water column were calculated for 0.5-m water layers as products of the layer volume and nutrient concentration. Concentrations between sampling depths were linearly interpolated. The internal mass balance (IB) that shows changes in total amounts of nutrients in the whole water column was also calculated on a daily basis but used only as an intermediate result for calculating the net balance:

$$\mathrm{IB} = C_n \cdot V_n - C_{n-1} \cdot V_{n-1} \tag{3}$$

in which

C – the nutrient concentration;
V – volume of the layer;
n – day number.

To enable an insight to internal processes, the effect of the external balance should be eliminated. The net balance (NB) which shows the intensity of resultant fluxes at the sediment-water and water-atmosphere interface was calculated as the difference between internal and external balances:

$$\mathrm{NB} = \mathrm{IB} - \mathrm{EB} \tag{4}$$

Results

Water balance

The annual amount of precipitation was 643 mm in 1991 and 663 mm in 1993. Despite the larger amount of precipitation in 1993, runoff and water exchange in the lake remained smaller compared to 1991 (Table 2). The first half of the year gave 50% of annual precipitation in 1991 while only 32% in 1993. The flood peak was smaller in 1993 and the outflow almost ceased by the end of July, and that was reflected also in modelling results (Fig. 2). Soon the outflow continued because of rainy July and August (42% of annual precipitation) and the water level nearly reached the height of the spring peak.

External mass balance of nutrients

Surface runoff water contained on average 183 ± 63 mg P m^{-3} (±standard deviation) and

Table 2. Annual water balance components and characteristics of water exchange in L. Verevi in the years 1991 and 1993

Balance component or characteristic	1991	1993
Inflow (surface + subsurface), (m^3)	328,048	224,884
Precipitation onto lake surface (m^3)	80,993	83,639
Outflow (m^3)	283,944	203,302
Evaporation from lake surface (m^3)	77,540	77,540
Accumulation (m^3)	47,557	27,681
Water level amplitude (m)	0.77	0.51
Mean specific runoff (l km^{-2} s^{-1})	10.7	7.3
Water exchange (y^{-1})	0.90	0.67

2845 ± 930 mg N m^{-3}. Groundwater measured in dug wells had a low phosphorus content (20 ± 8 mg P m^{-3}), while nitrogen concentrations (4240 ± 1984 mg N m^{-3}) were even higher than in ditches. Therefore the combination of surface and subsurface runoff in the model in different proportions yielded different nutrient loads to the lake.

The mass balance of nutrients showed that the influx of nitrogen exceeded the outflux at any supposed proportion of surface runoff (Table 3). The lake retained 45–90% of the nitrogen influx by sedimentation and/or denitrification. The calculated phosphorus retention remained positive only in the case of surface runoff being over 75% in 1991 and over 38% in 1993.

The seasonal dynamics of daily external balances were rather similar at any supposed surface runoff proportion (Fig. 3). The daily influx of P_{tot} was 0.01–0.17 kg and the outflux 0–0.38 kg d^{-1}. Nitrogen fluxes differed more: influx 1–21 kg d^{-1} and outflux 0–2 kg d^{-1}. More than half of the phosphorus EB variability was explained by the P_{tot} concentration in the outflow (R^2 = 0.65) and 22% by the P_{tot} concentration in the inflow. Water inflow and outflow had almost no effect on the dynamics of the phosphorus balance (R^2 = 0.002 and 0.004, correspondingly). The dynamics of the nitrogen balance, on the contrary, was mostly explained by the water inflow (R^2 = 0.39) but also by the N_{tot} concentration of the inflow (R^2 = 0.19) while the water outflow and nitrogen concentration in it explained only 1% and 6%, respectively.

Internal and net balance of substances

The total amount of nutrients contained in the epilimnion, in the non-mixed layer and in the whole water column exhibited remarkable changes during the 4 years of measurements (Table 4). Seasonally the depth of the mixed layer increased from 2–3 m in April to 4–6 m in October (Fig. 4). In most years the lake stratified shortly after ice break-up and was not mixed totally even during the short homothermal period (Ott et al., (a) present issue). Deepening of the mixed layer compensated the sedimentation and denitrification losses. As a result, the total amount of nutrients in the mixed layer remained rather stable during summer and increased rapidly in autumn (Fig. 5a, b). The seasonal decrease of the non-mixed volume

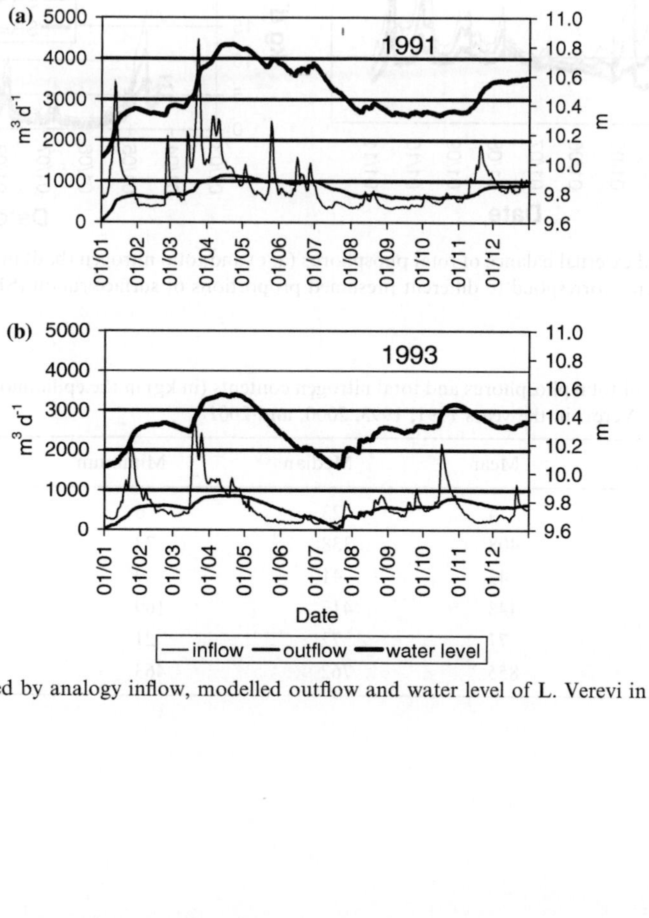

Figure 2. Calculated by analogy inflow, modelled outflow and water level of L. Verevi in 1991 and 1993.

Table 3. Mass balance of total nitrogen (N_{tot}) and total phosphorus (P_{tot}) of L. Verevi from April to October 1991 and 1993 in the case of different presumed proportions of surface inflow

Period	Surface inflow % of total	Influx		Outflux		Retention			
		P_{tot} kg	N_{tot} kg	P_{tot} kg	N_{tot} kg	P_{tot} kg	N_{tot} kg	P_{tot} %	N_{tot} %
Apr-Oct 1991	80	21	497	20	272	1	224	5	45
	50	14	573	20	272	−6	301	−40	52
	20	8	649	20	272	−12	377	−161	58
Apr-Oct 1993	80	16	1391	8	146	7	1245	47	90
	50	11	1062	8	146	2	916	22	86
	20	6	732	8	146	−3	587	−45	80

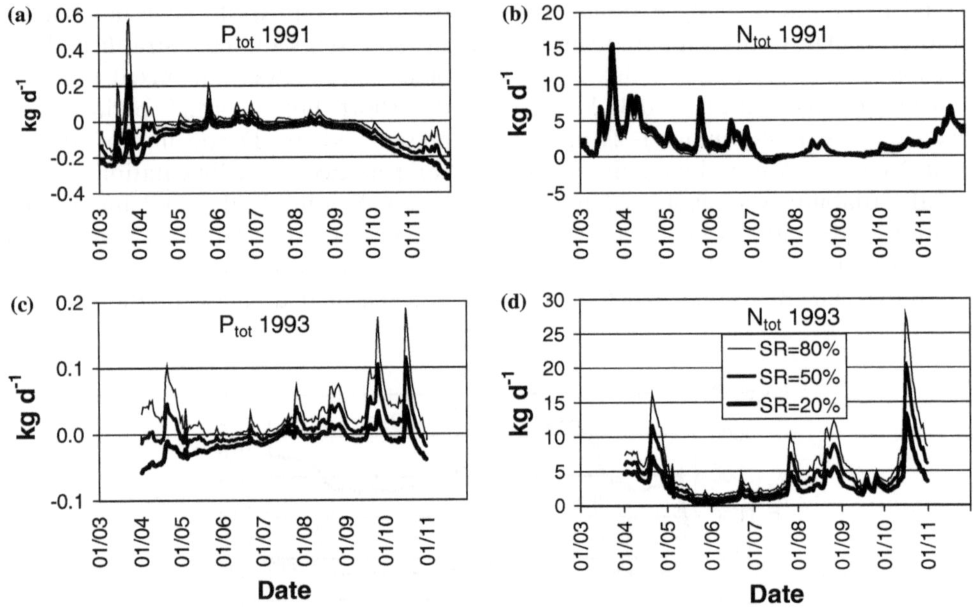

Figure 3. Daily step calculated external balance of total phosphorus (a, c) and total nitrogen (b, d) in L. Verevi for the years 1991 and 1993. The three curves in figures correspond to different presumed proportions of surface runoff (SR) in the inflow.

Table 4. Descriptive statistics of total phosphorus and total nitrogen contents (in kg) in the epilimnion, in the non-mixed layer, and in the whole water column of L. Verevi in the years 1991, 1993, 2000, and 2001

Layer	Valid N	Mean	Median	Minimum	Maximum	SD
$P_{tot\ epi}$	29	34	23	6	174	40
$N_{tot\ epi}$	29	408	338	72	1196	244
$P_{tot\ hypo}$	29	43	41	9	128	26
$N_{tot\ hypo}$	29	448	417	169	994	191
$P_{tot\ col}$	29	77	72	21	223	47
$N_{tot\ col}$	29	855	762	463	1365	278

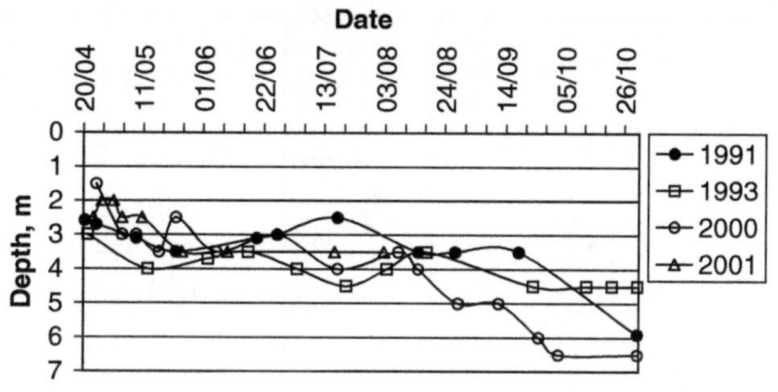

Figure 4. Seasonal changes in the thickness of mixed layer in L. Verevi.

was partly compensated by nutrient release from the sediment so that the absolute amount of nutrients was rather stable (in case of phosphorus; Fig. 5c) or showed a decreasing trend (in case of nitrogen; Fig. 5d). The total amount of phosphorus in the water column varied seasonally by a factor of 11 and was the smallest in June and the largest during autumn overturn in October–November (Fig. 5e). The nitrogen content was also the smallest in June, but the seasonal variation was smaller, only by a factor of 3. The nitrogen amount in water was nearly the same in March and in October–November (Fig. 5f).

The net balance of phosphorus (Fig. 6a, c) showed a vernal peak in April or May, followed by sedimentation during the following months (75 kg P in 1991 and 125 kg P in 1993). During a positive phase until November, phosphorus release from the sediment exceeded sedimentation by 205 kg in 1991 and by 79 kg in 1993. The net balance of nitrogen was mostly negative (Fig. 6b, d). The largest nitrogen losses with loss rates more than 10 kg N d^{-1} occurred in May and June and were followed by periods of positive net balance.

Differences between the 1990s and 2000s

A comparison of nutrient amounts in L. Verevi in the beginning of the 1990s and in the 2000s showed a statistically significant decrease in the epilimnetic nutrient pool (Table 5). The increase in the nutrient content of the non-mixed layer as well as the decrease in total amounts of nutrients in the lake remained statistically insignificant due to the small amount of data.

Discussion

The affluence of a year can be evaluated on the basis of specific runoff. In a wet year the average specific runoff for watersheds in Estonia is about 15 l km^{-2} s^{-1}, in a medium year about 10 l km^{-2} s^{-1}, and in a dry year around 5 l km^{-2} s^{-1} (Loopmann, 1984). According to this scale the year 1991 could be considered medium while 1993 was rather nearly dry despite the larger annual amount of precipitation.

Traditionally, most of the water in rivers and lakes has been considered to be a result of surface runoff. Research using environmental tracers has shown that non-channelised surface runoff is extremely rare in many environments and most of the water in rivers and lakes originates from subsurface runoff or groundwater (Vanek, 1987). Which of the sources, groundwater or surface water, dominates in lake hydrology is directly difficult to determine. An indirect way is to identify the hydrologic type of the lake according to the scheme proposed by Eilers et al. (1983):

Type I: Precipitation dominated lakes with no permanent inlets or outlets. Groundwater inflow is usually very small or moderate. Lakes are often of recharge type and exhibit a very low watershed-to-lake surface area ratio.

Type II: Precipitation/groundwater dominated lakes having intermittent outlet only. Groundwater inflow may be moderately strong, at least during some periods of the year. The switching between flow-through and recharge type may be common.

Type III: Mostly surface runoff dominated lakes with permanent inlet and outlet

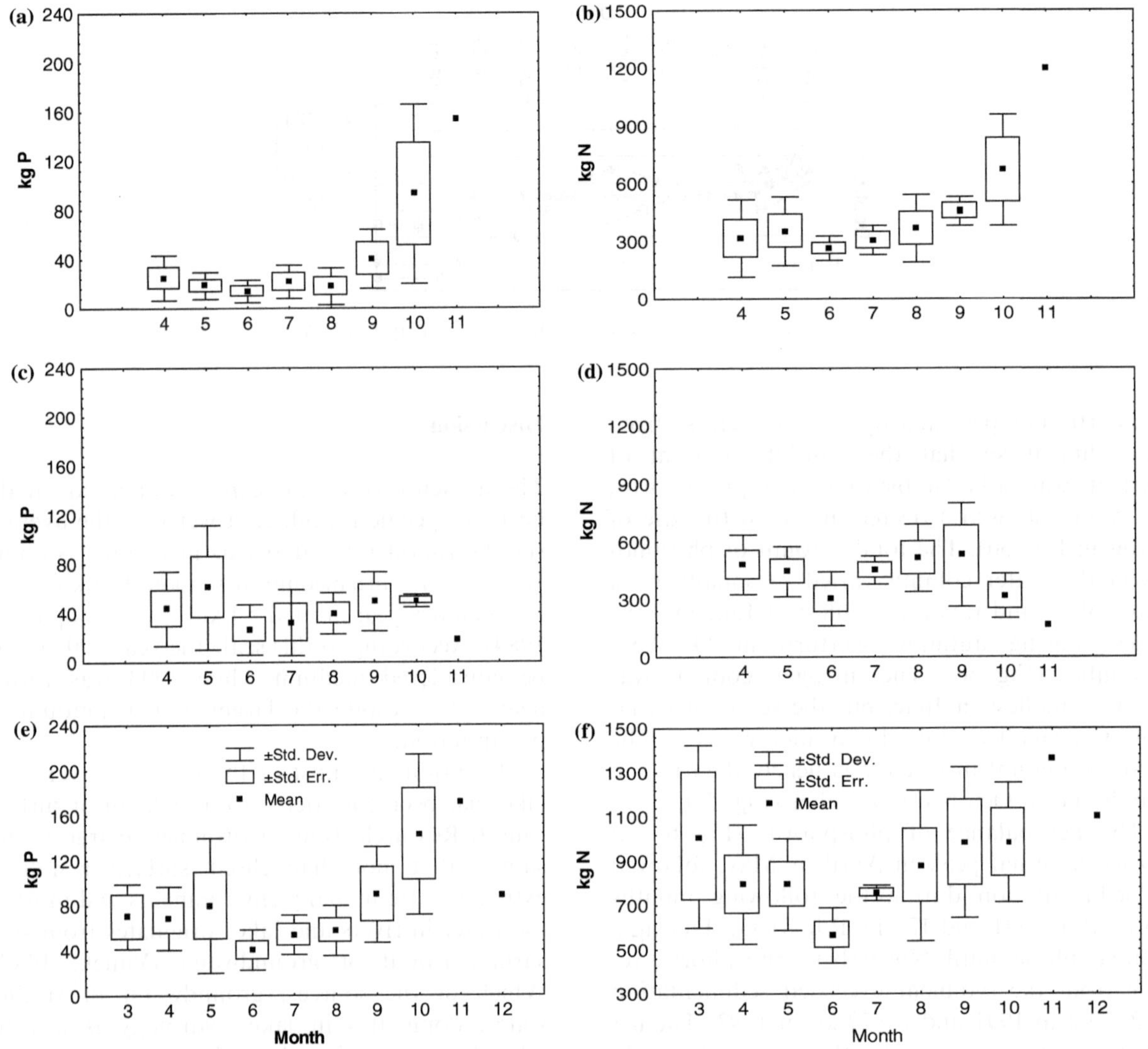

Figure 5. Seasonal changes in absolute amounts of total phosphorus and total nitrogen in the epilimnion (a, b), in the non-mixed layer (c, d) and in the whole water mass of L. Verevi in the years 1991, 1993, 2000, and 2001.

reflecting relatively strong surface water through-flow. Watershed-to-lake area ratio is usually high and the relative importance of groundwater flow is low to moderate.

Type IV: Groundwater-dominated lakes with permanent outlet but no inlet.

Lake Verevi, which is characterised by a changing number of small inlets (Mäemets & Ennok, 1991, Mäemets et al., 1991), absence of surface inflow during some periods (Riikoja, 1940), underwater and shoreline springs (Mäemets et al., 1991; Laugaste, 1994), and intermittent outflow (Riikoja, 1940; Mäemets et al., 1991), cannot be directly classified to any of these types but is closest to Type II. The existence of surface inflows add some features of Type III. Hence, L. Verevi can be considered to be intermediate between precipitation/groundwater dominated and surface runoff dominated lakes and the application of a 50:50 proportion of surface-to-groundwater inflow might be close to the reality.

It has been suggested by Boström et al. (1982) that over an annual cycle there is a net deposit of P in the sediment also in most eutrophic lakes. Supposing the 50:50 ratio between surface-

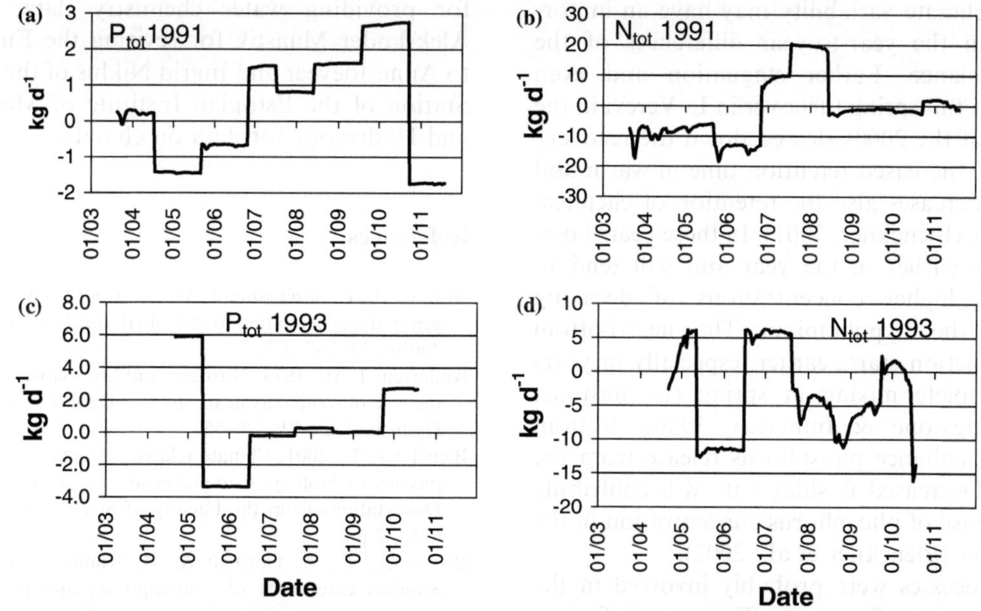

Figure 6. Daily step calculated net balance of total phosphorus (a, c) and total nitrogen (b, d) in L. Verevi for the years 1991 and 1993 assuming a 50:50% proportion of surface and groundwater inflow.

Table 5. Changes in the absolute content of total phosphorus and total nitrogen in the epilimnion, in the non-mixed layer, and in the whole water column of L. Verevi from 1991/93 to 2000/01 (April–June)

Layer	Mean 1991, 1993 (kg)	Mean 2000–2001 (kg)	Change	Significance *p*
$P_{tot\ epi}$	28	10	decrease 2.8 times	0.007
$N_{tot\ epi}$	412	208	decrease 2.0 times	0.009
$P_{tot\ non\text{-}mixhypo}$	39	48	increase .3 times	0.673
$N_{tot\ hyponon\text{-}mix}$	382	436	increase 1.1 times	0.562
$P_{tot\ col}$	67	58	decrease 1.2 times	0.717
$N_{tot\ col}$	793	644	decrease 1.2 times	0.260

to-groundwater inflows, the model showed a retention of 22% of P_{tot} influx in 1993 but in 1991 the balance was negative, i.e. more phosphorus was flushed out from the lake than discharged to the lake. Periods of negative external balance coincided clearly with mixing periods in the lake in spring and autumn when large amounts of phosphorus were transported to the surface layer where its concentration exceeded that in the inflow. As the balance was calculated only from April to October, the real annual retention of phosphorus might have been negative even in 1993. After onset of thermal stratification, phosphorus settles rapidly out of the epilimnion (Ott, et al., (b) present issue) and is no more flushed out of the lake. Consequently, the external balance turns positive during periods of stagnation. The mostly negative annual balance in the 1990s shows the recovery of the lake from an earlier larger phosphorus load. As shown by earlier investigators (Mäemets & Ennok, 1991; Laugaste, 1994), phosphorus in L. Verevi originates probably from the oxidation ponds of the town Elva, which were in operation between 1978 and 1987. During high floods, like in 1984 and 1989, overflow from the ponds was discharged to the lake.

Besides the decrease of anthropogenic nutrient loading resulting from the applied rehabilitation measures, climate variability may have an important role in the year-to-year differences of the nutrient balance. Earlier stagnation and even missing of a full spring turnover in L. Verevi in the beginning of the 2000s slowed down the recovery of the lake. Increased retention time in warm and dry years increases also the retention of chemical constituents (Blenckner, 2001). In these years lakes will stratify earlier in the year and will tend to accumulate higher concentrations of decaying matter in the hypolimnion. The near-bottom oxygen depletion starts earlier, especially in years with incomplete mixing in spring (Livingstone, 1997; Livingstone & Imboden, 1996). Bottom anoxia can enhance phosphorus release from the sediment. Decreased flushing rate will contribute to the increase of phosphorus concentration in the hypolimnion (Blenckner et al., 2002).

Two processes were probably involved in the efficient retention of nitrogen. The period of negative net budget of nitrogen in May and June coincided with periods of phosphorus sedimentation and could therefore be also attributed to sedimentation. However, lakes with anaerobic hypolimnion are also excellent places for bacterial denitrification that occurs even before oxygen is completely removed (Golterman, 1975). Low nitrite and nitrate levels measured in L. Verevi despite high nitrification rate, up to 308 mg N m^{-3} day^{-1} (Tõnno et al., present issue), support the idea of intensive denitrification in the lake. In lakes, a largely variable role of denitrification ranging from 0 to 10% (Andersen, 1974) up to 86–90% of the nitrogen loss rate (Jensen et al., 1992) has been observed. Chen et al. (1972) showed that up to 90% of nitrates added to lake sediments *in vitro* disappeared within 2 h. Denitrification may become especially important in lakes like Verevi, in which a part of nitrates received with groundwater are discharged directly to the anaerobic hypolimnion.

Acknowledgements

This work was supported by the target financed project No 0370208s98 of the Ministry of Education "Influence of stratification on biological matter circulation in lakes" and by grants No 3579 & 4835 of the Estonian Science Foundation. Special thanks to Dr. R. Laugaste and K. Ott for providing water chemistry data, to Prof. Aleksander Maastik for revising the English and to Anne Jõeveer and Ingrid Niklus of the Tõravere Station of the Estonian Institute of Meteorology and Hydrology for data on climate.

References

Ahlgren, I., F. Sörensson, T. Waara & K. Vrede, 1994. Nitrogen budgets in relation to microbial transformations in lakes. Ambio 23: 367–377.

Andersen, J. M., 1974. Nitrogen and phosphorus budgets and the role of sediments in six shallow Danish lakes. Archiv für Hydrobiologie 74: 528–550.

Blenckner, T., 2001. Climate related impacts on a lake. From physics to biology. Comprehensive Summaries of Uppsala Dissertations from the Faculty of Science and Technology 674, 37 pp.

Blenckner, T., A. Omstedt & M. Rummukainen, 2002. A Swedish case study of contemporary and possible future consequences of climate change on lake function. Aquatic Sciences 64: 171–184.

Boström, B., Jansson, M. & C. Forsberg, 1982. Phosphorus release from lake sediments. Archiv für Hydrobiologie Beiheft Ergebnisse der Limnologie 18: 5–59.

Chen, R. L., D. R. Keeney, D. A. Graetz & A. J. Holding, 1972. Denitrification and nitrate reduction in Wisconsin lake sediments. Journal of Environmental Quality 1: 158–162.

Dudel, G. & J.-G. Kohl, 1991. Contribution of dinitrogen fixation and denitrification to the N-budget of a shallow lake. Verhandlungen der Internationalen Vereinigung für Theoretische und Angewandte Limnologie 24: 884–888.

Eilers, J. M., G. E. Glass, K. E. Webster & J. A. Rogalla, 1983. Hydrologic control of lake susceptibility to acidification. Canadian Journal of Fisheries and Aquatic Sciences 40: 1896–1904.

Eloranta, P., 1992. Limnologian perusteet. Luentorunko, Helsinkin yliopisto, Helsinki, 190 pp.

Golterman, H. L., 1975. Physiological limnology. An approach to the physiology of lake ecosystems. Elsevier, Amsterdam, Oxford, New York, 49 pp.

Jensen, J. P., P. Kristensen & E. Jeppesen, 1990. Relationships between nitrogen loading and in-lake nitrogen concentrations in shallow Danish lakes. Verhandlungen der Internationalen Vereinigung für Theoretische und Angewandte Limnologie 24: 201–204.

Jensen, J. P., E. Jeppesen, P. Kristensen, P. B. Christensen & M. Søndergaard, 1992. Nitrogen loss and denitrification as studied in relation to reductions in nitrogen loading in a shallow, hypertrophic lake (Lake Søbygård, Denmark). Internationale Revue der gesamten Hydrobiologie 77: 29–42.

Jørgensen, S. E. & R. A. Vollenweider (eds), 1988. Principles of lake management. Guidelines of lake management. Vol. 1. ILEC, UNEP: 199 pp.

Laugaste, R., 1994. Verevi järve seisund, biogeensete ainete päritolu ja tervistamise abinõud. In Järvekülg, A. (ed.)Eesti jõgede ja järvede seisund ning kaitse. Teaduste Akadeemia

Kirjastus, Tallinn, 47–64 [The state, the origin of nutrients and the measures necessary for recovering Lake Verevi. In Estonian].

Livingstone, D. M., 1997. An example of the simultaneous occurrence of climate-driven 'sawtoth' deep-water warming/ cooling episodes in several Swiss lakes. Verhandlungen der Internationalen Vereinigung für Theoretische und Angewandte Limnologie 26: 822–826.

Livingstone, D. M. & D. M. Imboden, 1996. The prediction of hypolimnetic oxygen profiles: a plea for a deductive approach. Canadian Journal of Fisheries and Aquatic Sciences 53: 924–932.

Loopmann, A., 1984. Suuremate Eesti järvede morfomeetrilised and med ja veevahetus. Tallinn, 150 lk. [Morphometrical data and water exchange of larger Estonian lakes. In Estonian].

Löfgren, S., 1987. Phosphorus retention in sediments – implications for aerobic phosphorus release in shallow lakes. Acta Universitatis Upsaliensis. Comprehensive Summaries of Uppsala Dissertations from the Faculty of Science 100. 24 pp.

Mäemets, A., 1991. Füüsilis-geograafiline iseloomustus. In Timm, T. (ed.). Verevi järve seisund, Tartu, 13–19 [Physical-geographical characterization. In Estonian].

Mäemets, A. & K. Ennok, 1991. Valgla iseloom, sissevoolude vee keemiline koostis ja järve reostuskoormus. In Timm, T. (ed.) Verevi järve seisund, Tartu: 34–43 [Catchment features, chemical composition of water of inflows and pollution loading of the lake. In Estonian].

Mäemets, A. & E. L. Rembel Ainsalu, 1991. Sissevoolud ja veevahetus. In Timm, T. (ed.) Verevi järve seisund, Tartu: 19–22 [Inflows and water exchange. In Estonian].

Nõges, P. & A. Kisand, 1999. Forms and mobility of sediment phosphorus in shallow eutrophic Lake Võrtsjärv (Estonia). Internationale Revue der gesamten Hydrobiologie 84: 255–270.

Nõges, T. & K. Kangro, 2005. Primary production of phytoplankton in a strongly stratified temperate lake. Hydrobiologia 547: 105–122.

Ott, I., T. Kõiv, P. Nõges, A. Kisand, A. Järvalt & E. Kirt, 2005a. General description of partly meromictic hypertrophic Lake Verevi, its ecological status, changes during the past eight decades and restoration problems. Hydrobiologia 547: 1–20.

Ott, I., A. Rakko, D. Sarik, P. Nõges & K. Ott, 2005b. Sedimentation rate of seston during the formation of temperature stratification after ice break-up in the partly meromictic Lake Verevi. Hydrobiologia 547: 51–61.

Pärn & Eipre, 1983. Klimaticheskie usloviya ozera. In Sokolov, A. A. (ed.)Chudsko-Pskovkoe ozero. Gidrometeoizdat, Leningrad: 27–41 [Climatic conditions of the lake. In Russian].

Riikoja, H., 1940. Zur Kenntnis einiger Seen Ost-Eestis, insbesondere ihrer Wasserchemie. Publications of the Limnological Laboratory of the Natural Resources Research Institute of Estonia 2: 1–167.

Tõnno, I., K. Ott & T. Nõges, 2005. Nitrogen dynamics in steeply stratified temperate Lake Verevi, Estonia. Hydrobiologia 547: 63–71.

Vanek, V., 1987. The interactions between lake and groundwater and their ecological significance. Stygologia 3: 1–23.

Hydrobiologia (2005) 547:33–39
I. Ott & T. Kõiv (eds), Lake Verevi, Estonia – A Highly Stratified Hypertrophic Lake
DOI 10.1007/s10750-005-4141-2

Distribution of sediment phosphorus fractions in hypertrophic strongly stratified Lake Verevi

Anu Kisand
Võrtsjärv Limnological Station, Institute of Zoology and Botany, Estonian Agricultural University, 61101 Rannu, Tartu County, Estonia
E-mail: akisand@zbi.ee

Key words: sediment, phosphorus fractionation, internal loading, Estonia

Abstract

Lake Verevi is a hypertrophic and strongly stratified (partly meromictic) small temperate lake. Vertical distribution of sediment phosphorus fractions as well as iron, manganese, organic matter and calcium carbonate of the deep bottom sediment was determined. The study focused on the ecologically important layer of the sediment [<20(45) cm]. In the uppermost layers of the sediment, NaOH-NRP (organic P) dominated while HCl-RP (apatite-P) became dominant in some deeper layers below 7 cm. Extremely high concentrations of labile phosphorus fraction (NH_4Cl-RP) indicated the low binding capacity of phosphorus by lake sediment. Due to sediment and hypolimnion anoxia, the internal load of phosphorus in this lake is most likely. Potentially mobile phosphorus fractions (NH_4Cl-RP, BD-RP, NaOH-NRP) formed 301 kg in upper 10 cm thick sediment layer of hypolimnetic bottom sediment (40% of lake bottom area).

Introduction

Phosphorus is the key element of eutrophication processes in many lakes. Phosphorus supply to the euphotic zone depends on the external load and also on the tendency of sediments to retain or release phosphorus. The intensity of water mixing and stratification play an important role in this process. Stratification may lead to anoxic conditions, lowered pH and accumulation of nutrients in both hypolimnion and surface sediments of eutrophic lakes.

Total phosphorus concentration is a poor measure of the potential phosphorus release from sediments. Chemical fractionation of sedimentary phosphorus has served as a tool to predict the phosphorus binding capacity of sediments under different environmental conditions. In most cases, the potential internal load is the matter of interest.

The aim of present study was to determine the distribution of phosphorus fractions and other chemical properties of surface sediments in hypertrophic strongly stratified and partly meromictic Lake Verevi. The special emphasis of the study was focused on the upper 20 cm of sediment, which could take part in nutrient exchange between the water column and bottom.

Study site

Hypertrophic hard-water Lake Verevi is situated in South Estonia, within the borders of town Elva. General description of the ecosystem as well as detailed information on the location and bathymetry of the lake is presented by Ott et al., (2005a, 2005b). The lake has a surface area of 12.6 ha, maximum depth of 11 m, mean depth of 3.6 m and theoretical water residence time on an average 0.5 times per year. The deepest part of the lake is situated in the middle of the southern part of the lake while the narrow northern part is shallow and largely covered by macrophytes. The lake is strongly stratified and during winter covered by ice. In early warm springs, the stratification may be formed so rapidly that the spring turnover is absent. The lake is sheltered from winds and no marked water mixing due to wave action occurs.

The lake, especially its shallow northern part is rich in springs. The upper layer of sediment in the northern part is soft greyish-green calcareous mud. In the southern part, the upper 30–50 cm thick sediment layer consists of H_2S-rich greenish-black mud, apparently being formed in anoxic conditions. In 1989, the phosphorus content of the surface sediment of Lake Verevi was determined with the depth resolution of 20 cm (Rummi et al., 1991). The lowest phosphorus value of 0.4 mg g^{-1} of dry sediment (DW) has been reported from the northern part of the lake, the highest concentration from the middle area of the southern part was 1.4 mg g^{-1} DW.

Materials and methods

Field work

Sediment samples were collected using Willner core sampler in four occasions: August 1994 (one core, length 45 cm), March 2001 (two cores, 15 and 25 cm long), June 2001 (two cores, both 4 cm long) and August 2001 (both cores 15 cm long). In March and June 2001, samples were collected from the eastern part of the lake's deepest bottom area at a depth of 6–8 m. In 1994 and August 2001, samples were collected close to the deepest point of the lake at the water depth of 9 m. Parallel cores were collected within the distance of 20 m.

The cores were sliced immediately after sampling. The slicing intervals were following: 0–2 cm, 2–5 cm, 5–10 cm in 1994, 0–2 cm, 2–5 cm, 5–7 cm, 7–10 cm in March 2001 and 0–1 cm, 1–2 cm, 2–3 cm, 3–4 cm, 4–5 cm, 5–7 cm, 7–10 cm in June and August 2001; deeper than 10 cm, all cores were sliced with 5 cm intervals. The samples were kept in closed plastic bags in an icebox until they could be stored at 4 °C (normally within 1 h after sampling). The laboratory analyses began 1 week after sampling in 1994 and on the following day in 2001.

Laboratory analyses

The concentrations of dry matter and phosphorus fractions in the sediment were analysed. Additionally, in 1994, the concentrations of organic matter, carbonates, iron, manganese and total phosphorus were measured.

The concentration of dry matter of the sediment was calculated from weight difference before and after drying of triplicate samples at 105 °C for 24 h. The bulk density of the sediment was calculated according to Håkanson & Jansson (1983), enabling to determine the phosphorus amount in certain sediment volume.

The concentration of organic matter of the sediment was determined by the loss of weight during ignition at 550 °C for 2 h. Ignition residue was further ignited at 825 °C for 4 h. The loss of weight was ascribed to the emission of carbon dioxide serving as a basis for calculation of the carbonate concentration.

Manganese and iron concentrations were measured using Atomic Adsorption/ Flame Emission Spectroscopy (AA/FES, Shimadzu AA-670) after digestion of dry sediment in 7 M HNO_3. The content of total phosphorus was determined spectrophotometrically according to Murphy & Riley (1962) after boiling of dry sediment in mixed acids (conc. H_2SO_4, HNO_3, $HClO_4$).

Sediment phosphorus fractionation was performed in triplicates according to Hieltjes & Lijklema (1980) at the first three sampling cases. Sediment was extracted in four following steps with three solutions: NH_4Cl, NaOH and HCl. From these solutions, the concentration of soluble reactive phosphorus was measured according to Murphy & Riley (1962). The following fractions were gained:

1. NH_4Cl-RP (RP = reactive phosphate) – labile, loosely bound or adsorbed P.
2. NaOH-RP represents the phosphates adsorbed to metal (Fe, Al) oxides and other surfaces, exchangeable against OH^-, and phosphorus compounds soluble in bases.
3. NaOH-NRP (NRP = non-reactive phosphate) is calculated as the differences between total P in the NaOH extract, measured by peroxosulphate digestion, and NaOH-RP. It is assumed to represent the major part of organic and humic P.

4. HCl-RP represents phosphorus bound to carbonates, apatite-P and P released by the dissolution of oxides (not adsorbed to the surface).

In August 2001, the modified scheme of Psenner et al. (1988) was used for phosphorus fractionation. The advantage of the latter is the separation of iron-bound (BD-RP) and aluminium-bound phosphorus (NaOH-RP) by including an additional extraction step with reducing agent (buffered dithionite). The original scheme of Psenner et al. (1988) was modified by using 0.1 M NaOH instead of 1.0 M NaOH in the second extraction step for better comparison with the earlier results gained by the scheme of Hieltjes & Lijklema (1980). Results of phosphorus fractions are given as means of three or, in the case of two independent parallel cores, as six values.

Results

Dry and organic matter, carbonates

The content of dry matter (DM) of sediment samples varied from 2.8 to 30% of wet weight (WW). DM was lower at the sampling depth of 9 m and higher in shallower water (at 6–8 m). The maximum values of DM occurred around 2–10 cm below the sediment surface. In all cores, a gradual decrease was observed below peak values in the sediment depth between 7 and 15 cm. The lowest values always appeared in the surface layers of the cores differing up to 3.7 times between the cores (from 2.8 to 10.4% WW). The concentration of organic matter (OM) determined from the core of August 1994, showed increasing values towards the bottom of the core (Fig. 1), while the content of carbonates had the opposite vertical distribution.

Metals and total phosphorus

The concentration of Fe and Mn in the upper 40 cm of the sediment was 865–1448 and 110–520 mg g^{-1} DW, respectively, showing more than three-fold decrease from the surface layer towards the bottom of the core (Fig. 1). The concentration of total phosphorus reached the value of 1.5 mg g^{-1} DW in the sediment depth of 5–10 cm and declined more than 1.5 times by the depth of 40 cm. The Fe/P mass ratio was 7.9–18.1 in the core from August 1994 and remained below 10 in the layer of 5–10 cm and deeper than 30 cm.

Phosphorus fractions

Except of June 2001, NaOH-NRP was the largest phosphorus fraction in the surface sediment

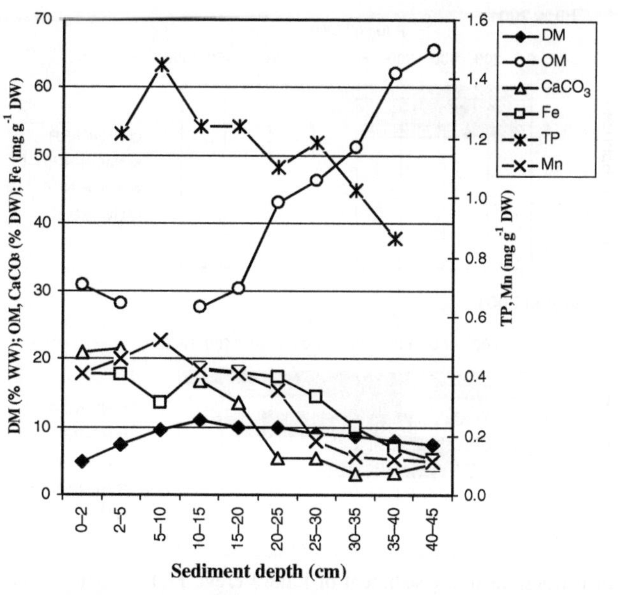

Figure 1. Vertical profiles of total phosphorus (TP), $CaCO_3$, dry weight (DW), organic matter (OM), total iron (Fe) and manganese (Mn) in the sediment of Lake Verevi in August 1994.

(Fig. 2). HCl-RP became dominant in some deeper (>7 cm) sediment layers. NaOH-RP, determined according to the fractionation scheme of Hieltjes & Lijklema, (1980) exceeded the concentration of NH_4Cl-RP. Analysed by the method of Psenner et al. (1988), NaOH-RP was the smallest fraction in upper 7 cm. Its concentration was 1.0–2.6 (average 1.9) times lower than that of BD-RP. Except of August 2001, NH_4Cl-RP was the smallest of all fractions. In August 2001, the extreme concentration of NH_4Cl-RP (526 $\mu g\ g^{-1}$ DW) appeared in the depth of 4–5 cm constituting 39% of the sum of all phosphorus fractions in that sediment layer.

The cores of June 2001, which were sampled from 6–8 m water depth, differed from other cores by comparatively high concentration of dry matter and low concentration of NaOH-NRP. In contrast to other cores, HCl-RP exceeded the NaOH-NRP concentration in surface layers.

The phosphorus content in upper 40 cm of the hypolimnetic bottom sediment was calculated on

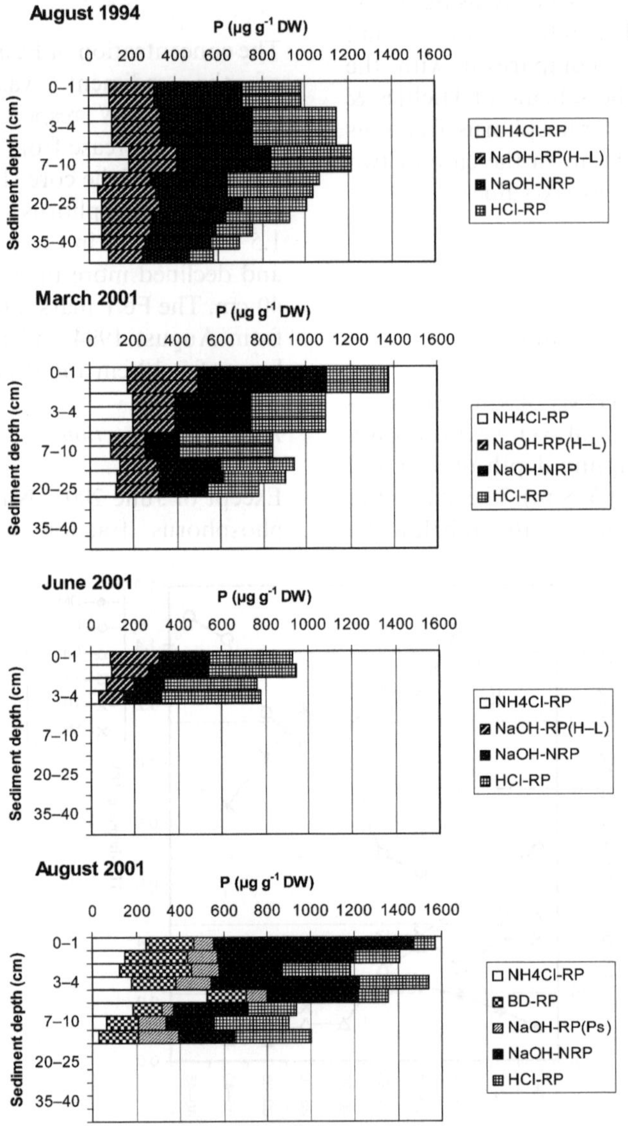

Figure 2. Distribution of phosphorus fractions in the sediment of Lake Verevi. H–L = determined according to Hieltjes & Lijklema (1980); Ps = determined according to modified scheme of Psenner et al. (1988). Intervals of the sediment slicing were different for different cores but fitted together with the smallest possible interval for comparison (see Materials and methods).

the basis of the data from August 1994 (Table 1). Due to poor sediment compaction, 5 cm thick surficial sediment layer contained only half of the phosphorus amount present in the 5–10 cm sediment layer. A gradual decrease of phosphorus amount appeared below 10 cm. Residual phosphorus, i.e. the calculated difference between the sum of measured phosphorus fractions and total phosphorus, constituted 0–10.7% (average 3.8%) of total phosphorus.

Discussion

Lake Verevi is highly eutrophic despite of the moderate external phosphorus load. The significant rise in phosphorus concentrations in hypolimnion during stagnation periods indicates that such a high content of that nutrient in the lake water is probably maintained by the phosphorus supply from the sediments (Nõges, 2005).

The total phosphorus content of the sediment of Lake Verevi was relatively low, compared to some other stratified and eutrophic Estonian lakes such as Lake Arbi (average for upper 35 cm 4.04 mg g^{-1}), Lake Martiska (average 2.6 mg g^{-1}) (unpublished data) and Lake Ruusmäe (average 5.2 mg g^{-1}) (Kruusement & Punning, 2000). Sediment phosphorus concentration of Lake Verevi was similar to shallow and eutrophic lakes such as Lake Võrtsjärv (average for upper 50 cm 0.78 mg g^{-1}) (Nõges & Kisand, 1999).

Total phosphorus content of lake sediment is not well correlated with the amount of released phosphorus, and the potentially mobile phosphorus in sediments is poorly defined. However, easily degradable organic phosphorus can be regarded as potentially mobile, and phosphorus adsorbed to iron surfaces is highly available during anaerobic conditions (Pettersson, 1998). In anoxic conditions, Fe^{3+} is reduced to Fe^{2+}. As a result, both iron and adsorbed phosphorus are transferred into soluble form (Boström et al., 1982). Using the fractionation scheme of Hieltjes & Lijklema (1980), phosphorus adsorbed to iron is included in NaOH-RP. The scheme of Psenner et al. (1988) separates this fraction into BD-RP (phosphorus adsorbed to iron) and NaOH-RP (Al-bound P). In the sediment of Lake Verevi BD-RP formed larger fraction than NaOH-RP.

The anoxic conditions and presence of H_2S in the hypolimnion of Lake Verevi suggest iron to occur in reduced form since sulphate reduction takes place at lower redox potential than the reduction of iron (Boström et al., 1982). In addition, in a reaction between Fe^{2+} and S^{2-} FeS can be formed which is highly a insoluble compound in natural conditions and hence inactivates iron. The mass ratio between active iron and phosphorus may be critical for phosphorus release if the sediment is oxidized or shifts from reduced conditions to oxidized status. This can be important for the sediment with changing oxidized or reduced conditions at the border of the hypolimnetic bottom area. Jensen et al. (1992) suggested, that a Fe/P ratio by weight of 15 in oxidized sediment would be able to control release of SRP (soluble reactive phosphorus) to the overlying water column if the pH of the sediment surface is less than 8.0. In the sediment core sampled from

Table 1. The contents of total phosphorus and phosphorus fractions (kg) in different sediment layers of the hypolimnetic bottom (40% of the lake bottom area)

	NH_4Cl-RP	NaOH-RP	NaOH-NRP	HCl-RP	TP	TP, cumulative
0–2 cm	5	10	20	14	49	49
2–5 cm	12	26	49	45	140	189
5–10 cm	46	55	109	93	362	551
10–15 cm	20	63	103	124	361	912
15–20 cm	12	69	87	105	327	1239
20–25 cm	15	68	99	79	285	1524
25–30 cm	14	54	80	68	277	1802
30–35 cm	14	50	67	38	228	2030
35–40 cm	12	41	63	27	178	2207

hypolimnetic bottom of Lake Verevi in August 1994 most Fe/P values remained below 15.

NaOH-NRP has been shown to be a potentially mobile phosphorus fraction (Rydin, 2000). This fraction is traditionally considered to consist of mainly organically bound phosphorus. Hupfer & Rübe (2004) showed in their study with many different lake sediments that up to 46% of the NaOH-NRP consisted of intracellular inorganic polyphosphates of microorganisms. The transformation of organic phosphorus compounds and polyphosphates can contribute significantly to the release of phosphorus during early diagenesis.

NH_4Cl-RP includes dissolved phosphates of sediment pore water. This phosphorus pool is maintained by the dissolution of particulate sediment phosphorus while diffusion into lake water takes place simultaneously to compensate the concentration gradient. High concentrations of NH_4Cl-RP at the depth of 4–5 cm in cores of August 2001 could refer to the extensive phosphorus saturation of the sediment. This stage could be favoured by seasonal changes such as enhanced sedimentation of organic matter and its intensive degradation due to high summer temperatures. However, somewhat higher concentration variations of all phosphorus fractions could probably be detected in previous sampling cases, too, if a more detailed depth resolution was used in surficial sediment layers.

NH_4Cl-RP, NaOH-NRP and BD-RP can be considered potentially available phosphorus fractions in Lake Verevi, i.e. a part of these fractions can be released into lake water under appropriate conditions. Referring to Table 1 and considering, that approximately two third of NaOH-RP (measured according to Hieltjes & Lijklema) consists of BD-RP, potentially mobile fractions form 301 kg in upper 10 cm thick sediment layer of hypolimnetic bottom sediment (40% of lake bottom area). The actual share of phosphorus to be released cannot be determined on the basis of phosphorus fractions.

Lake Verevi is stratified during most of the year. Phosphorus, released from sediments, accumulates in hypolimnion, often resulting in more than ten-fold concentration of this nutrient compared to that in epilimnion (Ott et al., 2005a). Since water mixing by wave action does not play an important role in Lake Verevi, phosphorus can spread into upper water column only during water circulation in spring and autumn. These events are followed by algal blooms (Kõiv & Kangro, 2005). Some of the phosphorus is likely to be removed from water column by macrophytes as Lake Verevi is relatively small by area, and, therefore, relatively significant phosphorus uptake by littoral zone vegetation could be suspected. After decaying of planktonic organisms, phosphorus is released both in water column and in sediments. Pettersson (1998) studied the phosphorus content in suspended matter, settling particles and sediment of stratified eutrophic Lake Erken, and showed that phosphorus rich material in the water releases phosphorus on its way down to the sediment, and that a further release takes place from the sediment. Probably it was the case for Lake Verevi too, the most part of organic material trapped into epi-, meta- and hypolimnion and was degraded before it ever reached to the sediments (Ott et al., 2005b). The remaining material reaches sediments and partially returns to the water column again.

Acknowledgements

This work was supported by target financed project of the Ministry of Education 'The influence of the stratification to the biological matter circulation of the lakes' No 0370208s98 and by Grants of Estonian Science Foundation No 3579 & 4835. Author appreciates assistance of Meeli Galuzo, Atko Heinsalu and Ingmar Ott.

References

Boström, B., M. Jansson & C. Forsberg, 1982. Phosphorus release from lake sediments. Archiv für Hydrobiologie – Advances in Limnology 18: 5–59.

Håkanson, L. & M. Jansson, 1983. Principles of Lake Sedimentology. Springer-Verlag, Berlin.

Hieltjes, A. H. M. & L. Lijklema, 1980. Fractionation of inorganic phosphates in calcareous sediments. Journal of Environmental Quality 9: 405–407.

Hupfer, M & B. Rübe, 2004. Origin and diagenesis of polyphosphate in lake sediments: a ^{31}P-NMR study. Limnology and Oceanography 49: 1–10.

Jensen, H. S., P. Kristensen, E. Jeppesen & A. Skytthe, 1992. Iron:phosphorus ratio in surface sediment as an indicator of phosphate release from aerobic sediments in shallow lakes. Hydrobiologia 235/236: 731–743.

Kruusement, K. & J.-M. Punning, 2000. Distribution of phosphorus in the sediment core of hypertrophic Lake Ruusmäe and some palaeoecological conclusions. Proceedings of the Estonian Academy of Sciences, Biology Ecology 49: 163–176.

Kõiv, T. & K. Kangro, 2005. Resource ratios and phytoplankton species composition in a strongly stratified lake. Hydrobiologia 547: 123–135.

Murphy, J. & J. P. Riley, 1962. A modified single solution method for the determination of phosphate in natural waters. Analytica Chimica Acta 27: 31–36.

Nõges, P., 2005. Water and nutrient mass balance of the partly meromictic temperate Lake Verevi. Hydrobiologia 547: 21–31.

Nõges, P & A. Kisand, 1999. Forms and mobility of sediment phosphorus in shallow eutrophic Lake Võrtsjärv (Estonia). International Review of Hydrobiology 84: 255–270.

Ott, I., T. Kõiv, P. Nõges, A. Kisand, A. Järvalt & E. Kirt, 2005a. General description of partly meromictic hypertrophic Lake Verevi, its ecological status, changes during the past eight decades and restoration problems. Hydrobiologia 547: 1–20.

Ott, I., A. Rakko, D. Sarik, P. Nõges & K. Ott, 2005b. Sedimentation rate of seston during the formation of temperature stratification after ice break-up in the partly meromictic Lake Verevi. Hydrobiologia 547: 51–61.

Pettersson, K., 1998. Mechanisms for internal loading of phosphorus in lakes. Hydrobiologia 373/374: 21–25.

Psenner, R., B. Boström, M. Dinka, K. Pettersson, R. Pucsko & M. Sager, 1988. Fractionation of phosphorus in suspended matter and sediment. Archiv für Hydrobiologie – Advances in Limnology 30: 98–103.

Rummi, P., T. Mägi, J. Ütsi, H. Mäemets, A. Lindpere & A. Mäemets, 1991. Põhjasetted. In: Timm, H. (ed.), State of Lake Verevi. Hydrobiological Researches 17: 22–33. (Bottom sediments. In Estonian).

Rydin, E., 2000. Potentially mobile phosphorus in Lake Erken sediment. Water Research 34: 2037–2042.

Hydrobiologia (2005) 547:41–49
I. Ott & T. Kõiv (eds), Lake Verevi, Estonia – A Highly Stratified Hypertrophic Lake
DOI 10.1007/s10750-005-4142-1

Optical properties and light climate in Lake Verevi

Anu Reinart[1,3,]*, Helgi Arst[2] & Donald C. Pierson[3]
[1]*Tartu Observatory, Tõravere, Tartu County 61602, Estonia*
[2]*Estonian Marine Institute, Tartu University, Mäealuse 10A, 12618, Tallinn, Estonia*
[3]*Department of Limnology, Norbyvägen 20, S-752 36 Uppsala, Sweden*
*(*Author for correspondence: E-mail: anu.reinart@aai.ee)*

Key words: light climate, optical properties, stratified lake

Abstract

The optical properties and light climate during the ice-free period in the highly stratified Lake Verevi (Estonia) have been studied together with other lakes in same region since 1994. The upper water layer above the thermocline belongs to class "moderate" by optical classification of Estonian lakes but can turn "turbid" (concentration of chlorophyll *a* up to 73 mg m^{-3} and total suspended matter up to 13.2 g m^{-3}) during late summer blooms. In the blue part of the spectrum, light is mainly attenuated by dissolved organic matter and in red part notably scattering but also absorption by phytoplanktonic pigments effect the spectral distribution of underwater light. Consequently, the underwater light is of greenish-yellow color (550–650 nm). Rapid change in optical properties occurs with an increase of all optically active substances close to thermocline (2.5–6 m). Optical measurements are often hampered beneath this layer so that modeling of the depth distribution of the diffuse attenuation coefficient is an useful compliment to field measurements. $K_{d,PAR}$ ranges from 0.8 to 2.9 m^{-1} in the surface layer, and model results suggest that it may be up to 5.8 m^{-1} in the optically dense layer. This forms a barrier for light penetration into the hypolimnion.

Introduction

The amount of photosynthetically active radiation (PAR) in water bodies is important in many aspects of aquatic ecology. Light is the principal determinant of depth distribution of living organisms and photochemical processes (Kirk, 1994). The spectral compositions, amount and angular distribution of incident light, scattering and absorbing properties govern the light field in lakes.

Following Secchi depth, PAR is one of the easiest optical properties to measure in the light field. Optical properties describing attenuation and scattering of incident solar light with depth are diffuse attenuation coefficient, K_d and diffuse reflectance, R. They are a composite measure of the effect of suspended particles, phytoplankton pigments and yellow substance on the underwater PAR (Dera, 1992).

In this paper data and results from investigations of the optical properties and light field characteristics in Lake Verevi are summarized. From the optical point of view Lake Verevi is very interesting as there is a strong vertical gradient of optical properties between 2.5 and 6 m depths, depending on the season.

We present here the limits of observed concentrations of optically active substances (OAS) and optical properties. Also the spectral behavior of K_d and R are estimated and the optical characteristics of Lake Verevi are compared with other investigated lakes in the same region. The seasonal change of euphotic depth and percent of irradiance penetrating to water is calculated using modeled diffuse attenuation coefficients.

Material and methods

Measurements of OAS

The marine optics workgroup in the Estonian Marine Institute has visited Lake Verevi among other lakes in Estonia and Finland during the ice-free period of 1994–2001, on 19 occasions. Optical measurements were performed together with researchers from the Department of Geophysics, University of Helsinki. Water samples were taken from upper water layer ~20 cm below surface and 3–4 additional samples were taken from the deeper layers. In years 1994–1997, only surface water was sampled. Additional measurements of the concentrations of optically active substances (22 days, analyses from about 7 different depths) were collected by studies carried out by researchers from the Võrtsjärv Limnological Station of the Institute of Zoology and Botany in years 1999–2000 (Kangro et al., 2005; Nõges & Solovjova, 2005; Ott et al., 2005a).

The concentration of chlorophyll *a* (C_{chl}) was measured filtering the water samples through Whatman GF/F filters and applying a standard method based on measuring the absorption of chlorophyll dissolved in ethanol at the wavelength of 665 nm (ISO 10260, 1996). The concentration of suspended matter (C_s), was obtained by its dry weight after filtration of the water through pre-weighted Whatman GF/F filters. Amounts of optically active dissolved organic matter is expressed by the absorption coefficient $a_y(380)$ obtained from spectrophotometric measurements of filtered water (filters CF/C with pore size 0.45 μm) relative to a reference of distilled water. Attenuation coefficient, $c^*(\lambda)$, was determined from non-filtered samples of water using spectrophotometer Hitachi U1000, with distilled water as the reference. No correction for the small-angle forward scattering was applied. However, the previous analysis of lakes data show, that $c^*(\lambda)$ spectra are rather good indicators of water transparency and quality (Arst et al., 1996, 1999).

Irradiance measurements

Measurements of underwater downward irradiance in PAR region (400–700 nm) at different depths in water were carried out using a LI-192 SA (units μmol m^{-2} s^{-1}). The first measurement of irradiance was always made at 0.01 m, then after that at 0.3–0.5 m intervals down to the depth, where light could no longer be detected.

To determine the mean value of the diffuse attenuation coefficient over the PAR region and depth, $K_{d,PAR}$ was found as the slope of the regression line through a plot of logarithmic irradiance vs. depth. According to Bowling & Tyler (1985), this method can lead to considerable errors in very clear or strongly heterogeneous lakes. So in Lake Verevi, only the upper layer, where optically active substances are distributed relatively uniformly, was used for calculations of vertically averaged $K_{d,PAR}$.

Our measurement system enables to determine the values of $K_{d,PAR}$ also for separate water layers with a thickness of 0.5 m. Two LI-192 SA sensors were fixed on a frame at a distance of 0.5 m from each other. Lowering the frame with the sensors, the vertical profile of $K_{d,PAR}$ can be obtained on the basis of its values for 0.5 m thick layers. Using these simultaneously measured irradiances, the values of $K_{d,PAR}$ (Δ0.5) were calculated:

$$K_{d,PAR}(\Delta 0.5) = \frac{1}{0.5}\ln\left[\frac{E_{d,PAR}(z_1)}{E_{d,PAR}(z_2)}\right], \tag{1}$$

where, $E_{d,PAR}(z_1;\ z_2)$ are underwater downward irradiance values measured at depths z_1 and z_2.

The spectral distribution of underwater downward and upward irradiances were measured between the depths of 0.5 and 2–4 m at 0.5 m intervals. A spectroradiometer LI-COR, LI-1800 UW (measuring in the range 300–800 nm, with a resolution of 2 nm) was used. K_d was calculated by the same method as described above for $K_{d,PAR}$. The irradiance reflectance R is calculated as a ratio of upward irradiance (E_u) to downward irradiance (E_d):

$$R(z) = \frac{E_u(z)}{E_d(z)}. \tag{2}$$

Equation 2 is valid for all wavelengths (λ).

Model used for estimating K_d, from concentrations of OAS

During year 2000, Lake Verevi was sampled intensively but without any optical *in situ* measurements. As there was not enough previous data to derive optical model for Lake Verevi, we used another lake optical model, described in detail in papers by Pierson & Strömbeck, (2000, 2001) and Pierson et al. (2002). Model inputs include the concentration of phytoplankton pigments, suspended inorganic material and dissolved organic matter. These substances are linked to absorption and scattering coefficients through a series of empirical relationships. Parameterization of the model is based on *in situ* measurements and laboratory analysis of Lake Mälaren's (Sweden) water. This model was chosen as it has been shown to reproduce variation in radiance reflectance with a good degree of accuracy and therefore could be reasonably applied to the lakes having similar variability of OAS. Concentration of inorganic suspended matter varied in Lake Mälaren from 0.1 to 31 g m^{-3}, C_{chl} from 0.1 to 79 mg m^{-3} and $a_y(380)$ = 0.6–13 m^{-1}. As the specific optical properties of Lake Mäleren and Lake Verevi are probably not the same it may cause some uncertainty. Clearly, more accurate estimation could be obtained if lake specific parameterization for Verevi were used. K_d is calculated from modeled inherent optical properties by Kirk (1994):

$$K_d(z) = (a^2 + 0.256ab)^{0.5} \quad (3)$$

where, a and b are spectral absorption and scattering coefficients accordingly. For every sampled depth spectral K_d was modelled and then irradiance in water was calculated according to equation:

$$E_d(z_2) = E_d(z_1)\exp[-K_d(z_1)\Delta z] \quad (4)$$

where, z_1 upper and z_2 lower sampled depth. To get values for PAR region values are integrated over wavelengths 400–700 nm.

As data set collected in Lake Verevi did not contain exactly the same parameters as the original model, we made some minor changes into it. (1) Instead of yellow substance absorption at 420 nm we used absorption at 380 nm. (2) We estimated the inorganic particular matter concentration (C_t) by assuming a 1:100 ratio between C_{chl} and dry weight of phytoplankton (Reynolds, 1984; Philips et al., 1995). So C_t is estimated as difference of total suspended matter and dry weight of phytoplankton.

This model is probably not valid in the layer where extremely high values of OAS occurs (Table 1). However, the point of this paper is to demonstrate change in optical properties accompanying with change in OAS and estimate light conditions in Lake Verevi overall.

Results

Optically active substances and optical properties

Lake Verevi water varies remarkably by its optical properties both seasonally and vertically. By the typical vertical distribution of OAS (examples in Fig. 1 (a–c)) C_s values ranged in the upper layer from 1 to 13.2 g m^{-3} but rise up to 310 g m^{-3} below 3 m (Table 1). The increase of C_{chl} in deep layer is even more pronounced, but yellow substance varies relatively less (from 1.5 m^{-1} upper layer to 8.6 m^{-1} in deeper layer) and higher values for it are associated not only with the same thin layer as C_s and C_{chl} but also for all the hypolimnion. Increase in the concentrations of all optical components causes the increase also in optical properties (Figs. 2d and 3d). Properties and formation of this thin layer are analyzed in more detail in Kangro et al. (2005); Ott et al. (2005a and 2005b), Kõiv & Kangro (2005).

Based on the vertical profiles of measured $E_{d,PAR}$ in Lake Verevi, we can assume an optically homogenous upper layer of Verevi and irradiance decrease is approximately described by exponential law. However, past the 2.5–4 m depth a rapid decrease in $E_{d,PAR}$ occurs (Fig. 2a). According to

Table 1. Minimum and maximum values of C_s, C_{chl}, $a_y(380)$, c^*_{PAR}, Secchi depth (SD) and $K_{d,PAR}$ based on the measurements in Lake Verevi in ice-free period of 1994–2001

	Layer 0–0.5 m (1994–2000)	Layer 3–6 m (1997–2000)
C_s, g m^{-3}	1.0–13.2	2.0–310
C_{chl}, mg m^{-3}	3.9–73.0	5.1–1012
$a_y(380)$, m^{-1}	1.5–6.4	3.3–8.6
c^*_{PAR}, m^{-1}	1.8–8.7	2.7–22.3
$K_{d,PAR}(\Delta z$ = 0.5 m)	0.8–2.9	3.5–4.6
SD, m	1.4–4.1	–

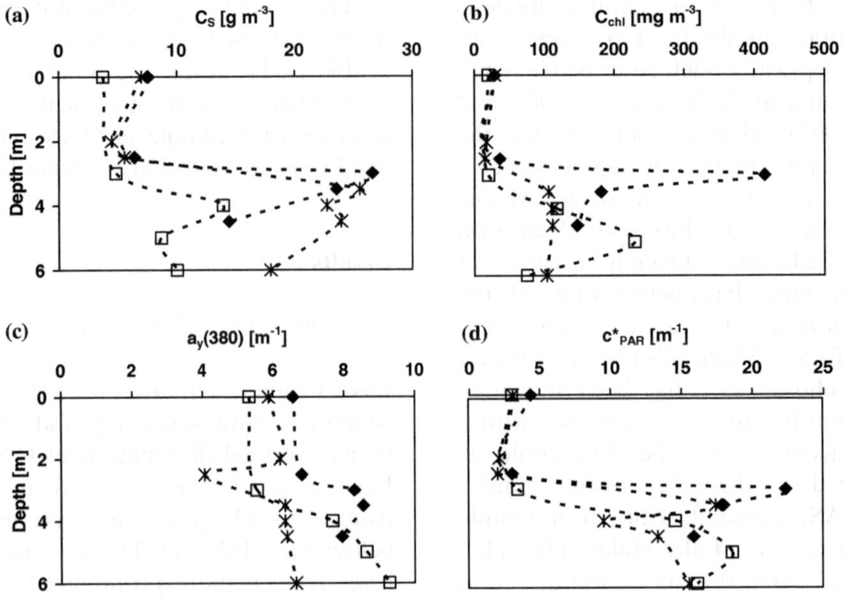

Figure 1. Examples of the vertical profiles of (a) C_s, (b) C_{chl}, (c) $a_y(380)$ and (d) c^*_{PAR} in Lake Verevi: 06 Aug, 1997 (stars); 17 June, 1999 (diamonds) and 15 Aug, 1999 (squares).

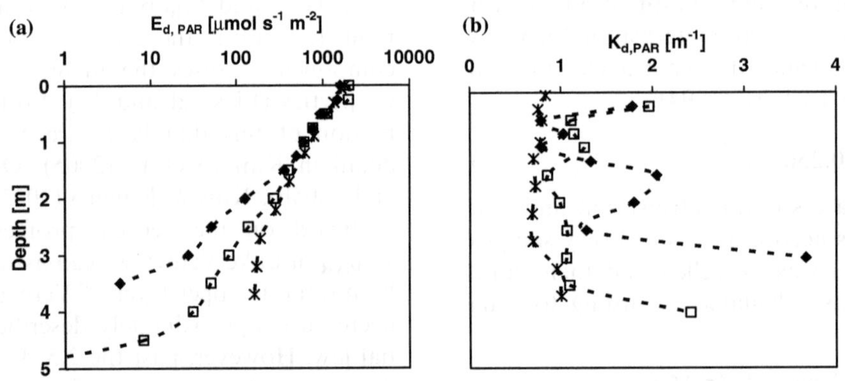

Figure 2. Examples for the measured downwelling irradiances $E_{d,PAR}$ (a) and according diffuse attenuation coefficients $K_{d,PAR}$ (0.5 m layers) (b) in Lake Verevi (same dates as in Figure 1).

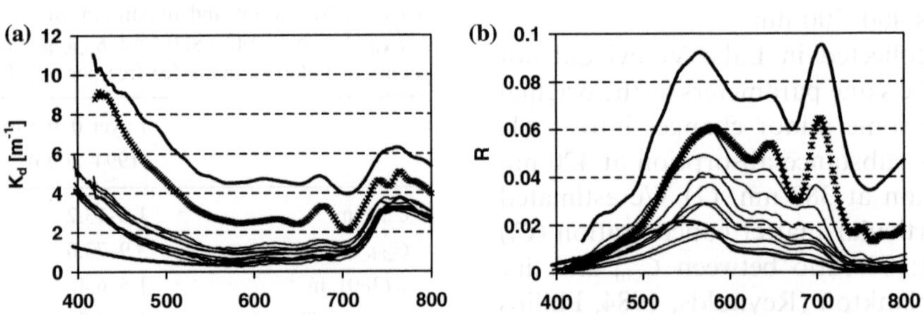

Figure 3. Spectra of K_d (a) and R (b) measured in Lake Verevi during 1994–2001 (thin lines). Stars show spectra from 12 Aug, 1998, when C_s = 13.2 g m^{-3}; C_{chl} = 73 mg m^{-3} and $a_y(380)$ = 4.6 m^{-1}. Solid lines show limits of the same parameters measured over same period in other Estonian lakes (lower line- Paukjärv, 15 June, 1997; upper line- Võrtsjärv, 16 Aug, 1999).

this, increase of light attenuation in the layer 2.5–4 m occurs notably. Measured $K_{d,PAR}$ (Δ0.5) can reach up to 4.6 m^{-1}, but real values may be even higher as measurements below the optically dense layer were often impossible due to too low irradiance values.

Secchi depth ranged from 0.65 to 4.1 m. When the highest depths were measured the white disc disappeared very rapidly, suggesting that it entered into the optically dense layer.

Correlation coefficients in Table 2 refer that relationships between OAS and apparent optical properties are often nonlinear. Secchi depth and $K_{d,PAR}$ are related ($p < 0.05$) as:

$$SD = 2.5(\pm 0.1) K_{d,PAR}{}^{-1.2(\pm 0.2)} \quad (5)$$

$r = 0.75$; $n = 17$, Std.err. = 0.4 m

Euphotic depth is the depth where 1% from subsurface irradiance reach is related to K_d:

$$z_{eu} = \frac{\ln(100)}{K_{d,PAR}} \cong \frac{4.6}{K_{d,PAR}}. \quad (6)$$

From Equation (5) and (6) we get an approximate relationship between euphotic depth and SD:

$$z_{eu} = 2.15 * SD^{0.83}. \quad (7)$$

Therewas also a rather good relationship between c^*_{PAR} and $K_{d,PAR}$ in Lake Verevi:

$$K_{d,PAR} = 0.24(\pm 0.06) c^*_{PAR} + 0.3(\pm 0.2) \quad (8)$$

$r = 0.74$; $p < 0.05$; $n = 17$, Std.err. = 0.3 m

Here, $K_{d,PAR}$ is the average for the whole upper water layer and c^*_{PAR} is measured from samples collected below surface. From this relationship it is possible to estimate values of the diffuse attenuation coefficient from laboratory measurements of c^*_{PAR}. This may be useful in some cases when field measurements are impossible due to bad weather. More complex model for such approximation for many Estonian lakes (including Verevi) is presented in Arst et al. (1997, 2002).

Attenuation of light in Lake Verevi is highest at both ends of the 400–800 nm region (below 450 nm and above 720 nm). As a result, green-yellow light (range 550–600 nm) dominates the sub-surface spectra in Lake Verevi (Fig. 3a). Spectral behavior of K_d is stable and measured spectra show rather small variation. Only one exceptional case was encountered – extremely high attenuation in the upper layer on Verevi was measured on 12th Aug, 1998, when water was very turbid (Secchi depth only 0.65 m) due do high concentration of chlorophyll (C_{chl} = 73 mg m^{-3}). Typically $K_d(490)$ values in this lake range from 1.1 to 2.4 m^{-1} but in the exceptional case it was 5.0 m^{-1}.

Reflectance has typically maximum between 580 and 600 nm. If there is more phytoplankton in the water an additional reflectance maximum is notable at 690-710 nm (Fig. 3b) and then there is also a remarkable decrease of the reflectance at 675–680 nm that corresponds to the phytoplankton absorption peak in this spectral region.

Light field in Lake Verevi in year 2000

Seasonal variation of the light field can be described by *in situ* measurements of OAS and estimates of $K_{d,PAR}$, based on the model introduced earlier. According to seasonal vertical change in the OAS, the modelled $K_{d,PAR}$ also changes (Fig. 4a). In the upper layer (0.5–2 m) $K_{d,PAR}$ varied from 0.64 to 1.4 m^{-1}, having higher values in spring. In the first half of August, $K_{d,PAR}$ increased for a short period due to a phytoplankton

Table 2. Correlation coefficients between OAS and c^*_{PAR}, Secchi depth (SD) and $K_{d,PAR}$ in the upper water layer. Correlations for logarithmic values are presented in brackets

	c^*_{PAR} ; $\ln(c^*_{PAR})$	SD; ln(SD)	$K_{d,PAR}$; $\ln(K_{d,PAR})$
C_s; $\ln(C_s)$	0.69 (0.84)	−0.49 (−0.71)	0.64 (0.78)
C_{chl}; $\ln(C_{chl})$	0.89 (0.84)	−0.83 (−0.90)	0.85 (0.79)
$a_y(380)$; $\ln(a_y(380))$	0.88 (0.77)	−0.53 (−0.56)	0.57 (0.41)
C^*_{PAR}; $\ln(c^*_{PAR})$	1	−0.84 (−0.86)	0.74 (0.68)
SD; ln(SD)	−0.84 (−0.89)	1	−0.75 (−0.83)
$K_{d,PAR}$; $\ln(K_{d,PAR})$	0.74 (0.68)	−0.75 (−0.83)	1

bloom, a similar increase was also observed in 1998 (data in Fig. 3). Concentrations of OAS in deeper layers, close to thermocline started to increase already from spring and had peak values in September (C_s = 122 g m^{-3}; C_{chl} = 1012 mg m^{-3}). This layer is located between 2.5 and 6 m, depending on local climatic conditions and season (Kõiv & Kangro, 2005). Those concentrations resulted in extremely high values of $K_{d,PAR}$ which for the 2–6 m layer were typically much higher than for upper layer and reached values of 5.8 m^{-1} (Fig. 4a).

As shown in Figure 4b, the irradiance penetrating the water, 32–47% reaches down to 0.5 m and 4–17% to 2 m depth, but at 6 m depth there is almost no PAR left (<0.005%), as it is attenuated in dense layer close to thermocline (Fig. 4b). These values became especially low after short bloom in the upper water layer at the beginning of August 2000. In October when the upper water column started to mix again, transparency increased.

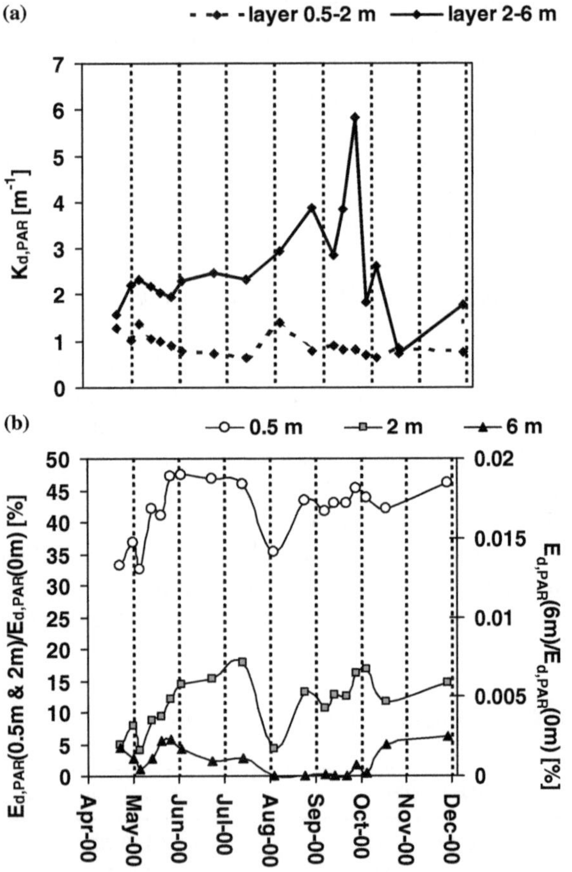

Figure 4. (a) Variation of modeled $K_{d,PAR}$ for upper layer where water is relatively clear (0.5–2 m) and lower layer, where strong stratification of OAS occurs (2–6 m) in Lake Verevi, 24 April–4 Dec, 2000. (b) For the same period variation of light penetrating to water in percent from incident irradiance $E_{d,PAR}(0)$. Depths 0.5, 2 m (left axis) and 6 m (right axis) are shown.

Discussion

Any change in the type and concentration of optically active substances (including temporal changes due to phytoplankton blooms, yellow substance increases at spring associated with snowmelt runoff or suspended particles associated with river discharge or resuspensions) change the optical properties of water. Estonian lakes can be divided into five optical types according to the optical properties and concentrations of optically active substances: clear, moderate, turbid, very turbid and brown (Reinart, 2000). Lake Verevi's upper layer (above thermocline, where strong stratification occurs) usually belongs to the moderate class of lakes but the water turns turbid toward autumn unlike other lakes investigated in Estonia and South Finland (Fig. 5) where C_s ranges from 1.5 to 145 g m^{-3}, C_{chl} 0.7–102 mg m^{-3} and $a_y(380)$ from 0.5 to more than 50 m^{-1}.

Every optical type of water has a rather characteristic spectral shape of attenuation coefficient and reflectance. The main factor influencing attenuation in the violet and blue parts of the spectrum is yellow substance and in the orange and red parts suspended particles. The Lake Verevi K_d spectra lie between limits measured in other lakes in our region (Fig. 3). The shape of reflectance spectra are similar to the types 3 and 5 defined by Vertucci & Likens (1989) and Reinart (2000). These are types where absorption in yellow substance and scattering by particles are the main factors affecting to PAR attenuation (about 95%).

In the relationship between SD and K_d (Equation 5) we estimated a slope parameter in the middle of the range found for 15 Estonian and Finnish lakes earlier: 1.09–4.11 and an exponent close to −1; similar values have been assumed by many previous investigations (for example Davies-Colley & Vant, 1988; Koenings & Edmundson, 1991).

There is a complicated feedback between water transparency and the amount of OAS in water:

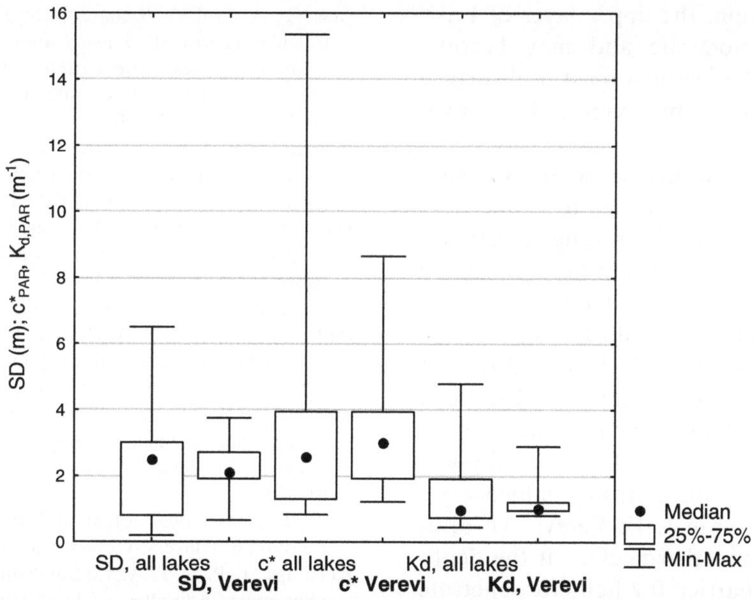

Figure 5. Optical parameters c^*_{PAR}, $K_{d,PAR}$ and Secchi depth (SD) measured in 15 lakes in Estonia and South-Finland in years 1994–2000 (67 cases) and same parameters measured in Lake Verevi (19 cases).

increasing PAR triggers photosynthesis but higher phytoplankton biomass and subsequently the amounts of dissolved organic matter and detrial material attenuate light more effectively, resulting in decrease of water transparency (Kirk, 1994). Modeling of underwater light climate can show the relative importance of the various components for the attenuation of PAR (Van Duin et al., 2001). However, final depth distribution of water properties is affected by many physical, chemical and biological processes. Comparing depth of the mixed layer as estimated by measurements of vertical distribution of OAS and euphotic depth (Equation 7), we can conclude that optical properties of the upper 2 m layer would allow PAR to penetrate down to as much as 7.3 m. However, light is blocked by optically dense water layer, which attenuates almost all of it (Fig. 6). This layer is not formed in response to a lack of PAR, but conversely it forms a barrier for light to penetrate to the deeper layer.

Conclusions

Lake Verevi is an optically complex water body because of high concentrations of optically active substances and strong vertical stratification of optical properties. The epilimnion is relatively well mixed and below this rapid change in all OAS results in rapid changes of optical properties. Concentrations of all optically active substances increase close to thermocline, chlorophyll *a* shows the highest variation (from 5.9 up to 1012 mg m^{-3}) while the amount of yellow substances vary in the smallest range. Comparing with other investigated

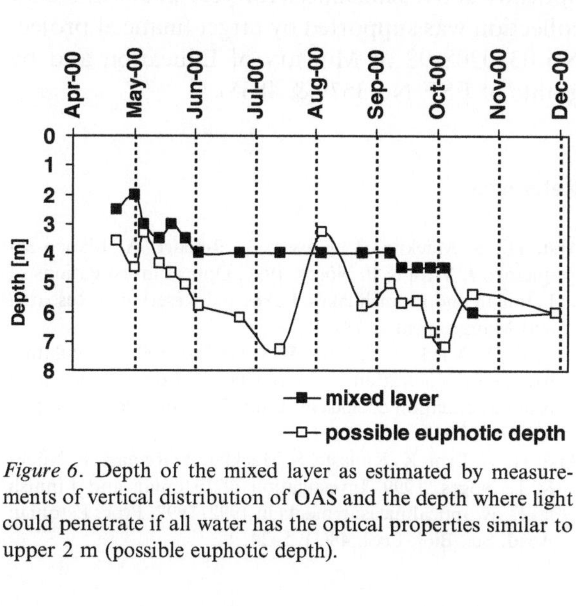

Figure 6. Depth of the mixed layer as estimated by measurements of vertical distribution of OAS and the depth where light could penetrate if all water has the optical properties similar to upper 2 m (possible euphotic depth).

lakes in the same region, the upper layer of Lake Verevi is optically moderate and may become turbid during algal blooms in August. Subsurface light fields are dominated by green–yellow wavelengths (580–600 nm).

From model calculations it is shown that 4–17% of surface PAR penetrates to a depth of 2 m. $K_{d,PAR}$ in upper 2 m layer ranged between 0.64 and 1.4 m^{-1} and in the 2–6 m layer it ranged between 0.9 and 5.3 m^{-1}. Based on the optical properties of upper layer it could be possible for PAR to penetrate down to a depth of 7.3 m, but due to the optically thick layer almost no PAR can reach this depth (less than 0.005%). These results show that biota and the gradient of physical properties in the water have strong influence on the overall light climate in Lake Verevi. An optically thick layer close to thermocline at the depth of 2.5–6 m forms a barrier for light to penetrate into the hypolimnion.

Acknowledgements

The authors are indebted to the Estonian Science Foundation (ESF grant No 1804). Water samples are analyzed by S. Mäekivi, A. Erm and L. Sipelgas in Estonian Marine Institute. Many thanks are to A. Herlevi (University of Helsinki) for carrying out the spectral measurements with LI-1800 UW, and to the working group of Võrtsjärv Limnological Station leaded by dr. I. Ott for providing data on optically active substances for year 2000 (this data collection was supported by target financed project No 0370208s98 of Ministry of Education and by grants of ESF No 3579 & 4835).

References

Arst, H., S. Mäekivi, T. Kutser, A. Reinart, A. Blanco-Sequeiros, J. Virta & P. Nõges, 1996. Optical investigations of Estonian and Finnish lakes. Lakes and Reservoirs: Research and Management 2: 187–198.

Arst, H., S. Mäekivi, T. Lukk & A. Herlevi, 1997. Calculating irradiance penetration into water bodies from the measured beam attenuation coefficient. Limnology and Oceanography 42: 379–385.

Arst, H., A. Erm, K. Kallaste, S. Mäekivi, A. Reinart, P. Nõges & T. Nõges, 1999. Investigation of Estonian and Finnish lakes by optical measurements in 1992–1997. Proc. Estonian Acad. Sci. Biol. Ecol. 48(1): 5–24.

Arst, H., A. Erm, A. Reinart, L. Sipelgas & A. Herlevi, 2002. Calculating irradiance penetration into water bodies from the measured beam attenuation coefficient, II: Application of the improved model to different types of lakes. Nordic Hydrology 33: 227–240.

Bowling, L. C. & P. A. Tyler, 1985. The underwater light-field of lakes with marked physiochemical and biotic diversity in the water column. Journal of Plankton Research 23: 69–77.

Davies-Colley, R. J. & W. N. Vant, 1988. Estimates of optical properties of water from Secchi disk depths. Water Resources Bulletin 24: 1329–35.

Dera, J., 1992. Marine Physics. Elsevier, Amsterdam 452 pp.

ISO, 10260, 1992. (E), Water quality – Measurement of biochemical parameters – Spectrophotometric determination of chlorophyll-*a* concentration. (Geneva, Switzerland: ISO), 1–6.

Kangro, K., R. Laugaste, P. Nõges & I. Ott, 2005. Long-term changes and seasonal development of phytoplankton in a strongly stratified, hypertrophic lake. Hydrobiologia 547: 91–103.

Kirk, J. T. O., 1994. Light and Photosynthesis in Aquatic Ecosystem. University Press, Cambridge 509 pp.

Koenings, J. P. & J. A. Edmundson, 1991. Secchi disk and photometer estimates of light regimes in Alaskan lakes: Effects of yellow colour and turbidity. Limnology and Oceanography 36: 91–105.

Kõiv, T. & K. Kangro, 2005. Resource ratios and phytoplankton species composition in a strongly stratified lake. Hydrobiologia 547: 123–135.

Nõges, T. & I. Solovjova, 2005. The formation and dynamics of deep bacteriochlorophyll maximum in the temperate and partly meromictic Lake Verevi. Hydrobiologia 547: 73–81.

Ott, I., T. Kõiv, P. Nõges, A. Kisand, A. Järvalt & E. Kirt, 2005a. General description of Lake Verevi, its ecological status, changes during the past eight decades and restoration problems. Hydrobiologia 547: 1–20.

Ott, I., A. Rakko, D. Sarik, P. Nõges & K. Ott, 2005b. Sedimentation rate of seston during the formation of temperature stratification after ice break-up in the partly meromictic Lake Verevi. Hydrobiologia 547: 51–61.

Philips, E. J., F. J. Aldridge & P. Hansen, 1995. Patterns of water chemistry, physical and biological parameters in shallow subtropical lake (Lake Okeechobee, Florida, USA). Ergebnisse der Limnologie 45: 117–135.

Pierson, D. C., H. Markensten & N. Strömbeck, 2002. Long and short term variations in sediment resuspension: the influence on light available to the phytoplankton community. Hydrobiologia. 494: 299–304.

Pierson, D. C. & N. Strömbeck, 2000. A modeling approach to evaluate preliminary remote sensing algorithms: use of water quality data from Swedish great lakes. Geophysica 36: 177–202.

Pierson, D. C. & N. Strömbeck, 2001. Estimation of radiance reflectance and the concentrations of optically active substances in Lake Mälaren, Sweden, based on direct and inverse solutions of a simple model. The Science of the Total Environment 268: 171–188.

Reinart, A., 2000. Light field characteristics in different types of Estonian and Finnish lakes, Ph.D. theses, Tartu University Press, Tartu, 195 pp.

Reynolds, C. S., 1984. The Ecology of Freshwater Phytoplankton. University Press, Cambridge, 390 pp.
Van Duin, E. H. S., G. Blom, F. J. Los, R. Maffione, R. Zimmerman, C. F. Cerco, M. Dortch & E. P. H. Best, 2001. Modeling underwater light climate in relation to sedimentation, resuspension, water quality and autotrophic growth. Hydrobiologia 444: 25–42.
Vertucci, F. A. & G. E. Likens, 1989. Spectral reflectance and water quality of Andirondack mountain region lakes. Limnology and Oceanography 34: 1656–1672.

Hydrobiologia (2005) 547:51–61
I. Ott & T. Kõiv (eds), Lake Verevi, Estonia – A Highly Stratified Hypertrophic Lake
DOI 10.1007/s10750-005-4143-0

Sedimentation rate of seston during the formation of temperature stratification after ice break-up in the partly meromictic Lake Verevi

Ingmar Ott*, Aimar Rakko, Diana Sarik, Peeter Nõges & Katrin Ott
Institute of Zoology and Botany, Estonian Agricultural University, Võrtsjärv Limnological Station, 61101, Rannu, Tartu County, Estonia
*(*Author for correspondence: E-mail: iott@zbi.ee)*

Key words: partly meromictic lake, spring sedimentation rate, primary production, export production, plankton dynamics

Abstract

The small strongly stratified hard-water hypertrophic lake Verevi (max. depth 11.0 m, surface area 12.6 ha, mean depth 3.6 m) was investigated in 2000 and in 2001. The lake is sheltered from winds, and the role of waves in mixing the water column is minimal. Eutrophication favours the strengthening of stratification. Early warm springs cause a fast stagnation of the water column forming partly meromictic conditions. Seston content of water and in sediment traps in 3 layers was measured several times during the formation of stratification. Besides measuring particulate matter, in 2001, the nutrient content of the trapped sediment was analysed. During the first 7 days of the investigation, 30% of the total particle sedimentation took place. The sedimentation rate of particulate matter was 0.4–6.3 g m^{-2} d^{-1} dry weight in different layers of the water column. Daily average sedimentation loss rate was 27% of the total amount of seston of the epilimnion, whilst from the meta- and hypolimnion the settling was much slower (9.6 and 7.3%, respectively). In our experiments with twin sediment traps, to one of which formaldehyde was added, the PO_4^{3-}-P concentration was 19% smaller in the trap without formaldehyde, probably due to planktonic uptake. The relationship between primary and export production is loop-like. The shape was irregular, indicating a high grazing rate of zooplankton.

Introduction

Sedimentation of particulate matter is one of the key processes in the circulation of matter in small stratified lakes. The effects of eutrophication in limnic environments are manifested as increased production and sedimentation (Heiskanen et al., 1999).

Under some circumstances the cycling of energy and particulate matter of lakes might not be so intensive, although the nutrient content is high. The strong competition between organisms for resources during the strengthening of stratification at the beginning of the vegetation period is crucial for the whole vegetation period. Large amounts of nutrients remaining in the aphotic layer cannot be used by autotrophic biota. A contradictory situation of starvation of plankton organisms, while nutrients are abundant, occurs. Early warm springs shift the beginning of thermal and chemical stratification, and resource limitation nearly by a month. If there is no spring overturn, several events unusual for the lake phenomena will occur, e.g., the changing of ordinary communities during the summer stagnation. Acceleration of limnological processes is characteristic of the spring season.

The aim of our investigation was to follow the dynamics of suspended solids, plankton, nutrients, primary production and sedimentation rate in spring in different water layers in order to measure the rate of the processes. The key questions were: which is the general sedimentation rate of suspended solids? which is the daily sedimentation loss of suspended matter from different layers of water? which is the dependence of export production from primary production? and how intensive could be the phosphorus uptake by plankton in the water column?. This study forms only a small part of the complex investigation of Lake Verevi that deals with ecosystem functioning during formation of stratification and changing of environmental conditions in the water column. Holistic study includes main trophic levels and biotic groups: bacterioplankton, protozoa, phytoplankton, metazooplankton, periphyton, macrophytes, meio- and macrozoobenthos, fishes (see: Tammert et al., 2005; Zingel, 2005; Nõges & Solovjova, 2005; Nõges & Kangro, 2005; Kõiv & Kangro, 2005; Kangro et al., 2005; Kübar et al., 2005; Laugaste & Reunanen, 2005; Mäemets & Freiberg, 2005; Timm & Möls, 2005; Järvalt et al., 2005).

Materials and methods

Description of lake (Ott et al., 2005), its environmental properties and peculiar partial meromixis in 2000 and 2001 is described in different articles of present issue (Nõges & Kangro, 2005; Reinart et al., 2005; Kõiv & Kangro, 2005; Nõges & Solovjova, 2005; Nõges, 2005).

Vertical flux measurements

In 2000 and 2001, in Lake Verevi, the seston content of water and its vertical flux in sediment traps was measured several times during the formation of stratification (Table 1). Besides measuring suspended matter, in 2001, the nutrient content of the trapped sediment was also analysed.

In the year 2000, five replicate plastic dark grey opaque cylinders (length 50 cm, inner diameter 5 cm, volume 0.98 l) were suspended at a depth of 5 m (at this spot the lake depth was 8 m). The cylinders were of the best height/diameter ratio of 10:1 (Rosa et al. 1994). On the basis of the great value of relative depth (2.7%; Wetzel, 1983), we assume that resuspension in L. Verevi was very small, and chose simple devices although much more sophisticated sampling techniques and methods exist (Bloesch, 1996). According to Huttula & Krogerus (1986), the material is not resuspended in these traps even in very dynamic conditions. Twin cylinders at depths of 1, 5 and 7 m were used in the water column also in 2001. Every time, a perforated 50 ml vial with concentrated formaldehyde was put at the bottom of one trap cylinder to preserve the settled material, while to the other no fixative was added.

Table 1. Sampling dates of water and trapped seston in 2000 and 2001

2000			2001		
Seston in water	Sedimentation	Exposition time	Seston in water	Sedimentation	Exposition time
24.04.00			29.03.01		
03.05.00			16.04.01		
16.05.00	16.05.00	13 days	19.04.01		
22.05.00			23.04.01		
29.05.00	29.05.00	13 days	26.04.01	26.04.01	3 days
05.06.00			30.04.01	30.04.01	4 days
13.06.00	13.06.00	15 days	03.05.01	03.05.01	3 days
			07.05.01	07.05.01	4 days
			10.05.01	10.05.01	3 days
			24.05.01	24.05.01	14 days
			05.06.01	05.06.01	12 days

Laboratory analyses

The suspended solid content was measured as 105 °C dry weight of suspended material filtered on Whatman GF/C filters after shaking the cylinder. Methods of studying weather conditions, hydrochemistry, phytoplankton, metazooplankton, and ciliates are described in other papers of this publication (Ott et al., 2005; Kõiv & Kangro, 2005; Kangro et al., 2005; Kübar et al., 2005; Zingel, 2005; Nõges & Solovjova, 2005; Nõges & Kangro, 2005). Total phosphorus (P_{tot}) was determined after persulfate oxidation as PO_4^{3-}. The content of $P{\cdot}PO_4^{3-}$ was determined with the molybdene blue method (Reports …, 1977). No filtration was carried out thus allowing the esimation also of P_{tot} in seston. The relative error of the analyses of total phosphorus and phosphates was ±5%.

Sediment rate calculations

All tubes were filled with water of the epilimnion, therefore for calculations we subtracted the amount of seston (also called suspended solids, particulate matter) in that water from the value measured in the traps. The sedimentation rate was calculated as the amount of total particulate matter in sediment traps per m^{-2} in different layers daily. The sedimentation loss rate was calculated as the vertical flux (mg m^{-2} d^{-1}) in the different layers divided by the standing stock (mg m^{-2}) and was given as percent.

Results

Particulate matter in lake water and sedimentation rate

After the ice break-up on 13 April 2001, despite a short period of homothermia (Fig. 1), the lake was stratified during the whole of spring (Figs 2–3). The seston content of lake water fluctuated in April (Fig. 4), but remained stable in May and June. In April, the content was the lowest in the hypolimnion and the highest in the epi- and metalimnions.

During the first 7 days of the investigation, (23–30 April; 16% of the total 43 days), 30% of the total particle sedimentation took place. The settling flux was the highest in the epilimnion

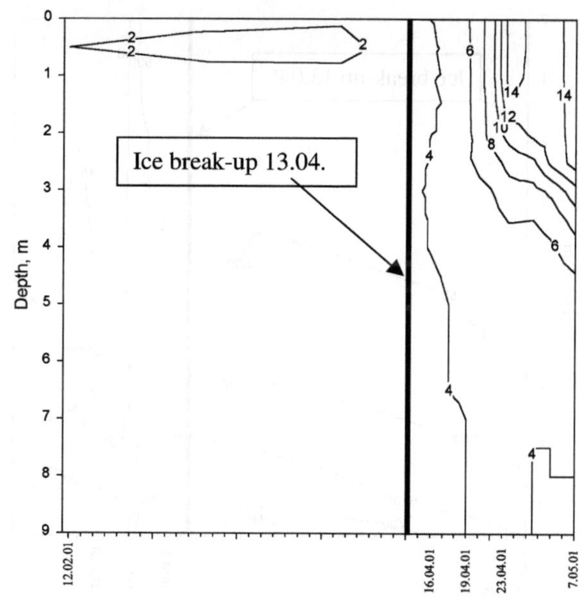

Figure 1. Vertical distribution of temperature in Lake Verevi between 12.02.–7.05.01.

(during the first 7 days 59% of the total settled matter at the same layer) while in the metalimnion this value was 50%.

The calculated daily settling flux (Fig. 5) in the 3 layers and in different traps was very different. The sedimentation rate was 0.4–6.3 g m^{-2} d^{-1} in different

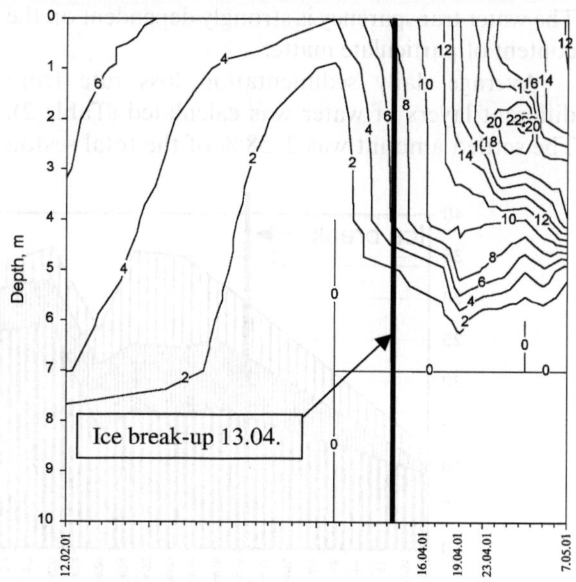

Figure 2. Vertical distribution of oxygen in Lake Verevi between 12.02.–7.05.01.

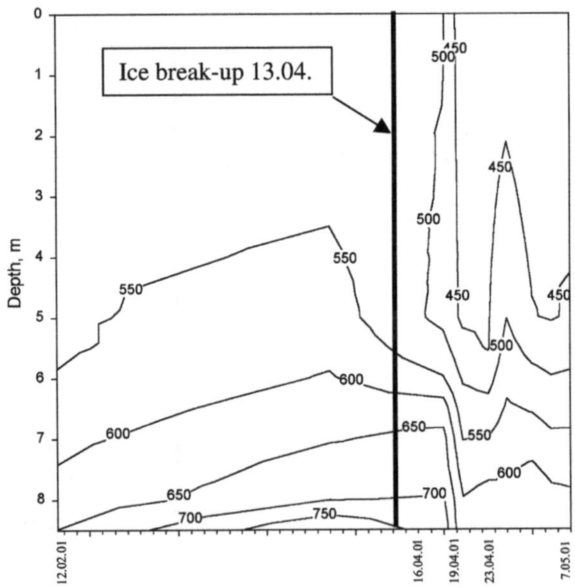

Figure 3. Vertical distribution of conductivity in Lake Verevi between 12.02.–7.05.01.

layers of the water column. The flux was the highest in the epilimnion in the beginning, and later in the meta- and hypolimnion. Then the sedimentation speed in the upper layers diminished, the decrease being slight in the hypolimnion. The same tendency was noticed in the dynamics of the Secchi disc transparency and the seston content of the epilimnetic water (Fig. 6). The water transparency is strongly dependent on the content of particulate matter.

Average daily sedimentation loss rate from different layers of water was calculated (Table 2). The settled amount was 2–58% of the total seston amount in the lake, and the settling was the greatest in the epilimnion.

Nutrients in samples from sediment traps

The total phosphorus (P_{tot}) and phosphates (PO_4^{3-}-P) content were analysed in samples from sediment traps in 2001. Generally, the concentration of PO_4^{3-}-P was 19% higher in the sediment trap with formaldehyde. This phenomenon occurred in the epilimnion and metalimnion in all cases, but in the hypolimnion only at the beginning of the experiment (Fig. 7).

The distribution of the P_{tot} content was of a different kind. During a short exposition time, P_{tot} had equal values in both upper cylinders (with and without formaldehyde), but when the exposition was long, the values in the traps with formaldehyde exceeded the others essentially. Obviously the great difference when the exposition time was longer, was caused by differences in the PO_4^{3-}-P content. The differences between the traps installed in the hypolimnion were smaller. Values were almost equal or those in the cylinders without preservative even higher (Fig. 8).

Sedimentation rate and plankton dynamics

The biggest values of phytoplankton biomass in the epilimnion were measured between 16–23 April 2001. The tiny centric diatom *Cyclostephanos dubius* (Fricke) Round dominated. The spring maximum of phytoplankton in whole water column (10.5 g m^{-3}) occurred a little later, on 26 April, at the depth of 5–6 m. In epi- and metalimnions

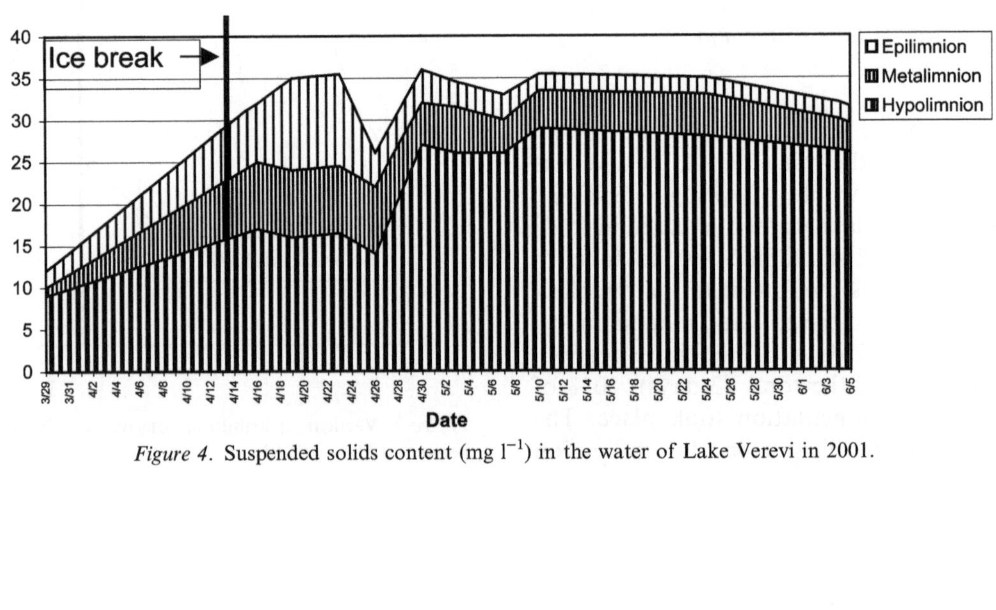

Figure 4. Suspended solids content (mg l^{-1}) in the water of Lake Verevi in 2001.

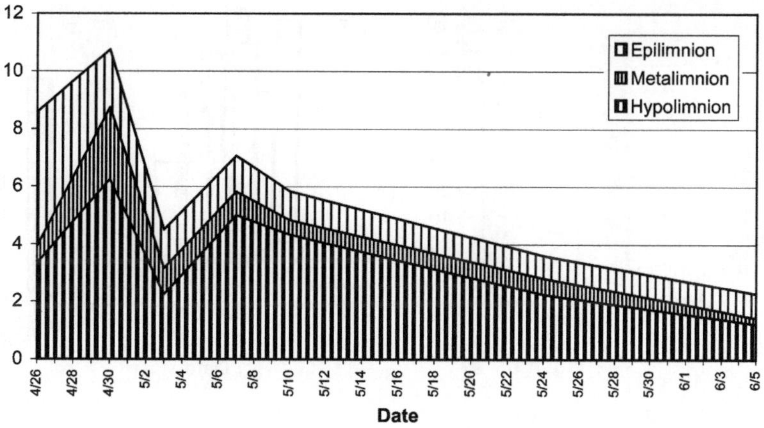

Figure 5. Calculated daily settling flux (dry weight g m^{-2} d^{-1}) in different layers of Lake Verevi.

(biomass 6–10 g m^{-3}) diatoms prevailed in short period, after that, already on 30 April the amount has drastically dropped (<0.5 g m^{-3}) and share of species with flagella, actively migrating in the water column, has increased staying since 3 May on the level 60–76% from the total biomass. The range of biomass of flagellated species was 0.4 –3.9 g m^{-3} (average 26% from the mean biomass during the investigation period). In early spring, dominating algae with flagella were *Chlorogonium* sp. and cryptophytes (Kangro et al., present issue). Later, number of species was increased including species from genera *Chlamydomonas*, *Trachelomonas*, *Euglena*, *Chroomonas* etc. In some cases also dinoflagellates were found.

Primary and export production

The daily primary production (PT) was measured in 2001 (Nõges et al., 2005) and the export production (PE) calculated from the Redfield's ratio (Redfield, 1958) on the basis of samples collected from traps with formaldehyde, in the metalimnion. The results are depicted in Fig. 9. The dynamics of PT had two peaks, at the beginning of the investigation cycle and on 5 June. PE was always lower than PT except 1 month after the ice breakup (10 May). The range of the proportion of PE to PT was wide (6–167%, average 73%).

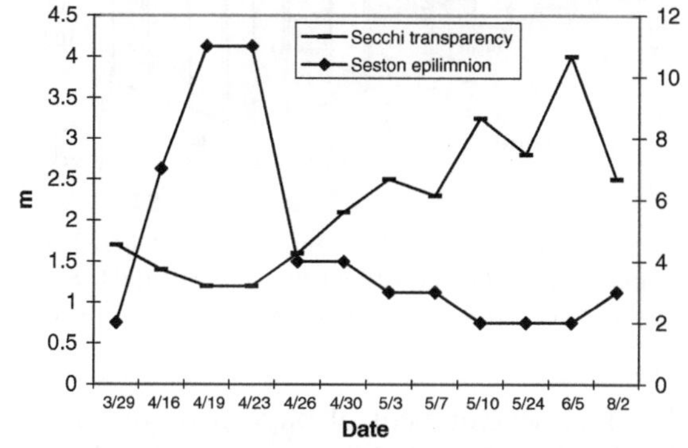

Figure 6. Dynamics of Secchi disc transparency (m) and suspended solids content (mg l^{-1}) in the epilimnetic water of Lake Verevi in 2001.

Discussion

Particulate matter in the lake water column and vertical flux

In order to describe the circulation of particulate matter in the lake, the investigation time was divided into subperiods, based on the suspended solid content of the water and in the traps:

Table 2. Daily sedimentation loss rate of L. Verevi in 2001

Period	Epilimnion	Metalimnion	Hypolimnion
23–26 April	58	4	12
26–30 April	25	25	12
30 April–3 May	22	6	4
3 May–7 May	21	10	10
7–10 May	25	6	7
10–24 May	20	6	4
24 May–5 June	21	10	2
Average	27.4	9.6	7.3

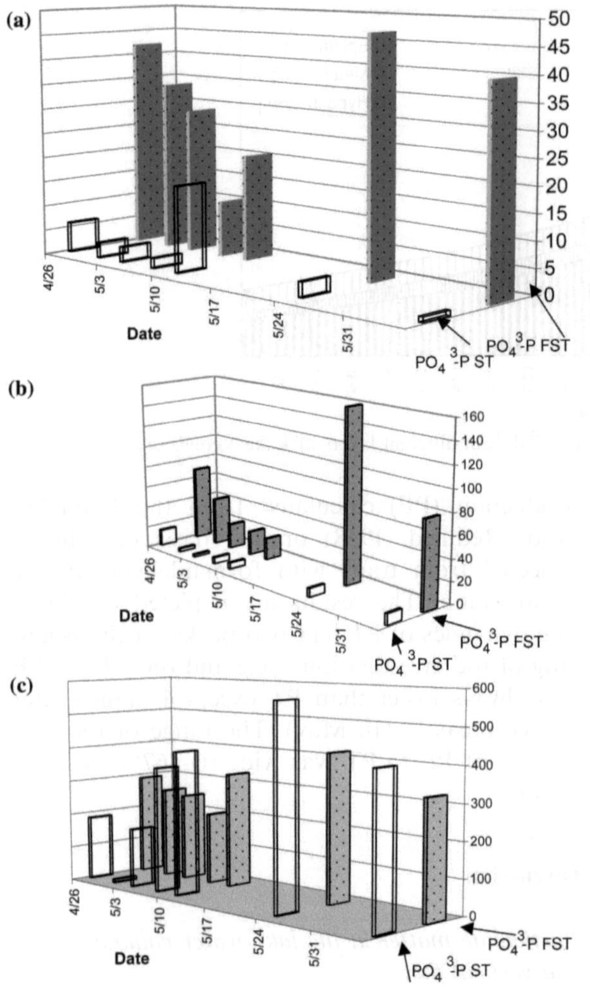

Figure 7. Phosphate (PO_4^{3-}-P) content (mg m^{-3}) in sediment trap water with (FST) and without preservative liquid (ST) in the epi- (a) meta- (b) and hypolimnion (c) of Lake Verevi.

a) rapid increase of the suspended solids content of the water column at the beginning of the vegetation period in 2001, presumably during ice break-up (13–16 April);
b) increase of the suspended solids content in the epilimnion, which lasted for 7 days (16–23 April);
c) increase of the sedimentation rate in the epilimnion (duration 3 days, 23–26 April);
d) increase of the sedimentation rate in the metalimnion (duration 4 days, 26–30 April);
e) stabilization and a slight permanent decrease of the seston content and sedimentation rate in the water column (duration 37 days, 30 April–5 June).

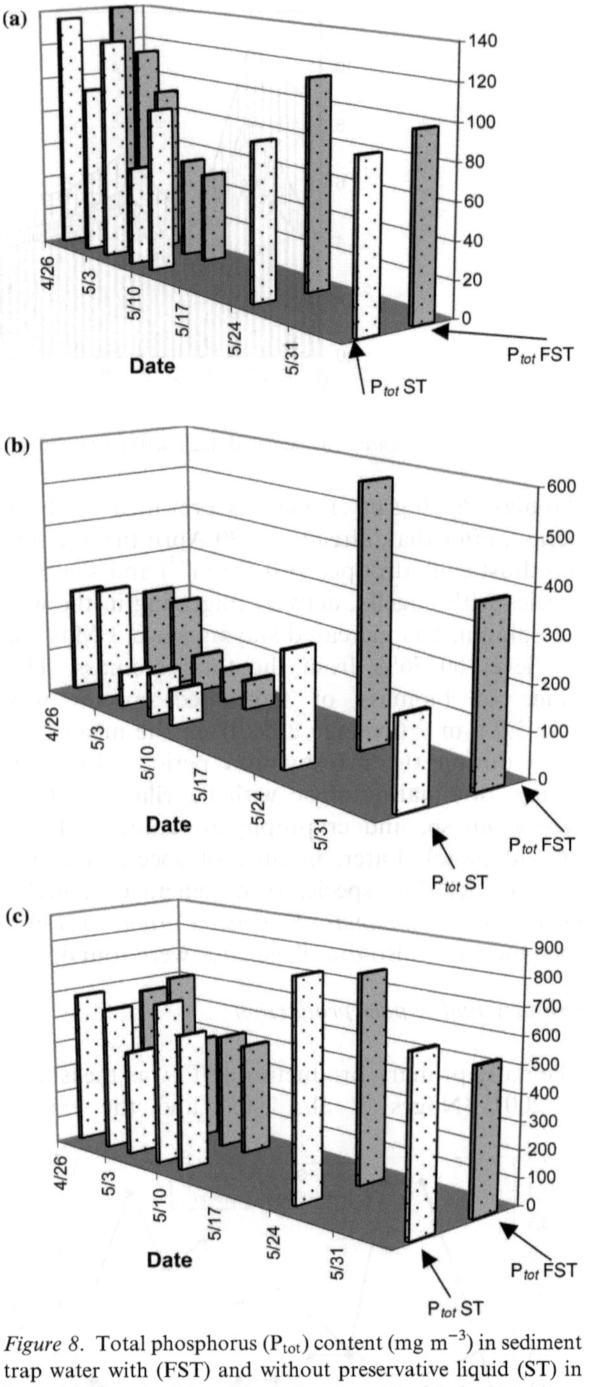

Figure 8. Total phosphorus (P_{tot}) content (mg m^{-3}) in sediment trap water with (FST) and without preservative liquid (ST) in the epi- (a) meta- (b) and hypolimnions (c) of Lake Verevi.

The fast development of phytoplankton in the epilimnion began probably already under the ice, but the greatest values were measured between 16 and 23 April. The particulate matter circulation in

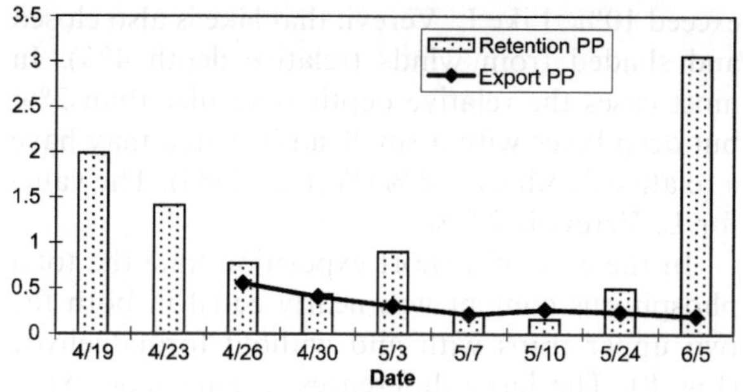

Figure 9. Dynamics of total primary production (PT; g C m^{-2} d^{-1}) and export production (PE) in the water column of Lake Verevi in 2001.

the upper layers of the water column was very fast for a short time after the ice break-up. Particulate matter cannot stay for a long period in the upper part of the water column, while the lower part stays homogenous.

An additional factor influencing the content of seston in the deeper sediment traps could be also resuspension, but in Lake Verevi it does not seem to be important. The lake is sheltered from the winds, and its relative depth is large. Also bioturbation seems to be not important since zoobenthos is scarce and bottom feeding fishes live only in the littoral (Järvalt et al., present issue). One of the most important benthophagous fish, bream, has disappeared from the lake (Ott et al., present issue). Steep stratification inhibits resuspension, and beside thermal stratification, the chemical stratification is very important (Fig. 3).

The suspended solids' content as well as the sedimentation rate, phytoplankton growth etc. are very dynamic at this first stage. After that, the resources become exhausted and a long period of plankton starvation follows. The strong competition for resources at the beginning of the vegetation period affects the species composition of plankton. This period is presumably crucial for development of the species dominating later. As this period is very short the result may be different in every year. This could be one of the reasons why entirely different communities dominate in different years (Ott et al., 2005; Kangro et al., 2005).

The very quick beginning of the vegetation period affects also the biota of deeper layers. Lack of nutrients during the sharp stratification of the epilimnion, lack of light in lower water layers as well as sharp gradients of chemical substances create a situation, in which different biotic groups are located in very narrow layers of water (Zingel & Ott, 2000; Solovjova & Nõges, 2005; Zingel, 2005).

The settling flux in Lake Verevi was generally high in comparison with data from the literature. In their overview Rosa *et al.* (1994) give a range of sedimentation rates for lakes from 0.1 to 30 g m^{-2} d^{-1} dry weight. Bloesch and Uehlinger (1986) investigated the settling flux in the deep eutrophic Lake Hallwil throughout the year and calculated the flux in midlake to be 887 g m^{-2} y^{-1} (2.4 g m^{-2} d^{-1}). The sedimentation rate in large shallow lakes strongly exceeds that of small stratified lakes. In the eutrophic Lake Võrtsjärv (area 270 km^2, mean depth 2.8 m) it varied between 26 and 700 g m^{-2} d^{-1} dry weight, mean value 170 g m−^{2} d^{-1} (T. Nõges et al., 1998), but in a such shallow lake resuspension is essential. This is much higher than was presented in an overview published by Rosa *et al.* (1994). Based on data of Punning (2002), we calculated sedimentation rate for the winter and spring period in Lake Jussi Pikkjärv to be 0.17 g m^{-2} d^{-1}.

Comparison of results for the years 2000 and 2001

The suspended solid content in the epilimnion of the lake was higher in 2000 than in 2001 (5.2 and 4.9 mg l^{-1}, respectively), and the same condition was observed with sedimentation rate in the metalimnion (2.1 and 0.9 g m^{-2} d^{-1}, respectively). In 2000, the ice broke up 2 days later (15 April) than in 2001, but the mean air temperature in April and May 2000 was higher by 1.6 °C and 0.5 °C, respectively. This coincided with higher values of suspended solids in the water and that of the sedimentation rate in 2000.

Nutrients in the sediment traps

The content of phosphorus in water samples from sediment traps reflects the biotic activity in water layers. Phosphates are often the limiting factor for the growth of photosynthetic organisms and they are also very quickly used in the euphotic zone. This explains why the PO_4^{3-}-P-content in sediment traps with formaldehyde was 19% larger than in water column water – that was caused by plankton uptake from the cylinder without formaldehyde. This phenomenon occurred only when the water column was illuminated. According to our measurements (see also Reinart et al., 2005), in L. Verevi only 1% of light penetrates generally down to 3.5–5.5 m and our intermediate trap was placed at the depth of 5 m. Despite the great variability of incident irradiance, the exponential light extinction was in all cases very similar.

We used different exposition periods (3–14 days). We considered that daily measurements could disturb the water column too often. Longer periods were chosen when settling flux was stabilised in the epi- and metalimnion. Migration of the organisms into and out of the trap seems to be obvious. Viner et al. (1999) stressed the role of fast migrating plankton species 'contaminating' traps. Fast swimming dinoflagellates are caught in poisoned traps. In Lake Verevi in 2001, dinoflagellates were rare, but lot of other flagellate species existed. The question raises, how accurate are the values of sedimentation, if an intensive migration into and out from the trap exists. May be algae just use cylinders for *in situ* growth and vertical flux is invalidated. We believe that primary production inside traps cannot be considerable, for many reasons. Our cylinders were dark and opaque with relatively small inner diameter and therefore illumination was very low. Duration of the exposition was different, but we did not notice any green biofilm on the trap walls. Because the dissolved nutrients do not sink gravitationally, phosphorus is trapped with seston and mineralised. The real amount of sedimented phosphorus resource is something between values in poisoned and non-poisoned traps. Part of the phosphorus in the traps without formaldehyde is removed by migrating organisms, while otherwise the traps with poison can catch more nutrients than were sunk naturally.

The rate of resuspension cannot be very important, because of the very strong thermal and chemical stratification (Figs. 1–3). This also means, that there was a difference in viscosity and density of the water. Nevertheless, Punning et al. (2002) claim that even 30% of the total annually discharged dry matter was resuspended in the steeply stratified small semidystrophic Lake Jussi Pikkjärv (Estonia). The resuspension of birch pollen grain and spherical fly ash particles did not exceed 10%. Like L. Verevi, that lake is also closed and shaded from winds (relative depth 4%). In most cases the relative depth is smaller than 2%, but deep lakes with a small surface area may have a relative depth over 4% (Wetzel, 1983). The value for L. Verevi is 2.7%.

In the case of a short exposition time the total phosphorus content was nearly equal in both the two upper traps with and without formaldehyde (Fig. 8). The large differences in long time exposition cases were caused by high PO_4^{3-}-P contents. A relatively large amount of organic phosphorus as well as organic matter is mineralized in the metalimnion. Suspended solids settle quickly from the epilimnion and stay in the metalimnion for a long period. There was no difference between P_{tot} in the hypolimnion traps. Besides the reason explained in the case of upper traps, this could also be explained with very high stable values of the total phosphorus as well as of phosphate content. That is why changes in the hypolimnion oversaturated by nutrients (P_{tot} = 533–808 mg m^{-3}) are not so clearly noticeable.

Sedimentation rate and plankton dynamics

The phytoplankton biomass dynamics (Fig. 10) was in a relatively good correlation (r = 0.73) with the sedimentation rate and seston content in the water column. A noticeable shift in time was observed. The peak of the phytoplankton content in the epilimnion was on 23 April and from this date on, began a fast development of metazooplankton (Fig. 10) first in the epi-, and then in the metalimnions. At the same time began the increase of ciliate abundance (Zingel, 2005). By 26 April, the biomass of phytoplankton had already decreased but the increase of metazooplankton and ciliates continued. This was especially remarkable in the metalimnion. The biomass of phytoplankton was the lowest on 5 June, and at the same time the Secchi disc transparency (4 m) and the biomass of metazooplankton were the highest.

Primary and export production

Export production is a function of primary production. Bottom up or top down control of primary production on the basis of similar calculations has

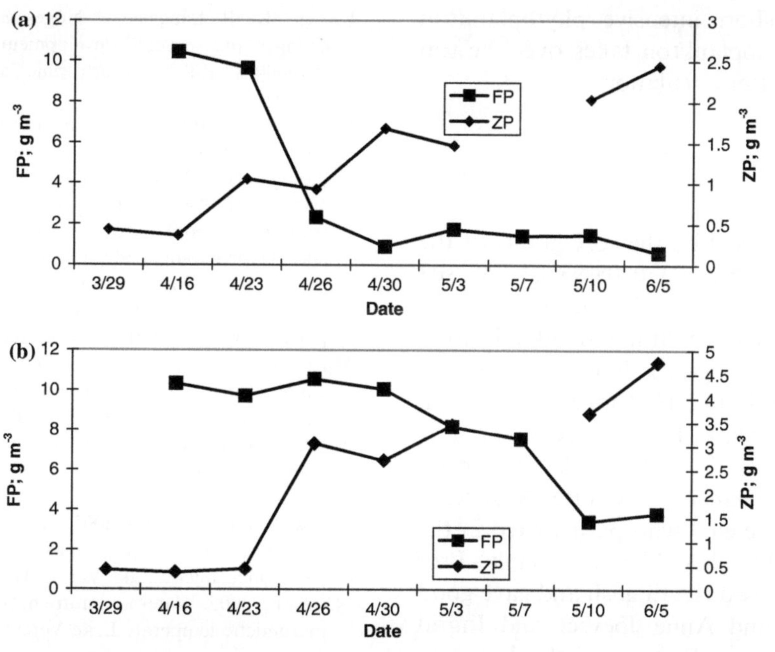

Figure 10. Dynamics of phytoplankton (FP) and zooplankton (ZP) biomass content (g m^{-3}), in the epi- (A) and metalimnion (B) of Lake Verevi.

mainly been discussed in the case of the marine environment (Eppley & Peterson, 1979; Heiskanen, 1999; Wassmann, 1999). We compiled a scatterplot between PT and PE daily data (Fig. 11) in order to check the hypothesis of very fast changes going on at the beginning of the vegetation period. Wassmann gives the most probable relationship between PT and PE as a regular loop. Our results allowed to construct only half of the loop, meaning that we were late, although we measured sedimentation already 10 days after the ice break-up. In our case, the shape of the loop was not regular and the latest values did not bring it to the theoretical midline (retention line) of the scatterplot. According to Wassmann, this situation corresponds with a high grazing rate of zooplankton. The increase of metazooplankton abundance in this time in L. Verevi (Fig. 10) shows that in spring, the biotic activity is high. The curve in Fig. 11 is similar to that found for marine environment where the trophic status is much lower. Lake Verevi functions similarly, due to the sharp stratification, in the case

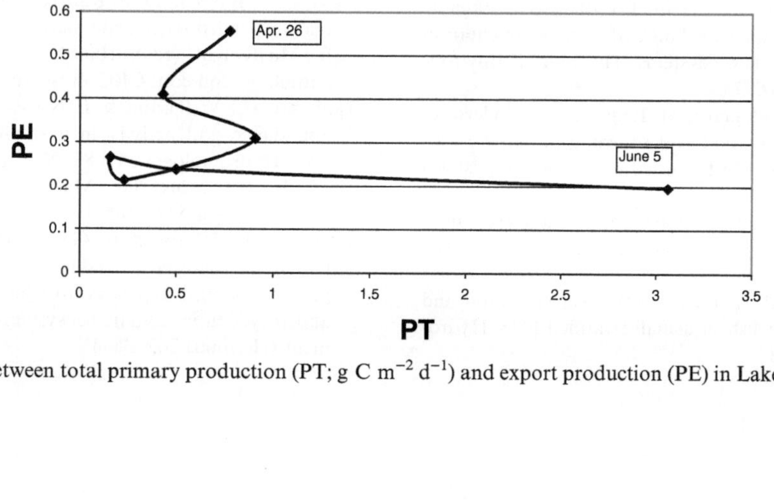

Figure 11. Scatterplot between total primary production (PT; g C m^{-2} d^{-1}) and export production (PE) in Lake Verevi in 2001 (– ♦ –).

of which after a short intensive phytoplankton activity in spring, zooplankton takes over the top down control of matter circulation.

Acknowledgements

The study was supported by the core grants of the Ministry of Education Nos. 0370208s98, 0362482s03 and by Grants of the Estonian Science Foundation No 3579 and 4835. We would like thank all group members (T. Kõiv, P. Nõges, T. Nõges, K. Kangro, P. Zingel, K. Kübar, E. Lill, H. Tammert, T. Timm, H. Timm, H. Mäemets, L. Freiberg, A. Kisand, V. Kisand, H. Starast, T. Möls, A. Lindpere, I. Solovjova, M. Reunanen, T. Krause, A. Järvalt, A. Palm, R. Laugaste etc.) who participated in the project. We would also like to thank Prof A. Maastik, who revised the English and gave good recommendations, and Anne Jõeveer and Ingrid Niklus of the Tõravere Station of the Estonian Institute of Meteorology and Hydrology for climatic data.

References

Bloesch, J., 1996. Towards a new generation of sediment traps and better measurement/understanding of settling particle flux in lakes and oceans: a hydrodynamical protocol. Aquatic Sciences 58/4: 283–296.

Bloesch, J. & U. Uehlinger, 1986. Horizontal sedimentation differences in a eutrophic Swiss lake. Limnology and Oceanography 31: 1094–1109.

Eppley, R. & B. J. Peterson, 1979. Particulate organic flux and planktonic new production in the deep ocean. Nature 282: 677–680.

Heiskanen, A.-S., 1999. Sedimentation in the open and enclosed water columns: the effect of algal blooms, planktonic food web, and resuspension on the quality of settling organic matter in the coastal northern Baltic Sea. In. Sedimentation and recycling in aquatic ecosystems. The Finnish Environment (Helsinki) 263: 69–78.

Heiskanen, A.-S., C. Lundsgaard, M. Reigstadt, S. Floderus & K. Olli, 1999. Sedimentation and recycling in aquatic ecosystems: Introduction. The Finnish Environment (Helsinki) 263: 3–4.

Huttula T. & K. Krogerus, 1986. Water currents and erosion of cellulose fibers in a short term regulated water course. Aqua Fennica (Helsinki) 16(2).

Järvalt, A., T. Krause & A. Palm, 2005. Diel migration and spatial distribution of fish in asmall stratified lake. Hydrobiologia 547: 197–203.

Kangro, K., R. Laugaste, P. Nõges & I. Ott, 2005. Long-term changes and seasonal development of phytoplankton in a strongly stratified, hypertrophic lake. Hydrobiologia 547: 91–103.

Kõiv, T. & K. Kangro, 2005. Resource ratios and phytoplankton species composition in a strongly stratified lake. Hydrobiologia 547: 123–135.

Kübar, K., H. Agasild, T. Virro. & I. Ott, 2005. Vertical distribution of zooplankton in a strongly stratified hypertrophic lake. Hydrobiologia 547: 151–162.

Laugaste, R. & M. Reunanen, 2005. The composition and density of epiphyton on some macrophyte species in the partly meromictic Lake Verevi. Hydrobiologia 547: 137–150.

Mäemets, H. & L. Freiberg, 2005. Long- and short-term changes of the macrophyte vegetation in strongly stratified hypertrophic Lake Verevi. Hydrobiologia 547: 175–184.

Nõges, T. & K. Kangro, 2005. Primary production of phytoplankton in a strongly stratified temperate lake. Hydrobiologia 547: 105–122.

Nõges, T. & I. Solovjova, 2005. The formation and dynamics of deep bacteriochlorophyll maximum in the temperate and partly meromictic Lake Verevi. Hydrobiologia 547: 73–81.

Nõges, P., 2005. Water and nutrient mass balance of the partly meromictic temperate Lake Verevi. Hydrobiologia 547: 21–31.

Nõges, T., P. Nõges, A. Kisand, V. Kisand, L. Tuvikene, P. Zingel, A. Põllumäe & J. Haberman, 1998. Ecological studies. Sedimentation rate. In Huttula, T. & T. Nõges (eds), Present state and future fate of Lake Võrtsjärv. Results from Finnish-Estonian joint project in 1993–1997. The Finnish Environment, 209: 98–104.

Ott, I., T. Kõiv, P. Nõges, A. Kisand, A. Järvalt & E. Kirt, 2005. General description of partly meromictic hypertrophic Lake Verevi, its ecological status, changes during the past eight decades and restoration problems. Hydrobiologia 547: 1–20.

Punning, J.-M., T. Koff, T. Alliksaar & J. Terasmaa, 2002. Tracing the pathways of settling particles into lake sediments. Proceedings of the Estonian Academy of Sciences, Biology, Ecology 51: 225–240.

Reinart, A., H. Arst & D. C. Pierson, 2005. Optical properties and light climate in Lake Verevi. Hydrobiologia 547: 41–49.

Redfield, A.S., 1958. The biological control of chemical factors in the environment. American Scientist 46: 205–211.

Reports of the Baltic Intercalibration Workshop, 1977, Kiel, 27–28.

Rosa, F., J. Bloesch, D. E. Rathke, 1994. Sampling the settling and suspended particulate matter (SPM). In Murdoch A. & S. MacKnight (eds), Handbook for Aquatic Sediments. Sampling, 2nd edn. CRC Press, Inc., 97–129.

Tammert, H., V. Kisand & T. Nõges, 2005. Bacterioplankton abundance and activity in a small hypertrophic stratified lake. Hydrobiologia 547: 83–90.

Timm, H. & T. Möls, 2005. Macrozoobenthos of Lake Verevi. Hydrobiologia 547: 185–195.

Viner, Y., T. Zohary. & A. Gasith, 1999. Dinoflagellate sedimentation followed by sediment traps with daily collection periods: Should a preservative be used? In Sedimentation and recycling in aquatic ecosystems. The Finnish Environment (Helsinki) 263: 38–43.

Wassmann, P., 1999. Primary and export production: problems in quantifying and predicting vertical export. In: Sedimentation and recycling in aquatic ecosystems. The Finnish Environment (Helsinki) 263: 10–21.

Wetzel, R., 1983, Limnology. Saunders College Publishing. 767 p.

Zingel, P., 2005. Vertical and seasonal dynamics of planktonic ciliates in a strongly stratified hypertrophic lake. Hydrobiologia 547: 163–174.

Zingel, P. & I. Ott, 2000. Vertical distribution of planktonic ciliates in strongly stratified temperate lakes. Hydrobiologia 435: 19–26.

Hydrobiologia (2005) 547:63–71
I. Ott & T. Kõiv (eds), Lake Verevi, Estonia – A Highly Stratified Hypertrophic Lake
DOI 10.1007/s10750-005-4145-y

Nitrogen dynamics in the steeply stratified, temperate Lake Verevi, Estonia

Ilmar Tõnno[1,2,*], Katrin Ott[1] & Tiina Nõges[1,2]
[1]*Institute of Zoology and Botany, Võrtsjärv Limnological Station, Estonian Agricultural University, 61101, Rannu, Tartu County, Estonia*
[2]*Institute of Zoology and Hydrobiology, University of Tartu, Vanemuise 46, 51014, Tartu, Estonia*
*(*Author for correspondence: E-mail: ilmar@zbi.ee)*

Key words: stratified lake, nitrogen dynamics, planktonic N_2-fixation, nitrification

Abstract

The dynamics of different nitrogen compounds and nitrification in diverse habitats of a stratified Lake Verevi (Estonia) was investigated in 2000–2001. Also planktonic N_2-fixation (N_2fix) was measured in August of the observed years. The nitrogen that accumulated in the hypolimnion was trapped in the non-mixed layer during most of the vegetation period causing a concentration of an order of magnitude higher than in the epilimnion. The ammonium level remained low in the epilimnion (maximum 577 mgN m^{-3}, average 115 mgN m^{-3}) in spite of high concentrations in the hypolimnion (maximum 12223 mgN m^{-3}, average 4807 mgN m^{-3}). The concentrations of NO_2^- and NO_3^- remained on a low level both in the epilimnion (average 0.94 and 9.09 mgN m^{-3}, respectively) and hypolimnion (average 0.47 and 5.05 mgN m^{-3}, respectively). N_2fix and nitrification ranged from 0.30 to 2.80 mgN m^{-3} day^{-1} and 6.0 to 107 mgN m^{-3} day^{-1}, respectively; the most intensive processes occurred in 07.08.00 at depths of 2 and 5 m, accordingly. The role of N_2fix in the total nitrogen budget of Lake Verevi (in August 2000 and 2001) was negligible while episodically in the nitrogen-depleted epilimnion the N_2fix could substantially contribute to the pool of mineral nitrogen. Nitrification was unable to influence nitrogen dynamics in the epilimnion while some temporary coupling with ammonium dynamics in the hypolimnion was documented.

Introduction

Nitrogen (N) is one of the main building blocks for the production of organic matter on the planet Earth and it is required in greatest quantities (Stolp, 1996; Williams et al., 2002). N, as many other elements in the world, is involved in cyclical transformations. Non-biological transformations have little importance in the nitrogen cycle by contrast with biological transformations, which are primarily controlled by microorganisms (Gorlenko et al., 1977; Sprent, 1987; Stolp, 1996; Voytek et al., 1999). A nitrogen cycle consists of four main components: molecular nitrogen (N_2) fixation (N_2fix), mineralization of organic N (ammonification), nitrification, and denitrification.

Only prokaryotes are capable of N_2fix. In aquatic ecosystems cyanobacteria appear responsible for most of the planktonic N_2fix while heterotrophic bacteria are most important N_2 fixers in lake sediments. The fixation of N_2 by microorganisms is the only process in nature that counteracts the nitrogen losses from the environment by denitrification. The central compound of the nitrogen cycle is ammonium (NH_4^+) which is released into the water by zooplankon and represents the main decomposition product of urea of other animals like fish. In anaerobic hypolimnion where animals are scarce, ammonium is formed at amino-acid degradation of proteins carried out by ammonificating bacteria, occurring in the water column and sediments (Gorlenko et al., 1977;

Howarth et al., 1988a; Stolp, 1996). Nitrification is a two-step oxidation of NH_4^+ through nitrite (NO_2^-) to nitrate (NO_3^-), carried out mainly by chemolithoautotrophic bacteria in aerobic conditions. The most important and intensive site in lakes for nitrification are aerobic sediments while planktonic nitrification could be also significant. Denitrification is an anaerobic heterotrophic process, which shares many of the same substrates and intermediates as nitrification. Denitrification leads to gaseous nitrogen (N_2, N_2O) losses counteracting N_2fix (Gorlenko et al., 1977; Hall, 1982; Henriksen et al., 1993; Stolp, 1996; Voytek et al., 1999). In stratified lakes phytoplankton takes up epilimnetic mineral nitrogen and transports it to the hypolimnion via sedimentation. N may accumulate in the hypolimnion during stratification period while in the epilimnion N deficiency may occur if resupply from the inflows is limited (Scheffer, 1998).

The aim of the present study was to investigate the dynamics of different nitrogen compounds as well as the rates of N_2-fixation and nitrification in diverse habitats of a stratified partly meromictic lake. The main processes of transformations (Fig. 1) were followed on the background of the dynamics of the physico-chemical stratification regime.

Materials and methods

Lake Verevi (0.126 km^2, mean depth 3.6 m, maximum depth 11 m) is a small stratified hypertrophic (see Ott et al., the present issue) lake in South Estonia. The lake is characterized by strong stratification from April to September and an anoxic hypolimnion. The main N_2fix cyanobacterium in Lake Verevi during the years 2000 and 2001 was *Aphanizomenon klebahnii* (Elenkin) Pechar *et* Kalina (Kangro et al., the present issue).

The water samples for nitrogen determination were collected from April to December 2000 and from March to August 2001. In the year 2000 eight and in 2001 three to eight vertical samples were taken at different depths of the epi-, meta- and hypolimnion. The water from the surface layer (0.5 m) was taken directly into the bottle, for other depths a Masterflex pump was used (for details see Zingel, the present issue). Total nitrogen (TN), NH_4^+, NO_2^- and NO_3^- was analysed at the laboratory of Võrtsjärv Limnological Station using the methods described by Grasshoff et al. (1983). For more detailed description of TN determination see Ott et al. (present issue). Ammonium was determined (detection error ± 5.5%) with indophenol blue method (Hansen & Koroleff, 1999). Nitrate was reduced to nitrite, and sulphanil-amide and

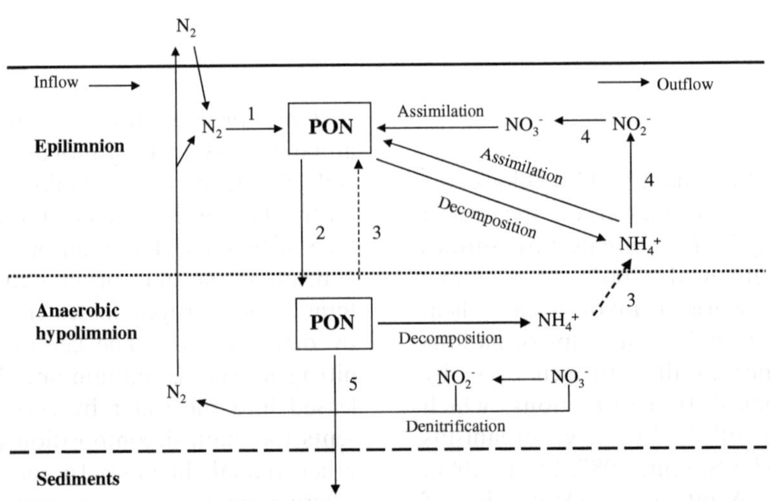

Figure 1. Conceptual scheme of the main processes of the nitrogen cycle in a stratified lake according to Lampert and Sommer (1997): 1 – planktic molecular nitrogen fixation (N_2fix); 2 – a part of the PON (particular organic nitrogen) sinks to the hypolimnion; 3 – due to water mixing some PON is carried from hypolimnion to epilimnion; 4 – in epilimnion ammonium (NH_4^+) is subject to nitrification; 5 – a part of PON is switched off from nitrogen cycle due to sedimentation.

N-(1-naphthyl)-ethylenediamine dihydrochloride was used (detection error ± 2%) for the determination of NO_2^- (Koroleff, 1982). The total amounts of measured nitrogen forms in the epilimnion, hypolimnion and in the whole water column for both years were calculated by integrating the concentrations of the compounds in different water layers. For a detailed description of the calculation method see Nõges et al. (in the present issue).

N_2-fixation in Lake Verevi was measured on August 7, 2000 at one, on August 28, 2000 at two, and on August 2, 2001 at four different depth horizons in the epilimnion (Table 1) applying the acetylene reduction method (Stewart et al., 1967; Présing et al., 1996). Three 60-ml glass bottles were filled with lake water from each depth horizons and exposed for 4 h in the incubator at a constant illumination of 120 W m^{-2} *in situ* temperature. For details see Tõnno & Nõges (2003).

Depths for the nitrification measurement were selected according to the supposed NH_4^+ and oxygen content (Table 1). At each depth four dark glass scintillation vials (two 'samples' and two 'blanks') with a capacity of 24 ml were filled with

Table 1. Dates, sampling depths and measured parameters in Lake Verevi in 2000–2001

Date	Depth (m)	Measured parameters					
		Nitrification rate (mgN m^{-3} day^{-1})	TN (mgN m^{-3})	NH_4^+ (mgN m^{-3})	NO_2^- mgN m^{-3})	NO_3^- (mgN m^{-3})	N_2-fixation (mgN m^{-3} day^{-1})
07.08.2000	2.0	–	+	+	+	+	+
07.08.2000	4.0	+	+	+	+	+	–
07.08.2000	5.0	+	+	+	+	+	–
28.08.2000	0.5	–	+	+	+	+	+
28.08.2000	1.0	–	–	–	–	–	+
28.08.2000	5.0	+	+	+	+	+	–
28.08.2000	5.5	+	+	+	+	+	–
23.04.2001	0.5	+	+	+	+	+	–
23.04.2001	5.5	+	+	+	+	+	–
23.04.2001	8.5	+	+	+	+	+	–
30.04.2001	0.5	+	+	+	+	+	–
30.04.2001	5.0	+	+	+	+	+	–
30.04.2001	5.25	+	+	+	+	+	–
30.04.2001	5.5	+	+	+	+	+	–
07.05.2001	0.5	+	+	+	+	+	–
07.05.2001	5.0	+	+	+	+	+	–
07.05.2001	5.25	+	+	+	+	+	–
07.05.2001	5.5	+	+	+	+	+	–
05.06.2001	0.5	+	+	+	+	+	–
05.06.2001	5.0	+	+	+	+	+	–
05.06.2001	5.25	+	+	+	+	+	–
05.06.2001	5.5	+	+	+	+	+	–
02.08.2001	0.5	+	+	+	+	+	+
02.08.2001	1.0	+	+	+	+	+	+
02.08.2001	2.0	+	+	+	+	+	+
02.08.2001	3.0	+	+	+	+	+	+
02.08.2001	4.0	+	+	+	+	+	–
02.08.2001	5.0	+	+	+	+	+	–
02.08.2001	6.0	+	+	+	+	+	–
02.08.2001	7.0	+	+	+	+	+	–

lake water. $NaH^{14}CO_3$ (VKI, Denmark) was added to each vial with a final activity of 0.07 μCi ml^{-1}. To 'blank' vials 100 μl of nitrification inhibitor 2-chloro-6-(trichloromethyl) pyridine (TCMP) was added (final concentration 10 mg l^{-1}). Thereafter the vials were incubated 24 h in thermos flasks containing the water from the same depth where the samples had been taken from. After the incubation 100 μl of water from each vial was taken and mixed with 0.5 ml of β-phenylethylamine (PEA) for the assessment of total radioactivity by using 5 ml of Optiphase solution and LSC RackBeta 1211 (Wallac, Finland). The rest of the water sample from the vials (23.9 ml) was filtered through membranes of 0.20 μm pore size (Millipore, HA). The filters were treated with concentrated HCl fumes for 5 min to remove the excess of inorganic ^{14}C, and air-dried for 24 h. Five milliliters of toluene-PPO-POPOP cocktail was added to filters and their radioactivity was assessed with LSC RackBeta 1211.

Chemosynthetic fixation of CO_2 was calculated by the formula:

$$R = [x * C * 1.05 * 1.06 * V1 * k]/y * V2 * t$$

where R, CO_2 fixation rate (mmole m^{-3} h^{-1}); x, difference of the radioactivities of the filter from the 'sample', and the filter from the 'blank'; C, concentration of HCO_3^- in water (mmole l^{-1}); 1.05, coefficient considering the difference of assimilation efficiencies of $^{12}CO_2$ and $^{14}CO_2$; 1.06, factor considering the respiration losses of the assimilated CO_2 during the exposition; $V1$, volume of the exposition vial (ml); k, 1000 coefficient from litres to cubic meters; y, radioactivity of $NaH^{14}CO_3$ solution added to the vial; $V2$, amount of the filtered water (ml); t, incubation time (h). To estimate nitrification, we used an average conversion factor of 8.3 moles of N oxidized per mole of carbon fixed (Owens, 1986; Joye et al., 1999).

Results

In Lake Verevi the concentration of total nitrogen in the hypolimnion (annual average 6646 mgN m^{-3}) was by an order of magnitude higher than in the epilimnion (annual average 948 mgN m^{-3}). TN concentration in the epilimnion and in the hypolimnion from April to September was 781 and 4007 mgN m^{-3}, respectively, increasing by the end of the vegetation period up to 1284 and 11922 mgN m^{-3}, respectively (Fig. 2a). Mean epilimnetic concentration of ammonium from May to August was 6.4 mgN m^{-3}, followed by a sharp increase. In the hypolimnion, the concentration of NH_4^+ was about 35 times higher (average from May to August 2118 mgN m^{-3}) than in the epilimnion but followed the same dynamics (Fig. 2b). The NO_2^- content in the epi- and hypolimnion stayed on a low level from April to October (average 0.21 and 0.16 mgN m^{-3}, respectively), increasing up to 6.8 and 2.9 mgN m^{-3}, respectively by November/December (Fig. 2c). The mean epilimnetic concentration of NO_3^- from April to October was 1.5 mgN m^{-3} increasing abruptly up to 130 mgN m^{-3} by November/December (Fig. 2d). Mean hypolimnetic nitrate concentration in spring was 3.4 mgN m^{-3}, by the end of June it decreased to undetectable values, and two peaks occurred in autumn: in September (4.1 mgN m^{-3}) and in December (58 mgN m^{-3}).

N_2-fixation was measured in 2000 on August 7th at a depth of 2 m (2.80 mgN m^{-3} day^{-1}), and on August 28th at a depth of 0.5 m (0.30 mgN m^{-3} day^{-1}) forming, respectively, 0.31 and 0.043% of TN, and 140 and 15% of the amount of mineral nitrogen in the investigated depth horizons. On August 2, 2001 the N_2fix occurred only at a depth of 3 m (0.38 mgN m^{-3} day^{-1}), taking up 0.05% of TN and 5.37% of mineral nitrogen in the investigated depth horizon.

Nitrification occurred in 2000 on August 7th at a depth of 5 m (107 mgN m^{-3} day^{-1}), and on August 28th at a depth of 5.5 m (54.4 mgN m^{-3} day^{-1}). In 2001 we were unable to detect any nitrification on April 23rd and on May 7th (Table 1) while on April 30th and June 5th nitrification occurred at a depth of 0.5 m (6.0 and 7.5 mgN m^{-3} day^{-1}, respectively) and 5.5 m (13.3 and 17.7 mgN m^{-3} day^{-1}, respectively). On August 2nd we detected nitrification at four depths: 0.5, 3, 6, and 7 m (26.1, 50.7, 79.3 and 308 mgNm^{-3} day^{-1}, respectively).

Discussion

As it is common to stratified lakes (Scheffer, 1998), in Lake Verevi the nitrogen that accumulated in

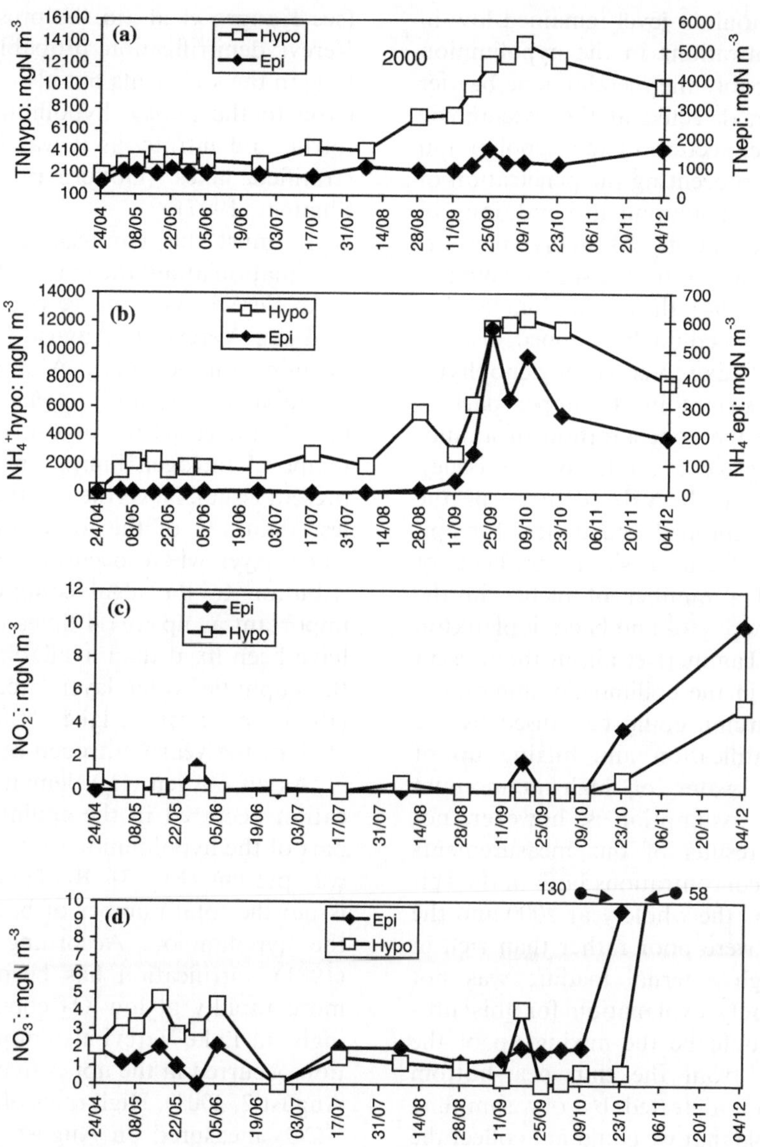

Figure 2. Seasonal course of the epilimnetic (Epi) and hypolimnetic (Hypo) (a) total nitrogen (TN) (b) ammonium (NH_4^+) (c) nitrite (NO_2^-), and (d) nitrate (NO_3^-) in Lake Verevi in 2000.

the hypolimnion was trapped in the non-mixed layer during most of the vegetation period remaining inaccessible to the epilimnetic community.

The concentration of ammonium in the hypolimnion of Lake Verevi was high (maximum 12223 mgN m^{-3}, average 4807 mgN m^{-3}) compared to the other Estonian stratified eutrophic lakes, where according to the database of Võrtsjärv Limnological Station (108 lakes), the average is 742 mgN m^{-3}, and four Canadian lakes, where the NH_4^+ content in the hypolimnion remained below 2000 mgN m^{-3} (Knowles et al., 1981). The high hypolimnetic ammonium concentration implies that most of the epilimnetically derived particulate organic matter was decomposed in this region (Priscu et al., 1986). According to Tammert et al. (present issue) bacteria are one of the most important pools of nutrients (nitrogen, phosphorus) in the hypolimnion of Lake Verevi. In the epilimnion, the ammonium level remained low in spite of high concentrations in the hypolimnion (Fig. 2b). Most probably the metalimnetic barrier but also nitrification detected at the oxic/anoxic interface of the upper section of the hypolimnion were responsible for preventing the penetration of ammonium into the epilimnion in conditions of stable summer stratification. As many microorganisms prefer NH_4^+ as a nitrogen source (Wetzel, 1983; Ahlgren et al., 1994), the ammonium leaking from the hypolimnion could be trapped also by phytoplankton. In Lake Verevi euglenophytes were numerous important in the upper part of hypolimnion, where the concentration of ammonium was high (Kangro et al., in the present issue). Although there was high NH_4^+ content in the hypolimnion, it was trapped and useless for epilimnetic organisms. Such a sharp gradient of nutrients composed a number of niches in the water column for the phyto- and bacterioplankton (see Kangro et al.; Tammert et al., in the present issue). An increase in the epilimnetic ammonium concentration in autumn could be caused by the disturbance of stratification and mixing up of some hypolimnetic water of high ammonium concentration. This assumption is, however, not supported by the results of our measurements showing increasing concentrations both in the epi- and hypolimnion. As the whole year 2000 and the September of 2001 were poor rather than rich in precipitation, a high external loading was not likely either. The only explanation for this pronounced increase could be the mixing up of the nutrient-rich water from the thin near-bottom layer which was not detected by our sampling strategy. It is possible that we could not collect the nearest to the bottom water layer by applied water sampler. As shown by Kõiv & Kangro (the present issue), the concentration of total phosphorus (TP) and SRP in the whole water column also increased in September, which supports the hypothesis that the near-bottom water layer had been mixed up. The hypothesis, however, remains speculative as we have no evidence of such near bottom nutrient rich layer. The concentrations of NO_2^- and NO_3^- remained on a low level both in the epi- and hypolimnion (Fig. 2c, d). The nitrogen mineralized in the epilimnion was probably quickly assimilated by the phytoplankton, causing temporal nitrogen limitation and increase of N_2-fixing cyanobacteria (see Kangro et al., in the present issue). In Lake Verevi denitrification probably could occur not only in the sediments but also in the water column close to the anoxic hypolimnion, and, thus, use nitrite and nitrate, as it was found also in other stratified lakes (Golterman, 1975; Lampert & Sommer, 1997).

It must be emphasized that there is no information about the seasonality of N_2-fixation in Lake Verevi. As a source of nitrogen, N_2-fixation in Lake Verevi in August 2000 and 2001 was of minor importance. Abundance of N_2-fixing cyanobacteria (mainly *Aphanizomenon klebahnii*) usually increased in August, during the period of temporal nitrogen limitation (Kangro et al., the present issue). Nevertheless, the daily input formed less than 1% of the total amount of TN in the water layer where occurred. The temporary contribution to the algal community could still be important as up to 1.4 times more nitrogen could have been fixed than available in mineral form in the euphotic water layer. According to Kostjaev (1986), in eutrophic lakes N_2fix could form up to 50% of the yearly nitrogen budget.

As an oxygen (O_2) demanding process, nitrification occurred in the epilimnion and the upper part of the hypolimnion (5–5.5 m), where some O_2 was present (Fig. 3). By Tammert et al. (present issue) the total number of bacteria was highest in the hypolimnion. According to Knowles et al. (1981), nitrification has been reported to occur more rapidly at low O_2 concentrations. Accordingly, in Lake Verevi also more intensive nitrification occurred in the upper hypolimnion (except on August 2, 2001). High rates of dark assimilation of $^{14}CO_2$ measured on August 2nd, 2001 in anoxic H_2S-rich water at depths of 6 and 7 m could probably indicate not the nitrification but rather the oxidation of H_2S by sulphur chemoautotrophes, which can also assimilate inorganic carbon in darkness (Gorlenko et al., 1977). Nitrification intensity in Lake Verevi (up to 308 mgN m^{-3} day^{-1}, Fig. 3) was much higher than recorded by Hall (1982) in the hypolimnion of mesotrophic L. Grasmere (in all cases less than 8 mgN m^{-3} day^{-1}). According to Hall (1982), planktonic nitrification in aerobic hypolimnetic water could be important in affecting changes in the water chemistry of this water layer. In Lake Verevi, however, a remarkable influence on nitrogen dynamics

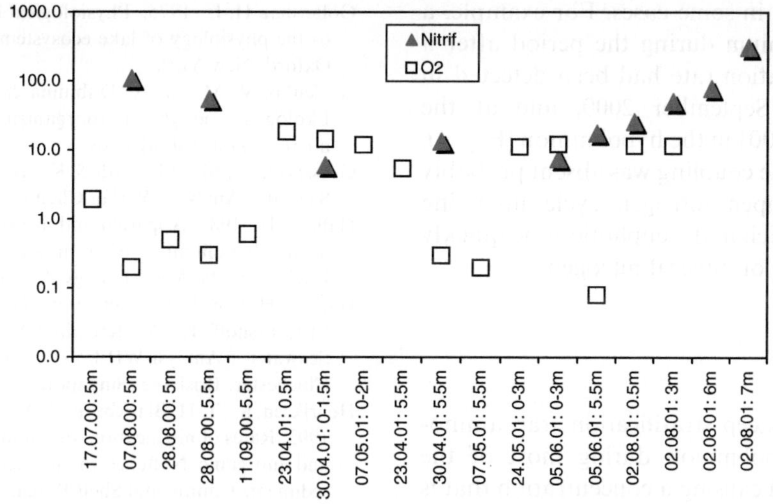

Figure 3. Dynamics of nitrification (Nitrif: mgN m^{-3} day^{-1}) and dissolved oxygen (O_2: mg l^{-1}) in Lake Verevi in 2000–2001 (depth-integrated water samples were collected from water layers 0–1.5, 0–2 and 0–3 m).

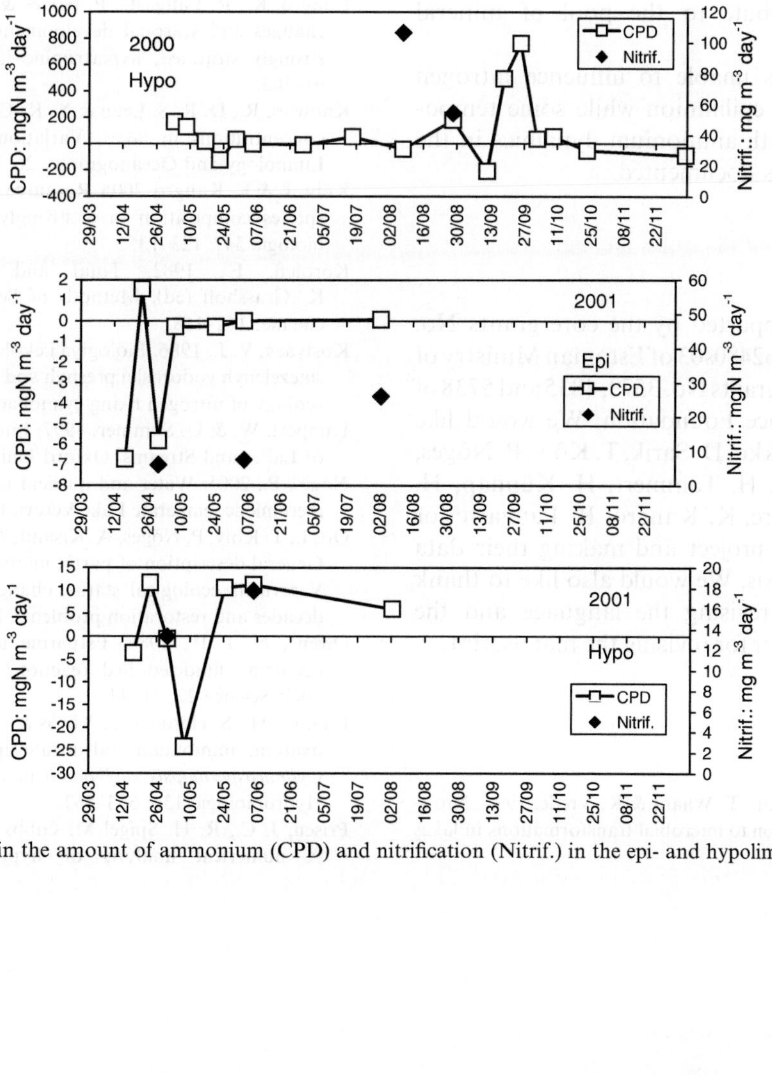

Figure 4. Daily changes in the amount of ammonium (CPD) and nitrification (Nitrif.) in the epi- and hypolimnion of Lake Verevi in 2000–2001.

could be quantified in some cases. For example, a decrease in ammonium during the period after a rather high nitrification rate had been detected in late August/early September 2000, and at the beginning of May 2001 in the hypolimnion (Fig. 4). In the epilimnion the coupling was absent probably because a more open nitrogen cycle involving phytoplankton, which in the euphotic zone quickly consumes all forms of mineral nitrogen.

Conclusions

- In Lake Verevi steep stratification trap ammonium in the hypolimnion during most of the vegetation period causing a concentration that is by an order of magnitude higher than in the epilimnion.
- The role of N_2-fixation in the total nitrogen budget of Lake Verevi in our investigation period was negligible while episodically in the nitrogen-depleted epilimnion N_2fix could substantially contribute to the pool of mineral nitrogen.
- Nitrification was unable to influence nitrogen dynamics in the epilimnion while some temporary coupling with ammonium dynamics in the hypolimnion was documented.

Acknowledgements

This work was supported by the core grants No. 0370208s98 and 0362480s03 of Estonian Ministry of Education, and by grants No. 3579, 4835 and 5738 of the Estonian Science Foundation. We would like thank I. Ott, A. Rakko, D. Sarik, T. Kõiv, P. Nõges, K. Kübar, E. Lill, H. Tammert, H. Künnap, H. Starast, A. Lindpere, K. Kangro, R. Laugaste for taking part in the project and making their data available for analysis. We would also like to thank Dr. E. Veldi for revising the language and the anonymous reviewer for revising the manuscript.

References

Ahlgren, I., F. Sörensson, T. Waara & K. Vrede, 1994. Nitrogen budgets in relation to microbial transformations in lakes Ambio 23: 367–377.

Golterman H. L., 1975. Physiological limnology. An approach to the physiology of lake ecosystems. Elsevier, Amsterdam, Oxford, New York.

Gorlenko, V. M., G. A. Dubinina & C. I. Kuznetsov, 1977. Ekologija vodnyh mikroorganizmov [Ecology of water micro-organisms]. Moscoy.

Grasshoff, K., M. Ehrhardt & K. Kremling, 1983. Methods of Seawater Analysis. Verlag Chemie, Weinheim.

Hall, G. H., 1982. Apparent and measured rates of nitrification in the hypolimnion of a mesotrophic lake Applied and Environmental Microbiology 43: 542–547.

Hansen, H. P. & F. Koroleff, 1999. Determination of nutrients. In Grasshoff, K., M. Kremling & K. EhrhardtMethods of Seawater AnalysisWILEY-VCHWeinheim New York Chichester. Brisbane. Singapore. Toronto,600 pp.

Henriksen, K., T. H. Blackburn, B. A. Lomstein & C. P. McRoy, 1993. Rates of nitrification, distribution of nitrifying bacteria and inorganic N fluxes in northern Bering-Chukchi shelf sediments Continental Shelf Research 13: 629–651.

Howarth, R. W., J. J. Cole, R. Marino & J. Lane, 1988. Nitrogen fixation in freshwater, estuarine and marine ecosystems. 1. Rates and importance Limnology and Oceanography 33: 669–687.

Joye, S. B., T. L. Connell, L. G. Miller & R. S. Oremland, 1999. Oxidation of ammonia and methane in an alkaline, saline lake Limnology and Oceanography 44: 178–188.

Kangro, K., R. Laugaste, P. Nõges & I. Ott, 2005. Long-term changes and seasonal development of phytoplankton in a strongly stratified, hypertrophic lake. Hydrobiologia 547: 91–103.

Knowles, R., D. R. S. Lean & Y. K. Chan, 1981. Nitrous oxide concentrations in lakes: Variations with depth and time Limnology and Oceanography 26: 855–866.

Kõiv, T. & K. Kangro, 2005. Resource ratios and phytoplankton species composition in a strongly stratified lake. Hydrobiologia 547: 123–135.

Koroleff, F., 1982. Total and organic nitrogen. In K. Grasshoff (ed), Methods of Seawater Analysis. Verlag chemie, 162–168.

Kostyaev, V. J., 1986. Biologiya i èkologiya azotfiksiruyushchih sinezelenyh vodoroslej presnyh vod. Leningrad. [Biology and ecology of nitrogen fixing cyanobacteria from inland water].

Lampert, W. & U. Sommer, 1997. Limnoecology: The Ecology of Lakes and Streams. Oxford University Press.

Nõges, P., 2005. Water and nutrient mass balance of the partly meromictic temperate Lake Verevi. Hydrobiologia 547: 21–31.

Ott, I., T. Kõiv, P. Nõges, A. Kisand, A. Järvalt & E. Kirt, 2005. General description of partly meromictic hypertrophic Lake Verevi, its ecological status, changes during the past eight decades and restoration problems. Hydrobiologia 547: 1–20.

Owens, N. J. P., 1986. Estuarine nitrification: A naturally occurring fluidized bed reaction? Estuarine, Coastal and Shelf Science 22: 31–44.

Présing, M., S. Herodek L. Vörös & I. Kobor, 1996. Nitrogen fixation, ammonium and nitrate uptake during a bloom of *Cylindrospermopsis raciborskii* in Lake Balaton Archiv für Hydrobiologie 136: 553–562.

Priscu, J. C., R. H. Spigel M. Gibbs & M. T. Downes, 1986. A numerical analysis of hypolimnetic nitrogen and phosphorus transformations in Lake Rotoiti, New Zealand: A geothermally influenced lake Limnology and Oceanography 31: 812–831.

Scheffer, M., 1998. Ecology of Shallow Lakes. Chapman & Hall, London.

Sprent, J. I., 1987. The Ecology of the Nitrogen Cycle. Cambridge University Press, USA.

Stewart, W. D. P., G. P. Fitzgerald & R. M. Burris, 1967. In situ studies on N_2 fixation using the acetylene reduction technique Proceedings of the Nnational Academy of Sciences of the U.S.A. 58: 2071–2078.

Stolp, H., 1996. Microbial Ecology: Organisms, Habitats, Activities. Cambridge University Press, USA.

Tammert, H., V. Kisand & T. Nõges, 2005. Bacterioplankton abundance and activity in a small hypertrophic stratified lake. Hydrobiologia 547: 83–90.

Tõnno, I. & T. Nõges, 2003. Nitrogen fixation in a large shallow lake: rates and initiation conditions Hydrobiologia 490: 23–30.

Voytek, M. A., J. C. Priscu & B. B. Ward, 1999. The distribution and relative abundance of ammonia-oxidizing bacteria in lakes of the McMurdo Dry Valley, Antarctica Hydrobiologia 401: 113–130.

Wetzel, R., 1983. Limnology. Saunders College Publishing.

Williams, P. J. le B., D. N. Thomas & C. S. Reynolds (eds), (2002) Phytoplankton Productivity. Carbon Assimilation in Marine and Freshwater Ecosystems. Blackwell Science, UK 109–141.

Zingel, P., 2005. Vertical and seasonal dynamics of planktonic ciliates in a strongly stratified hypertrophic lake. Hydrobiologia 547: 163–174.

Hydrobiologia (2005) 547:73–81
I. Ott & T. Kõiv (eds), Lake Verevi, Estonia – A Highly Stratified Hypertrophic Lake
DOI 10.1007/s10750-005-4147-9

The formation and dynamics of deep bacteriochlorophyll maximum in the temperate and partly meromictic Lake Verevi

Tiina Nõges[1,*] & Irena Solovjova[2]
[1]*Estonian Agricultural University, Institute of Agriculture and Environmental Sciences, Limnological Centre, 61101 Rannu, Tartu County, Estonia*
[2]*University of Tartu, Institute of Zoology and Hydrobiology, Vanemuise 46, 51014, Tartu, Estonia*
*(*Author for correspondence: E-mail: tnoges@zbi.ee)*

Key words: stratified lake, deep chlorophyll maximum, bacteriochlorophyll, photosynthetic bacteria

Abstract

Vertical distribution of phytoplankton and the formation of deep chlorophyll maximum (DCM) in the metalimnion of a small stratified and partly meromictic temperate lake was studied in 1999 and 2000. During summer DCM usually occurred on the borderline of H_2S and oxygen-containing waters. At the depths where the bacteriochlorophyll (Bchl) maxima were observed, the sulphide concentration was usually relatively low compared to the bottom layers, where its concentration reached as high as possible saturation level. In April 2000, DCM was formed at the depth of 3.5 m, and lowered thereafter slowly to 6.5 m by October. The concentration of Bchl *d* reached the highest values (over 1000 $\mu g\ l^{-1}$) just before the water column was mixed up in autumn. In December and April Bchl *d* was detectable only near the bottom of the lake. The concentration of chlorophyll *a* yielded by the spectrophotometric phaeopigment corrected method and by HPLC (high pressure liquid chromatography), fit rather well in the upper layers. In deeper water layers chlorophyll *a* concentration (Chl *a*) measured by spectrophotometry was overestimated about 47 times if compared to HPLC values because of the high Bchl *d* in that layer. In most cases vertical profiles of primary production (PP) did not coincide with the vertical distribution of the pigment content; the maximum values of PP were found in the epilimnion. In some cases PP had notably high values also at the depth of DCM. In the upper layers Chl *a* usually did not exceeded 20 $\mu g\ l^{-1}$ in spring and 10 $\mu g\ l^{-1}$ in summer. The moderately high Chl *a* in the epilimnion in spring was significantly reduced after the formation of thermocline most probably because of the establishment of the nutrient limitation in epilimnion. Decreasing Chl *a* concentration in the epilimnion led to increased water transparency and better light conditions for photosynthetic bacteria in metalimnion.

Introduction

Anoxygenic photosynthetic bacteria contain a single type of reaction centre with a pigment bacteriochlorophyll, which absorbs light of longer, less energy-rich wavelengths than plant chlorophylls (Brock et al., 1994). Green bacteria (Chlorobiaceae) and purple sulphur bacteria (Chromatiaceae) use elemental sulphur, sulphide, thiosulfate, or hydrogen gas as electrondonor, whereas the purple non-sulphur bacteria use electrons from hydrogen or organic substrates. All these bacteria require anaerobic conditions for photosynthetic activity and can grow at very low light intensities. They are present where light reaches anaerobic, sulphide-containing zones in lakes. In meromictic or stagnant holomictic lakes, a dense population of photosynthetic bacteria appears frequently in the contact layer between the oxidative and reductive zones (Takahashi & Ichimura, 1970). By dominating in anoxic regions of the water column they influence primary production (PP), elemental cycling, and trophic interactions (Hurley & Watras, 1991). These bacteria keep growing during the season and show a characteristic pattern of vertical distribution (Takahashi & Ichimura, 1968).

The main factors determining the growth of photosynthetic sulphur bacteria in lakes are H_2S concentration and light conditions (Takahashi & Ichimura, 1970; Steenbergen & Korthals, 1982; Rodrigo et al., 2000). In many aquatic ecosystems, selective attenuation determines that only certain wavelengths of light reach the water layer containing sulphide. For purple and green sulphur bacteria that possess pigments with a specific absorption spectrum, light quality and intensity are key factors in determining which types of bacteria can develop in certain conditions (Parkin & Brock, 1980; Guerrero et al., 1985).

Pigment concentrations are used as indirect measures of photosynthetic biomass in planktonic community. Chlorophyll *a* is the most common measure of phytoplankton biomass while bacteriochlorophylls act as markers for phototrophic bacteria (Hurley & Watras, 1991).

The present paper studied the formation of deep chlorophyll maximum (DCM) in the metalimnion of a small stratified and partly meromictic temperate lake with the aim to quantify the extent of DCM and to find out the regularities and controlling factors of its development.

Description of the study site

Lake Verevi is a hypertrophic but partly meromictic lake with an area of 12.6 ha, maximum depth of 11.0 m, and mean depth of 3.6 m (for more details see Ott et al., present issue). During the summer period the thermocline and chemocline are well established (Figs. 1 and 2).

Materials and methods

Lake Verevi was investigated 15 times from April to December in 2000. On 7 September 1999, samples were taken for performing the HPLC (high pressure liquid chromatography) analysis. Samples were collected at the deepest area of the lake from eight different depth horizons (Table 1), chosen depending on the vertical profiles of temperature and oxygen. Water transparency was measured by a Secchi disc. The profiles of temperature and dissolved oxygen were measured vertically using an Aqua-Check Water Analyzer (USA). Nutrient analyses were performed using the photometric methods described by Grasshoff et al. (1983): soluble reactive phosphorus (SRP) was measured by the molybdate blue method using ascorbic acid as reductant; nitrates were reduced to nitrites by reduction with a cadmium column; in order to determine total nitrogen and total phosphorus, organic compounds were mineralized into nitrite and phosphate, using persulphate. The concentration of sulphide was determined by the methylene blue method described by Parkin & Brock (1980) using an HACH DR/2000 spectrophotometer (USA).

Chlorophyll *a* and Bchl *d* were determined spectrophotometrically using their maximum absorption wavelengths in 90% acetone, 662–665 nm and 654 nm respectively. Pigment samples were filtered on Whatman GF/F filters and extracted with 90% acetone by soaking the filters in the solvent for 4 h at room temperature. The extracts were vortexed and centrifuged for 10 min at 3000 rpm min^{-1}. The absorption of the extract was determined in the region of 430–800 nm by the scanning UV–VIS spectrophotometer Cecil-3000 (Great Britain). Chl *a* concentration was calculated by the equations of Jeffrey & Humphrey (1975) and Lorenzen (1967). For bacteriochlorophyll *d* the equation of Takahashi & Ichimura (1970) was used.

For HPLC analysis in September 1999, the filters were soaked in 1 ml methanol and extracts were sonicated before injection into the column. The analysis of photosynthetic pigments was performed with an HPLC system consisting of an HP 79852A solvent delivery system, coupled to an HP1100 variable-wavelength adsorbance detector. The separating column was a 25 cm long 0.5 cm ID 5 μm ODS-Hypersil column. The injection volume was 100 μl and the flow rate was 1 ml min^{-1}. The samples were injected via a Rheodyne 7125 dosator. Absorption was measured at the wavelength of 435 nm. The solvents and the gradient are described by Mantoura & Llewellyn

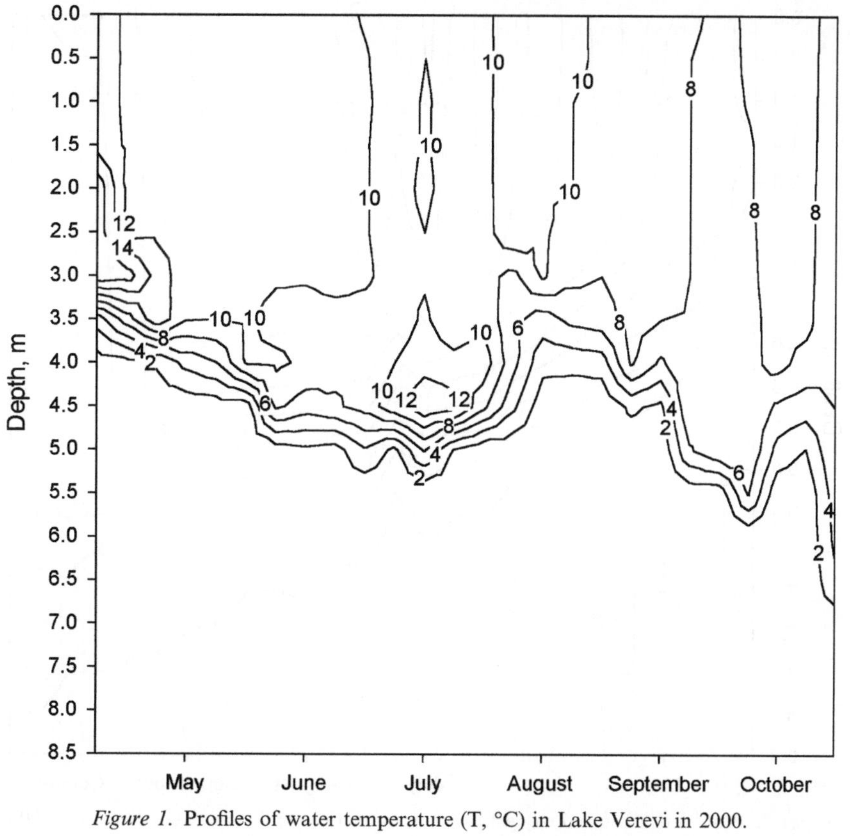

Figure 1. Profiles of water temperature (T, °C) in Lake Verevi in 2000.

(1983). The calibration of Chl *a* was performed by using external Chl *a* standard (Sigma Chemical Company) solution at a concentration of 25 μg ml^{-1}.

The primary production (PP) of phytoplankton was estimated *in situ* using the $^{14}CO_2$ assimilation technique (Steeman-Nielsen, 1952). Water from 5 to 6 different depth horizons in the epi- and metalimnion were poured into 24 ml glass scintillation vials. If handling the anaerobic water from the metalimnion, special attention was paid to keep anaerobic and low light conditions – vials were kept in dark box, the tube from the water sampler was put on the bottom of the vial and the water was let to overflow slowly from the vial for some seconds. 50 μl of sterile $NaH^{14}CO_3$ (VKI, Denmark) solution (1.7 μCi per vial) was added to achieve final activity 0.07 μCi ml^{-1}. The vials were incubated for 2 h at six depths in the lake. Subsequently the 100 μl of water from each bottle was mixed with 0.5 ml β-phenylethylamine (PEA) for the assessment of total radioactivity. Six millilitre of water from each sample was poured into a clean glass scintillation vial and acidified (pH < 2) by adding 150 μl of 0.5 N HCl. Inorganic ^{14}C was assumed to be removed during 24 h (Niemi et al., 1983; Hilmer & Bate, 1989; Lignell, 1992). Next, 5 ml subsamples were poured into new plastic vials. The radioactivity was assessed by LSC RackBeta 1211 (Wallac, Finland) using external standardization for DPM calculations. Scintillation cocktail OPTIPHASE (Wallac, Finland) was applied. Primary production was calculated according to the standard formula (Guidelines for the Baltic monitoring programme for the third stage, 1984). Non-photosynthetic carbon fixation was measured in dark vials and subtracted from light assimilation.

Results

In 2000, Lake Verevi was thermally stratified with a termocline at 3–5 m (Fig. 1). In September 1999, the thermocline reached from the depth of 4–6 m. The water column was not completely mixed in the

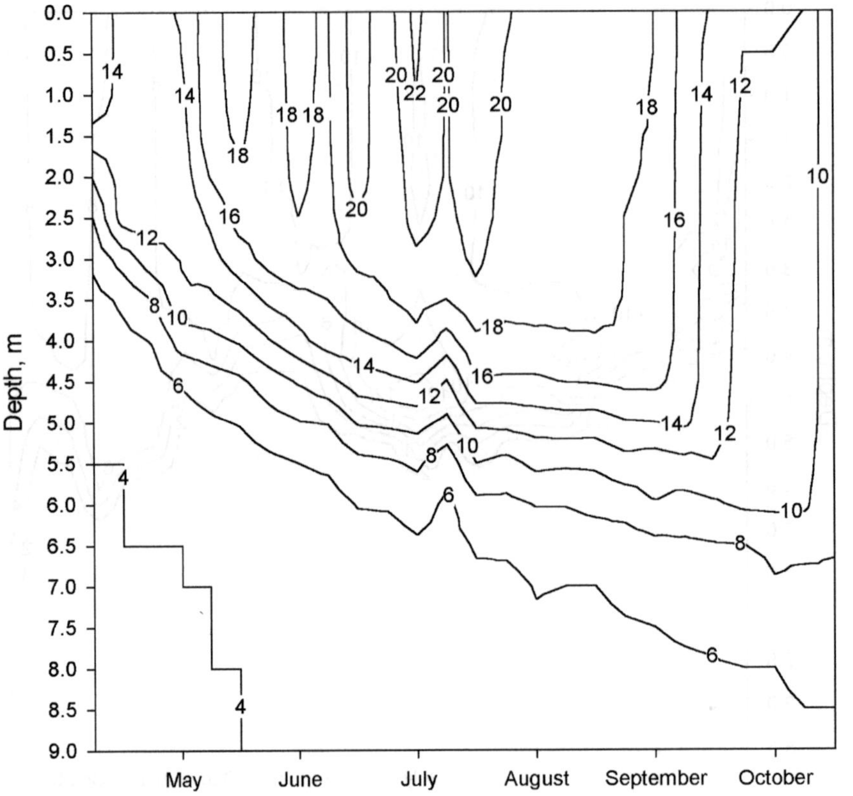

Figure 2. Profiles of dissolved oxygen concentration (O_2, mg l^{-1}) in Lake Verevi in 2000.

Table 1. Sampling depths (m) in Lake Verevi: E – epilimnion; M – metalimnion; H – hypolimnion

Date	E1	E2	M1	M2	M3	M4	H1	H2
07.09.99	0.5	1.5	3.0	3.5	4.0	4.5	6.5	8.0
24.04.00	0.5		1.5	2.0	2.5	3.5	6.0	9.0
03.05.00	0.5	2.0	3.0	3.5	4.0	4.5	6.5	9.0
08.05.00	0.5	2.0	3.0	3.5	4.0	4.5	6.5	9.0
16.05.00	0.5	2.0	3.5	4.0	4.5	5.0	7.0	9.0
22.05.00	0.5	1.5	2.5	3.0	4.0	4.5	7.0	9.0
29.05.00	0.5	1.5	3.0	3.5	4.0	4.5	7.0	9.0
05.06.00	0.5	2.0	3.5	4.0	4.5	5.5	7.0	9.0
26.06.00	0.5	1.5	3.0	4.0	4.5	5.5	7.0	9.0
17.07.00	0.5	2.0	4.0	4.5	5.0	5.5	7.0	9.0
07.08.00	0.5	2.0	3.5	4.0	5.0	6.0	7.0	9.0
28.08.00	0.5	3.0	5.0	5.5	6.0	6.5	7.0	9.0
11.09.00	0.5	2.5	5.0	5.5	6.0	6.5	7.0	9.0
17.09.00	0.5	3.0	5.5	6.0	6.5	7.0	8.0	9.0
25.09.00	0.5	3.0	6.0	6.5	7.0	7.5	8.0	9.0
02.10.00	0.5	4.5	6.5	7.0	7.5	8.0		9.0
09.10.00	0.5	4.5	6.5	7.0	7.5	8.0		9.0
23.10.00	0.5	2.0	4.0	5.0	6.0	7.0		9.0
04.12.00	0.5			5.0				9.0

spring of 2000 (Nõges & Kangro, the present issue).

In the spring and summer of 2000, the Secchi depth varied between 1.4 m in April and 3.6 m at the end of June. The highest transparency of 4.7 m was found in December. In September 1999 the transparency was 2.7 m. Oxygen concentration approached zero usually at a depth of 4.5–5.5 m (Fig. 2). In September 1999 the oxygen depletion started at a depth of 4.5 m.

During summer a DCM usually occurred at the borderline of H_2S- and oxygen-containing waters. The filters from anoxic water were dark green in colour and consisted mainly a bacteriochlorophyll *d*, which was manifested in the colour of the filters and the absorption spectra of the pigments. The position of the DCM changed seasonally. In April 2000, it was formed at a depth of 3.5 m; declined slowly to 6.5 m in October (Fig. 3). In September 1999, the DCM was located at a depth of 4.5 m.

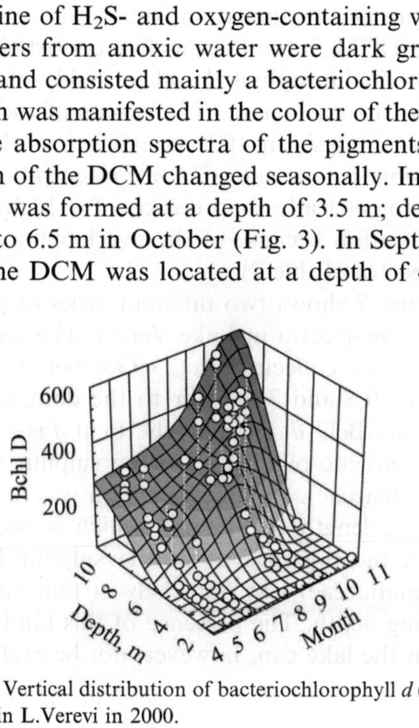

Figure 3. Vertical distribution of bacteriochlorophyll *d* (Bchl *d*, mg m^{-3}) in L.Verevi in 2000.

At the depths where the bacteriochlorophyll maxima were observed, the sulphide concentration was usually relatively low compared to the bottom layers, where the concentration reached as high as possible saturation level. In spring 2000, in the upper layers of Lake Verevi chlorophyll *a* concentration usually did not exceed 20 $\mu g\ l^{-1}$. In summer the concentration was usually below 10 $\mu g\ l^{-1}$ (Fig. 4). In 2000 the highest concentration of bacteriochlorophyll was measured on 25 September (Fig. 3). In December and April, Bchl *d* could be detected only near the bottom of the lake. In September 1999, an HPLC analysis of the pigments was performed. The concentration of chlorophyll *a* yielded by the spectrophotometric phaeopigment corrected method of Lorenzen (1967) and by HPLC fit rather well in the upper layers of Lake Verevi and never exceeded 10 $\mu g\ l^{-1}$. The spectrophotometric equation of Jeffrey & Humphrey (1975) overestimated chlorophyll *a* concentration about twice. In the deeper water layer the concentration of chlorophyll *a* measured by spectrophotometry (1324 $\mu g\ l^{-1}$) was overestimated about 47 times if compared to the HPLC values (28 $\mu g\ l^{-1}$) because of the high concentration of bacteriochlorophyll *d* in that layer. The concentration of bacteriochlorophyll *d* measured spectrophotometrically was 1237 $\mu g\ l^{-1}$.

In most cases the vertical profiles of PP did not coincide with the vertical distribution of the pigment content; the maximum PP values were found in the epilimnion. In some cases (e.g. in May, 16; September 11 & 25, 2000) PP had notably high values also at the depth of DCM (Fig. 5).

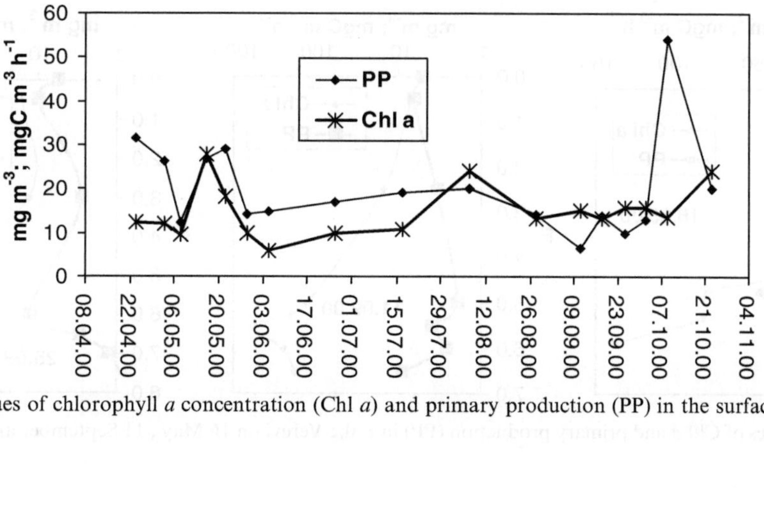

Figure 4. Maximum values of chlorophyll *a* concentration (Chl *a*) and primary production (PP) in the surface layer (0–2 m) of Lake Verevi in 2000.

Discussion

The concentration of bacteriochlorophyll in the metalimnion of L. Verevi (up to more than 1000 μg l^{-1}) was very high, exceeding the values reported by other authors: <20 μg l^{-1} by Steenbergen & Korthals (1982) in L. Vechten; 13 - > 500 μg l^{-1} by Hurley & Watras (1991) in some North-American lakes, and up to 825 μg l^{-1} by Takahashi & Ichimura (1970) in Japanese lakes and reservoirs.

During stratification, three different parts could be distinguished in the water column of Lake Verevi. The upper part can be described as a nutrient-limited and well-illuminated layer. A rather low phytoplankton concentration in the upper layer enables the light to penetrate into the thermocline and the growth of photosynthetic bacteria on the borderline of oxidative and reductive zones. In hypolimnion, the nutrient and sulphide concentration is high and represents a continuous supply for phototrophic bacteria forming chlorophyll maximum in the thermocline.

The moderately high Chl *a* concentration in the epilimnion in spring was significantly reduced after the formation of the thermocline (Fig. 4) most probably because the nutrient limitation was established (Fig. 6). The lower the Chl *a* concentration in the upper part of the water column, the higher became the water transparency and the more intensive growth of photosynthetic bacteria was enabled. The formation of the deep bacteriochlorophyll maximum in the middle of May was supported by the stratification of the water column. The sharpening of the thermocline supported the development of photosynthetic bacteria, the amount of them (according to the concentration of Bchl *d*) grow permanently until reaching the highest concentration immediately before the water column was mixed up at the end of September (Figs. 2 & 3). Overwintering of photosynthetic bacteria could occur only in the water layer near the bottom of the lake, which was not assumingly mixed up and remained anoxic.

The DCM in the thermocline consisted mostly of bacteriochlorophyll *d* and probably of chlorophyll *a* to some extent. The results of the HPLC pigment analysis in 1999 convinced us that it is impossible to estimate Chl *a* concentration spectrophotometrically in presence of a high Bchl *d* concentration because of the overlapping absorption spectra (Table 2).

Figure 7 shows two different types of pigment absorption spectra in Lake Verevi. The spectra of the samples collected on 23 October 2000 at a depth of 0.5 and 7 m refer to the domination of Chl *a* and Bchl *d*, respectively. Bchl *d* is a specific photosynthetic pigment of green sulphur bacteria (Chlorobiaceae). As no absorption was detectable at a wavelength of 772 nm, which is specific to Bchl *a*, the presence of purple sulphur bacteria (Chromatiaceae) was not likely at that particular sampling depth. The presence of this kind of bacteria in the lake can, however, not be excluded as

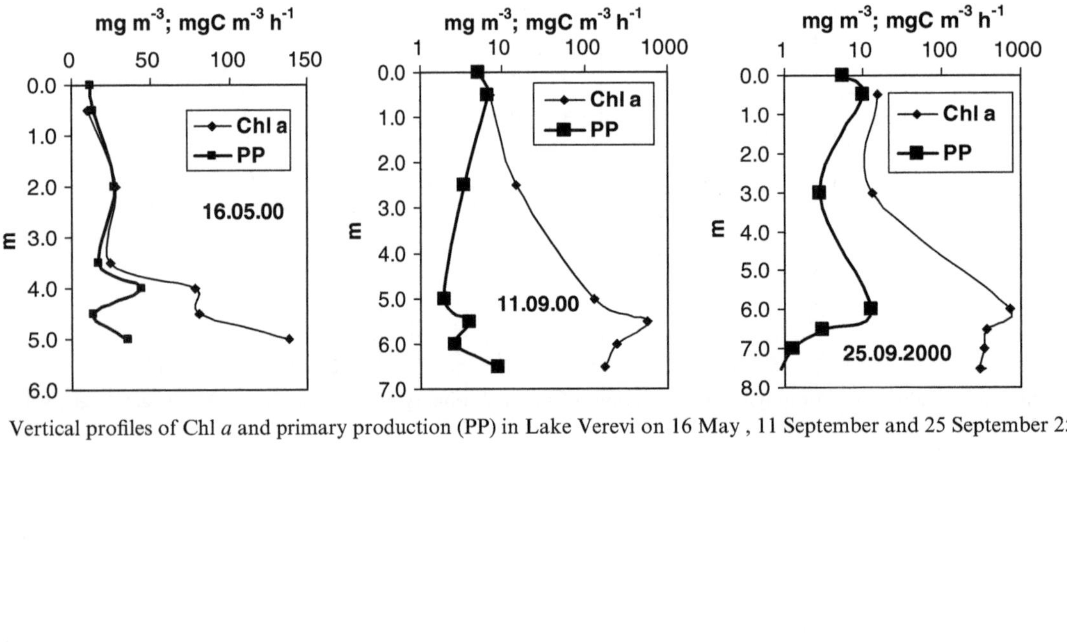

Figure 5. Vertical profiles of Chl *a* and primary production (PP) in Lake Verevi on 16 May , 11 September and 25 September 25 2000.

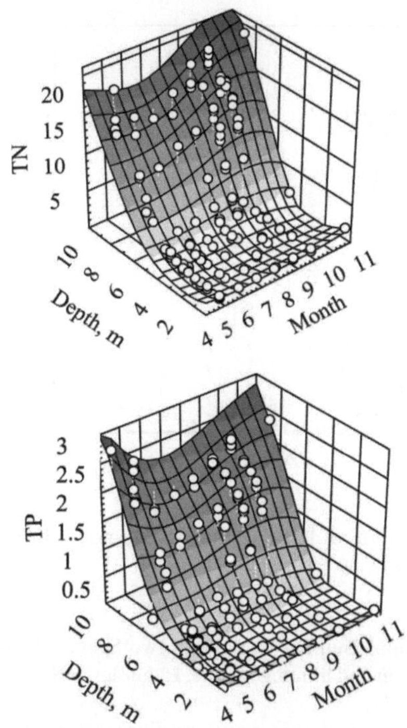

Figure 6. Vertical distribution of total nitrogen (TN, g m^{-3}) and total phosphorus (TP, g m^{-3}) concentration in Lake Verevi in 2000.

they frequently form very thin layers that are easily missed when sampling the water on a large scale (Hurley & Watras, 1991).

In most cases, the vertical profiles of PP and chlorophyll concentration did not coincide. This situation could be explained probably by the accumulation of photosynthetic bacterial biomass with low photosynthetic activity in the metalimnion of Lake Verevi. As a consequence of a sharp light gradient within the bacterial layer, metalimnetic plankton has to be light-limited. However, due to low temperature and reduced grazing pressure photosynthetic bacteria and phytoplankton accumulated in the metalimnion. It should also be taken into consideration that the specific chlorophyll content of the bacterial cells is higher than that of phytoplankton (Gemerden & Mas, 1995). The spectrophotometrically measured high concentration of phaeopigments in the lower part of the water column seems to be mostly a false- positive signal caused by a high Bchl *d* concentration. This pigment elevates the phaeopigment estimates upon acidification as described by Hurley & Watras (1991). A part of that high phaeopigment concentration in the aerobic water layer was probably caused by physiologically inactive desintegrating algal cells, which had settled into that layer.

As to conclude – L. Verevi represents an example of a sharply stratified lake with the metalimnion supporting DCM with the remarkably high amount of photosynthetic bacteria. In summer, a steep thermocline divides the water column in two rather isolated parts. High amount of nutrients are trapped in anoxic and H_2S-rich hypolimnion. Nutrient limitation in epilimnion keeps surface water clear and enables light to penetrate into the metalimnion where a dense population of photosynthetic bacteria develops. The composition and activity of these bacteria as well as the relations with other autotrophs like cyanobacteria and chemosynthetic bacteria requests further detailed studies.

Table 2. Main photosynthetic pigments of phytoplankton and bacteria (Brock et al., 1994; Jeffrey et al., 1997; Airs & Keeley, 2003; Hoogewerf et al., 2003)

Pigment	Main organisms	Main absorption max; λ (nm)
Chl *a*	All photosynthetic microalgae except prochlorophytes	428–431, 660–664
Chl *b*	Green algae	453, 643
Chl c_1, c_2, c_3	Diatoms	446–452, 579–586, 626–630
Bchl *a*	Purple sulphur bacteria	363, 606, 767
Bchl *b*	Some purple sulphur and non-sulphur bacteria	1012–1016
Bchl *c*	Green sulphur bacteria	435, 663–665
Bchl *d*	Green sulphur bacteria	427–430, 660

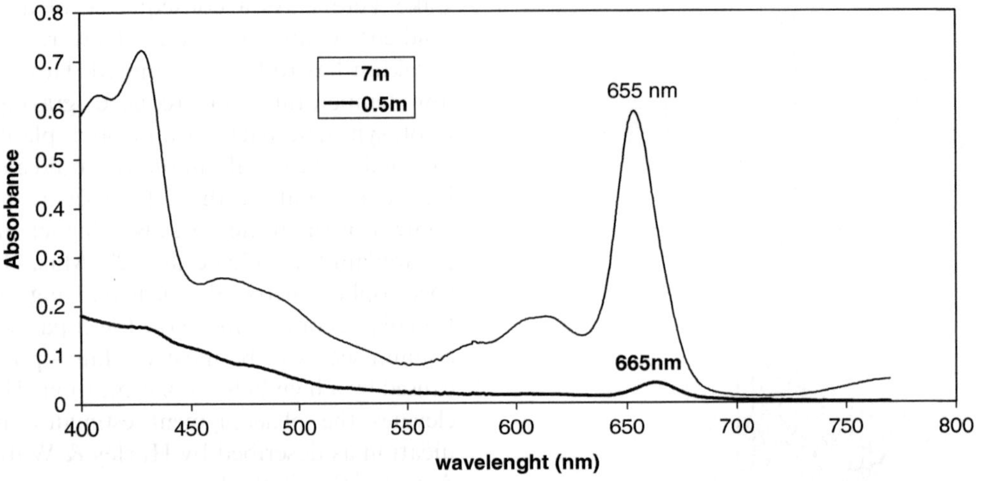

Figure 7. Absorption spectra of pigments in 90% acetone extract at two different depths of Lake Verevi on 23 October 2000. Thin and bold lines represent the content of Bchl *d* and Chl *a* at depths 7 and 0.5 m, respectively.

Acknowledgements

This work was supported by the target-financed project of Ministry of Education "The influence of the stratification to the biological matter circulation of the lakes" No 0370208s98 and by grants of the Estonian Science Foundation No 3579, 4080, 4835. We would like thank all the members of the research group (I. Ott, K. Ott, A. Rakko, D. Sarik, T. Kõiv, P. Nõges, E. Lill, H. Tammert, H. Künnap, V. Kisand, H. Starast, A. Lindpere, K. Kangro, R. Laugaste etc.) who participated in the project. We would also like to thank Dr Enn Veldi, who revised the English.

References

Airs, R. L. & B. J. Keely, 2003. A high resolution study of the chlorophyll and bacteriochlorophyll pigment distributions in a calcite/gypsum microbial mat. Organic Geochemistry. 34(4): 539–551.

Brock, T. D., Madigan, M. T., Martinko, J. M. & Parker, J., 1994. The Bacteria. Biology of Microorganisms. Prentice-Hall International, pp. 718–814.

Gemerden, H. V. & Mas, J. 1995. Ecology of phototrophic sulphur bacteria. Anoxygenic photosynthetic bacteria. In Blankenship, R. E., Madigan, M. T., Bauer & C. E., (eds). Kluwer Academic Publishers, 49–85.

Grasshoff, K., Ehrhardt, M., Kremling, K., (eds), 1983. Methods of Seawater Analysis. ISBN (Verlag Chemie), 3–527, 25998–8.

Guerrero, R., E. Montesinos, C. Pedros-Alio, J. Esteve, J. Mas, H. van Gemerden, P. A. G. Hofman & J. F. Bakker, 1985. Phototrophic sulphur bacteria in two Spanish lakes: vertical distribution and limiting factors. Limnolgy and Oceanography 30: 919–931.

Guidelines for the Baltic monitoring programme for the third stage, 1984. The Baltic Marine Biologists. Publ. 1. 2nd Edition.

Hilmer, T. & G. C. Bate, 1989. Filter types, filtration and post-filtration treatment in phytoplankton production studies. Journal of Plankton Research. 11: 49–63.

Hoogewerf, G. J., D. O. Jung & T. Michael, 2003. Madigan Evidence for limited species diversity of bacteriochlorophyll b-containing purple nonsulfur anoxygenic phototrophs in freshwater habitats. FEMS Microbiology Letters 218(2): 359–364.

Hurley, J. P. & C. J. Watras, 1991. Identification of bacteriochlorophylls in lakes via reverse-phase HPLC. Limnology and Oceanography 32: 307–315.

Jeffrey, S. W. & G. F. Humphrey, 1975. New spectrophotometric equations for determining chlorophylls *a*, *b*, c_1 and c_2 in higher plants, algae and natural phytoplankton. Biochem. Physiol. Pflanzen (BPP) 167: 191–194.

Lorenzen, C. J., 1967. Determination of chlorophyll and pheopigments: spectrophotometric equations. Limnology and Oceanography 12: 343–346.

Niemi, M., J. Kuparinen, A. Uusi-Rauva & K. Korhonen, 1983. Preparation of algal samples for liquid scintillation counting. Hydrobiologia 106: 149–159.

Nõges, T. & Kangro. K., 2005. Primary production of phytoplankton in a strongly stratified temperate lake. Hydrobiologia 547: 105–122.

Ott, I., Kõiv, T., Nõges, P., Kisand, A., Järvalt A. & Kirt. E., 2005. General description of partly meromictic hypertrophic Lake Verevi, its ecological status, changes during the last eight decades and restoration problems. Hydrobiologia 547: 1–20.

Parkin, T. B. & T. D. Brock, 1980. Photosynthetic bacterial production in lakes: the effects of light intensity. Limnology and Oceanography 25: 711–718.

Rodrigo, M. A., E. Vicente & M. R. Miracle, 2000. The role of light and concentration gradients in the vertical stratification and seasonal development of phototrophic bacteria in a meromictic lake. Archiv für Hydrobiologie 148: 533–548.

Steeman-Nielsen, E., 1952. The use of radioactive carbon (^{14}C) for measuring primary production in the sea. Journal du Conseil permanent international pour l' exploration del la mer 18: 117–140.

Steenbergen, C. L. M. & H. J. Korthals, 1982. Distribution of phototrophic microorganisms in the anaerobic and microaerophilic strata of Lake Vechten (The Netherlands). Pigment analysis and role in primary production. Limnology and Oceanography 27: 883–895.

Takahashi, M. & S. Ichimura, 1968. Vertical distribution and organic matter production of photosynthetic sulphur bacteria in Japanese lakes. Limnology and Oceanography 13: 644–655.

Takahashi, M. & S. Ichimura, 1970. Photosynthetic properties and growth of photosynthetic sulphur bacteria in lakes. Limnology and Oceanography 15: 929–944.

Hydrobiologia (2005) 547:83–90
I. Ott & T. Kõiv (eds), Lake Verevi, Estonia – A Highly Stratified Hypertrophic Lake
DOI 10.1007/s10750-005-4148-8

Bacterioplankton abundance and activity in a small hypertrophic stratified lake

Helen Tammert*, Veljo Kisand & Tiina Nõges
Institute of Zoology and Botany, Estonian Agricultural University, Võrtsjärv Limnological Station, 61101, Rannu, Tartu County, Estonia
*(*Author for correspondence: E-mail: helen@zbi.ee)*

Key words: total number of bacteria, bacterioplankton production, stratified hypertrophic lake

Abstract

Bacterioplankton abundance and production were followed during one decade (1991–2001) in the hypertrophic and steeply stratified small Lake Verevi (Estonia). The lake is generally dimictic. However, a partly meromictic status could be formed in specific meteorological conditions as occurred in springs of 2000 and 2001. The abundance of bacteria in Lake Verevi is highly variable (0.70 to 22 × 10^6 cells ml^{-1}) and generally the highest in anoxic hypolimnetic water. In 2000–2001, the bacterial abundance in the hypolimnion increased probably due to meromixis. During a productive season, heterotrophic bacteria were able to consume about 10–40% of primary production in the epilimnion. Our study showed that bacterioplankton in the epilimnion was top-down controlled by predators, while in metalimnion bacteria were dependent on energy and carbon sources (bottom-up regulated). Below the thermocline hypolimnetic bacteria mineralized organic matter what led to the depletion of oxygen and created anoxic hypolimnion where rich mineral nutrient and sulphide concentrations coexisted with high bacterial numbers.

Introduction

Bacteria are the most numerous planktonic organisms in freshwater lakes, they can be responsible for transformation of all net primary production. In lakes bacteria contribute ~10–90% of the total respiration rate (Biddanda et al., 2001) and their importance seems to increase toward more oligotrophic systems (e.g., Baines & Pace, 1991). Large vertical heterogeneity and steep gradients are characteristic to thermally stratified lakes facilitating the sequence of specific ecophysiologically different microbial populations over small depth intervals. Changes in the concentrations of key environmental factors such as oxygen, sulphides, nitrogen and phosphorus compounds as well as in light intensity and quality lead to the differences in food web structure and in abundance of its major players.

In present study the inter-annual, seasonal and vertical distribution of the total abundance and activity of non-photosynthetic planktonic bacteria was followed in a small steeply stratified hypertrophic lake. This a temperate region lake (Lake Verevi, area 12.6 ha, mean depth 3.6 m, maximum depth 11 m) is a partly meromictic and strongly stratified hypertrophic freshwater lake in South Estonia protected from wind and has the average water exchange of 0.63 $year^{-1}$ (Loopmann, 1984). Summer stratification develops quickly after the ice-break in April leading to fast oxygen depletion in the hypolimnion. In 1991–2001 the average Secchi depth was ~2 m, chlorophyll *a* concentration varied from 3.5 to 128 μgChl l^{-1} in the mixed layer (Nõges & Kangro, 2005). Concentrations of total nitrogen and phosphorus were 980 mg N m^{-3} and 55 mg P m^{-3} in the surface layers (<2.5 m) and 6322 mg N m^{-3} and 830 mg P m^{-3} in the bottom layers (>5 m) in 1984–2001 (Ott et al., 2005). The aim of our study was to follow and understand the reasons of bacterial abundance and activity distribution on the background of formation of the lake water stratification, on vertical gradients of environmental factors, and also the food web interactions.

Materials and methods

Sampling

Water samples were taken from 3 to 8 layers at the deepest point of the lake. In 1991, 1993, 1994 and 1998 sampling was carried out by Ruttner or van Dorn sampler. In 2000 and 2001 a water pump (Masterflex N 7533–60) with "easy-load" pump-head (model 7518–12) connected to a tube (diameter 8 mm), designed for study of thin (20–25 cm) water layers, was used for sampling. Temperature and oxygen concentration were measured before sampling. In a diurnal study the samples were taken at 1 m intervals from the layer of 0.5 to 7 m at 12:00 and 16:00 in August 2, and at 8:00 and 12:00 in August 3, 2001.

Analytical methods

Water temperature and the concentration of dissolved oxygen were measured by thermooxymeter Landorem 200 (Tartu University, Estonia). In 2000 and 2001, the parallel measurements were done with Aqua-Check Water Analyzer (O.I. Analytical Corporation). Chemical analyses were performed as described by Ott et al. (2005).

Total number of bacteria (TNB) was determined by DAPI staining (Porter & Feig, 1980). Formaldehyde or glutaraldehyde preserved samples (final concentration 2%) were incubated with DAPI (final concentration 10 $\mu g\ ml^{-1}$) for 5 min in the dark. Samples were filtered onto black 0.22-μm-pore-size polycarbonate filters (Poretics) and stored at −21 °C until counting with epifluorescence microscope (Leica DM RB) at 1000× magnification.

Bacterial activity and production was estimated by the tritiated thymidine incorporation method (Bell et al., 1983). Triplicate 10 ml subsamples of each sample (+3 formaldehyde killed blanks) were treated with 10 nM ^{3}H-thymidine (Amersham; specific activity 26 Ci $mmol^{-1}$). The subsamples were incubated 30 min at room temperature. Cold base–acid–ethanol extraction was used for purification of DNA as described by Wicks & Robarts (1987). The uptake of thymidine was converted to the number of produced cells by using conversion factor of 2×10^{18} cells per mole of incorporated thymidine.

For the estimations of chlorophyll *a* concentration (Chl *a*) plankton was filtered on Whatman GF/F filters. In 1991 and 1993 the pigments were extracted by 90% acetone (Edler, 1979), in 1998–2001 in parallel by 90% acetone and 96% ethanol (Jespersen & Christoffersen, 1987). The absorption of the extract between 430 and 750 nm was determined with a scanning UV-VIS spectrophotometer (Cecil-3000). When applying extraction both with acetone and with ethanol the maximal concentration of Chl *a* was used in further analysis as recommended by Nõges & Solovjova (2000). Primary production (PP) of phytoplankton was estimated *in situ* using $^{14}CO_2$ assimilation technique (Steeman-Nielsen, 1952). Detailed description of PP method is given by Nõges & Kangro (2005).

Results

Stratification

Morphometrical characteristics of the lake result in strong gradients of temperature and oxygen (Figs. 1, 2, and 4). In mid-summer aerobic epilimnion expanded only to the upper ~1–1.5 m, temperature in this layer was the highest. Thickness of the metalimnion was usually ~2 m, and an extensive anoxic zone developed in hypolimnion with increased H_2S concentration during the productive season (May–October). Total number of bacteria (TNB) was statistically significantly different between different layers (Tukey's *post-hoc* test ANOVA, $p < 0.001$). In average highest TNB was recorded in the hypolimnion ($9.5 \pm 0.5 \times 10^6$ cells ml^{-1}) and lowest in the epilimnion ($5.7 \pm 0.3 \times 10^6$ cells ml^{-1}). At the end of July, 1998, bacterial activity increased with depth in the epilimnion and peaked at the aerobic/anaerobic interface (Fig. 1). The average values of bacterioplankton production (BP) were not statistically different between epi-, meta- and hypolimnion, however the variation of BP was the highest in epilimnion (mean 5.8 ± 3.6 $\mu gC\ l^{-1}h^{-1}$). The lowest

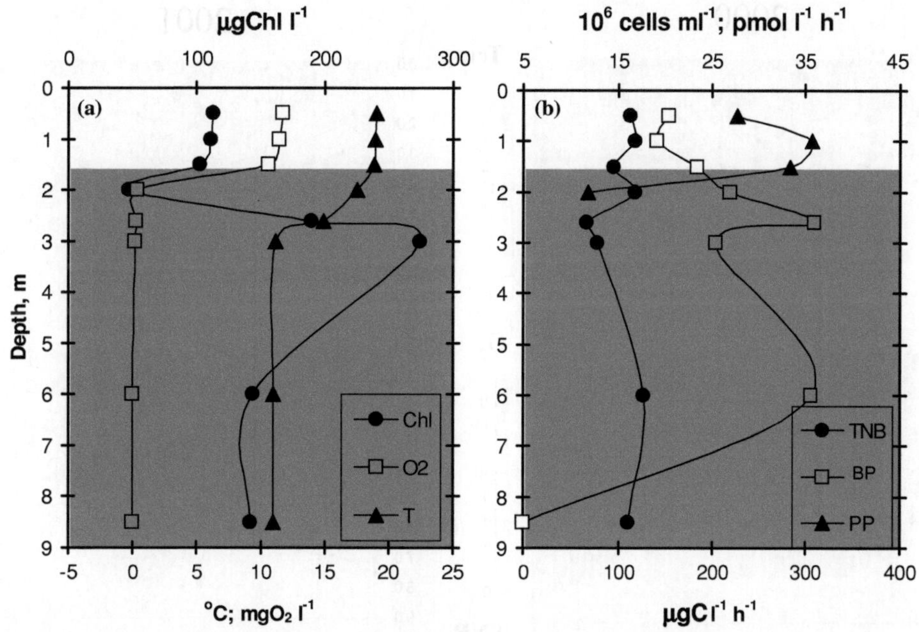

Figure 1. Vertical distribution of (a) temperature (*T*), oxygen (O_2), chlorophyll *a* concentration (Chl), and (b) primary production (PP), abundance (TNB), production of bacteria (BP) in Lake Verevi during strong summer stratification (in July 31, 1998). Shaded area on the plot shows the part of water-column below the light attenuation of 1% ($z_{1\%}$) (Reinart et al., 2005).

values and variability of activity were found in the hypolimnion (0.8 ± 0.3 μgC $l^{-1}h^{-1}$).

Seasonal dynamics and long term trends

TNB increased slightly during the productive season in all water layers, this increase was the most pronounced in metalimnion (Fig. 2). Hypolimnetic bacteria achieved the highest numbers when anoxic and sulphide rich waters spread further into water column. The increase of TNB in the hypolimnion was more pronounced compared between epi- or metalimnion. Typically in spring and late summer/autumn the additional peaks of TNB developed in the upper or lower metalimnion. Thymidine incorporation (TTI) was variable between the years, and no clear seasonal dynamics could be observed.

In 1990s, TNB did not change in the epi- and metalimnion but slightly increased in hypolimnion (Fig. 3, MANOVA of differences between co-effect of years and layers $p = 0.069$). Bacterial production data were scattered, however, a decrease of activity could be noticed in 1998, and 2001 as compared to the beginning of 1990s (Fig. 3, ANOVA $p = 0.001$, $n = 28$).

Diel dynamics of bacterioplankton

In August 2–3, 2001 when diurnal dynamics of bacteria (24 h cycle) was followed, the temperature and oxygen profiles were stable (Fig. 4b). TNB in epi- and metalimnion ($\sim$4–5 $\times 10^6$ cells ml^{-1}) had low variability and did not change over time. In deeper layers of the hypolimnion TNB was significantly higher ($\sim$10–14 $\times 10^6$ cells ml^{-1}) than in upper water layers. BP fluctuated highly in depth and during the diurnal cycle (Fig. 4a). The highest bacterial activity was estimated at the epilimnion/hypolimnion interface (average 0.218 μgC l^{-1} h^{-1}), in addition the quite high BP values (average 0.214 μgC l^{-1} h^{-1}) occurred also at the surface. Depth profiles of bacterial activity did not change during the diurnal cycle, though highest values were recorded in the afternoon (average 0.208 μgC l^{-1} h^{-1}), and lowest at noon (average 0.130 μgC l^{-1} h^{-1}).

Carbon flux

The average integrated bacterioplankton production in epilimnion was 354 mgC m^{-2} day^{-1} (assuming equal productivity over 24 h) while integral primary

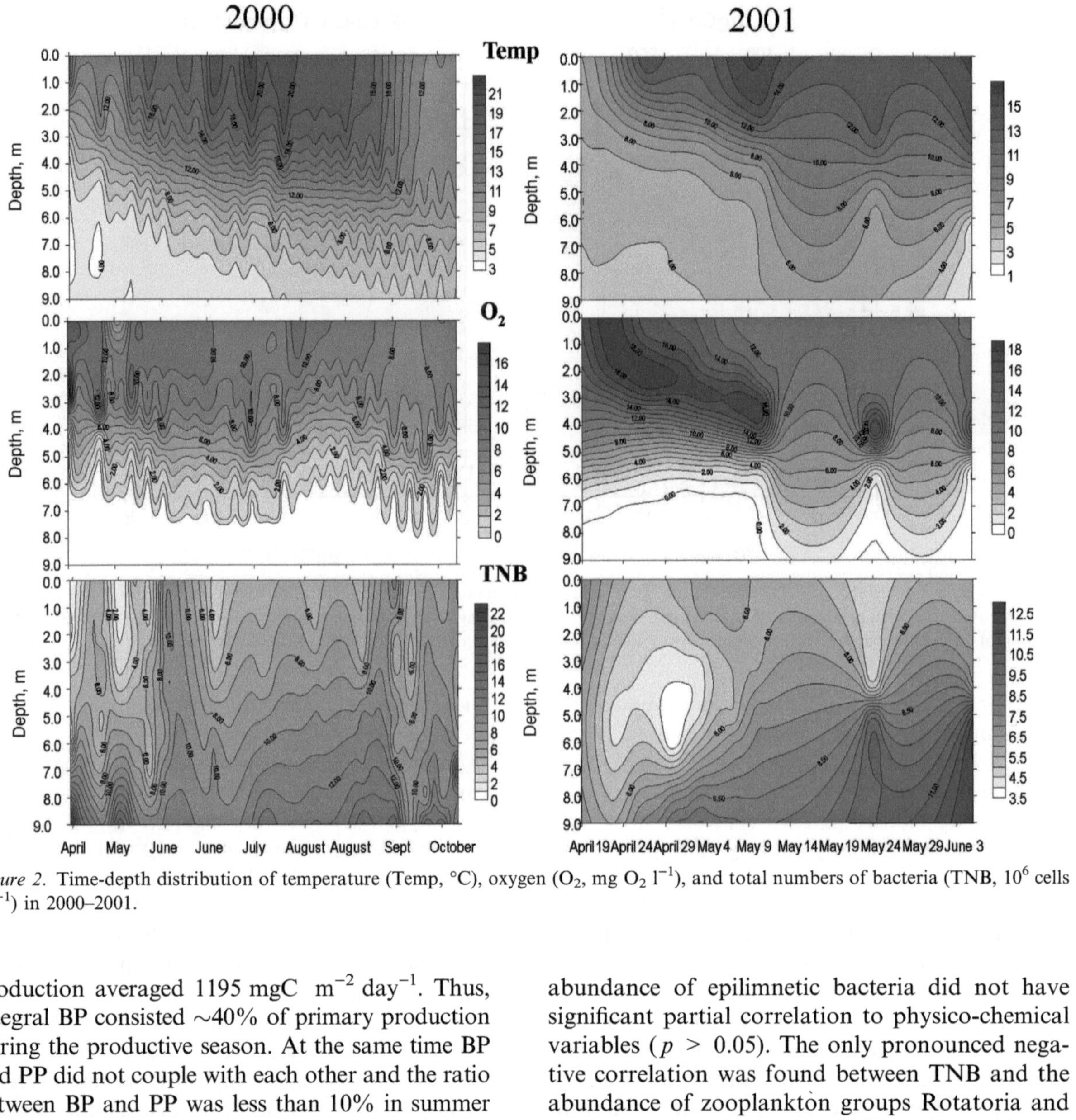

Figure 2. Time-depth distribution of temperature (Temp, °C), oxygen (O_2, mg O_2 l^{-1}), and total numbers of bacteria (TNB, 10^6 cells ml^{-1}) in 2000–2001.

production averaged 1195 mgC m^{-2} day^{-1}. Thus, integral BP consisted ~40% of primary production during the productive season. At the same time BP and PP did not couple with each other and the ratio between BP and PP was less than 10% in summer (June–August) and in most cases reached highest values in late autumn (up to 80% of integral PP).

Relationships between bacterioplankton with other biota and physico-chemical environment

Partial correlation analysis (interrelation between variables is adjusted) was used to estimate relationships between TNB and other available variables. Data of different layers (i.e. epi-, meta- and hypolimnion) were analyzed separately. The abundance of epilimnetic bacteria did not have significant partial correlation to physico-chemical variables ($p > 0.05$). The only pronounced negative correlation was found between TNB and the abundance of zooplankton groups Rotatoria and Cladocera (partial $r = -0.63$ and -0.56, $p < 0.001$, respectively). In the environment of steep gradients of temperature and oxygen (metalimnion), TNB showed significant positive correlation (partial $r > 0.40$, $p < 0.05$) with total and inorganic phosphorus, temperature and Chl *a*. In the mainly anaerobic hypolimnetic layer the bacterial number was strongly correlated (partial $r > 0.80$, $p < 0.01$) to the concentration of total nutrients (N, P) and phosphate, but not with the inorganic nitrogen. A strong correlation was also found

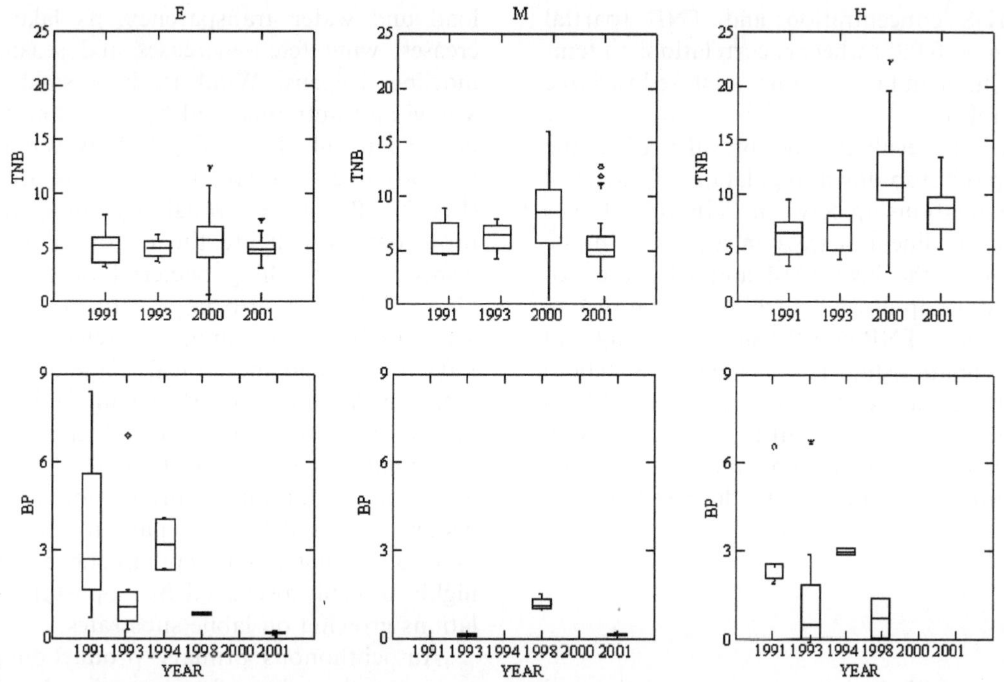

Figure 3. Summarized box-plot (with mean, standard deviation and absolute range) of abundance and activity of bacterioplankton in different layers in Lake Verevi through 1991–2001: (a) total number of bacteria (TNB, 10^6 cells ml^{-1}); (b) bacterioplankton production (BP, μg C l^{-1} h^{-1}); E: epilimnion M: metalimnion, H: hypolimnion.

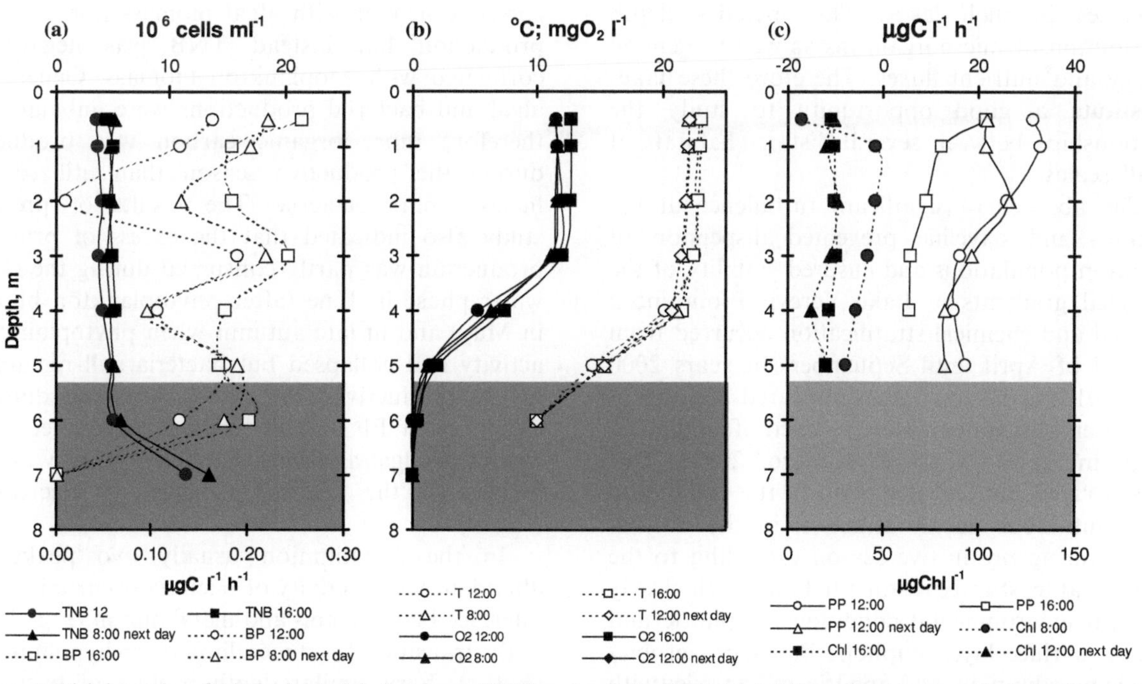

Figure 4. Diel dynamics of (a) the bacterial abundance (TNB) and activity (TTI-tritiated thymidine incorporation), (b) temperature and oxygen concentration and (c) chlorophyll *a* concenctration (Chl*a*) and primary production (PP). Shaded area on the plot shows the part of water-column below the light attenuation of 1% ($z_{1\%}$) (Reinart et al., 2005) at mid-day (12:00).

between H_2S concentration and TNB (partial $r = 0.61$, $p = 0.035$) whereas correlations to temperature, Chl *a* and to any other analyzed variable were insignificant.

In order to evaluate the prevalence of the 'bottom-up' or 'top-down' regulation of bacterial growth the relationship between TNB and BP was analyzed using linear regression analysis. In the upper layer of the lake TNB and BP were not related (strong top-down control by predation), in the metalimnion TNB and BP showed strong and significant relationship ($R^2 = 0.90$, $p < 0.000$) demonstrating the bottom-up regulation of bacterial growth. An insignificant negative regression ($p > 0.05$) was found in hypolimnion indicating weak N and P limitation in the deepest layers of the lake.

Discussion

Strongly stratified dimictic and monomictic, and in particular meromictic lakes produce extreme types of aquatic environments at comparatively small scales. Thermal stratification influences a wide variety of biological, chemical and physical processes in such lakes. This includes depth distribution of microorganisms as well as general energy and nutrient fluxes. Therefore these lakes constitute a good opportunity to study the relationships between several distinct habitats at small scales.

The absence of significant turbulence at the thermo- and oxycline prevented dispersion of plankton populations and ensured stability of the chemical gradients in Lake Verevi. Pronounced thermal and chemical stratification occurred from the end of April until September. In years 2000 and 2001, partial meromixis appeared because of the water was not completely mixed after the ice-break in April (Nõges & Kangro, 2005). This caused rapid nutrient depletion in the epilimnion and continuous anoxic situation in the bottom layers during productive season providing to the bacteria at least three distinct habitats with clearly different environmental conditions: euphotic and aerobic surface layer, euphotic but microaerobic/anoxic metalimnion, and aphotic and anoxic (with H_2S) hypolimnion. Generally, the seasonal thermocline depth is influenced by lake size, nutrients load and water transparency, as lake area increases, wind fetch increases and seasonal thermocline deepens. Wind fetch is small in Lake Verevi and nutrients load high, leading to a thermocline at only 1–2 m depth, however some fluctuations due to weather conditions are possible (Fig. 2). Resource availability and grazing by protozoans which are the major known mechanisms for controlling bacterial production (e.g., Gasol, 1994) are also important in Lake Verevi. The number of epilimnetic bacterioplankton in Lake Verevi had the typical range of hyper- and eutrophic lakes ($\sim 5 \times 10^6$ cells ml^{-1}). Productivity of bacteria was the highest and most variable in this compartment (0.001–64 $\mu gC\ l^{-1}\ h^{-1}$, mean 3.5). The epilimnion usually is more subjected to disturbances and bacteria have to grow irregular or erratic bursts, thus, epilimnetic bacteria were highly possibly prevailed by opportunistic populations growing on labile substrates.

Autochthonous primary production provided energy and carbon for heterotrophic bacteria, however, the growth of bacteria was instead controlled by predators and therefore no relationship between abundance and activity of bacteria was found. Also TNB and BP did not correlate neither with algal biomass nor primary production but instead TNB was negatively correlated with zooplankton biomass. Generally, algal and bacterial productions were unbalanced, therefore more organic carbon was produced during the productive season than utilized by heterotrophic bacteria. The results of present study also indicated that the excess of primary production was partly consumed during the clear water phase in June (after phytoplankton bloom in May) and in late autumn when phytoplankton activity had collapsed but bacteria still remained highly productive afterwards. Also the diurnal dynamics of PP and BP showed uncoupled variations of algal and bacterial activity: BP was the highest in the afternoon when PP decreased (Fig. 4).

In the metalimnion usually two peaks of abundance and activity of bacteria occurred at the interface between epi- and metalimnion (Fig. 4) or at the transition from metalimnion to hypolimnion (Fig. 1). Very similar depth profiles of bacterial abundance were found in L. Plußsee (Weinbauer & Höfle, 1998). As typical to eutrophic lakes the productive epilmnion was dominated by production of particulate organic matter (Biddanda et al., 2001), and in the process of sedimentation these particles were trapped in the upper part of the thermocline (Ott et al., 2005) and were possibly utilized by aerobic heterotrophic bacteria. Another peak of BP associated with microaerobic/anoxic conditions and with bacteriochlorphyll maximum. This zone had better access to H_2S together with light favouring development of phototrophic sulfur bacteria (e.g. Camacho et al., 2001). Bacteria in the thermocline could depend more on abiotic environment at the same time remineralizing organic matter and releasing inorganic nutrients, as correlation of bacterial abundance and BP with inorganic nutrients and temperature were strong ($p > 0.40$, $p < 0.05$). Water temperature was below the range (10–14 °C) what is reported to limit growth rate of bacteria (Hoch & Kirchman, 1993; Carlsson & Caron, 2001). Thus, temperature could also be an important factor controlling the development of bacteria in thermocline. Also BP and TNB were positively related to each other indicating bottom-up regulation of bacterioplankton activity.

Similar to other lakes (Weinbauer & Höfle, 1998; Kasprzak et al., 2000), TNB reached the highest values ($15–20 \times 10^6$ cells ml^{-1}) in the physically most homogenous hypolimnion. Abundance of bacteria in deep layers increased from 1990s to 2000s because of more rapid oxygen depletion in deep layers, caused most probably because of spring meromixis in warmer springs (Nõges & Kangro, 2005). At the same time anoxic and rich in H_2S environment was created by bacteria itself. This was expressed by a good correlation between number of bacteria and H_2S concentration. However, main energy and carbon still originated mostly from the upper highly productive layers. Hypolimnetic bacteria were not highly active (or measurements of BP failed in anoxic waters) and their high numbers were supported rather by specific conditions (lack of most eukaryotic organisms, therefore no grazing, undisturbed environment, etc) than high productivity. Such growth is typical to equilibrium populations (K-strategists) growing in stable environments (Andrews & Harris, 1986). However, the bacterial activity in binding of nutrients in hypolimnion is important as the concentrations of total phosphorus and nitrogen were strongly correlated with TNB. Probably bacteria were one of the most important nutrient pools in hypolimnion. At the same bacteria also re-mineralized organic compounds and released excesses of nutrients as the concentrations of inorganic N and P were very high in hypolimnion.

In conclusion, small steeply stratified water bodies such as Lake Verevi provide a promising environment for studying bacterial physiological and species diversity. Very strong stratification is stabilized by the small size of the lake ensuring that certain populations of bacteria develop under the specific environmental conditions. Ecophysiological studies of these bacteria would provide a deeper insight into energy and matter fluxes of the ecosystems of such kind of lakes.

Acknowledgements

We would like to thank and Dr. Hans-Peter Großart from the University of Oldenburg (ICBM) for useful comments on the manuscript. The study was supported by the core grants of the Ministry of Education Nos. 0370208s98, 0362482s03 and by grants of Estonian Science Foundation Nos. 3579, 4080 & 4835.

References

Andrews, J. H. & R. F. Harris, 1986. r- and K-selection and microbial ecology. Adv. Microbial Ecol. 9: 99–147.

Baines, S. B. & M. L. Pace, 1991. The production of dissolved organic matter by phytoplankton and its importance to bacteria–patterns across marine and freshwater systems. Limnology and Oceanography 36: 1078–1090.

Bell, R. T., G. M. Ahlgren & I. Ahlgren, 1983. Estimating bacterioplankton production by measuring (3H)thymidine incorporation in a eutrophic Swedish lake. Applied and Environmental Microbiology 45: 1709–1721.

Biddanda, B., M. Ogdahl & J. Cotner, 2001. Dominance of bacterial metabolism in oligotrophic relative to eutrophic waters. Limnology and Oceanography 46: 730–739.

Camacho, A., J. Erez, A. Chicote, M. Florin, M. M. Squires, C. Lehmann & R. Bachofen, 2001. Microbial microstratification, inorganic carbon photoassimilation and dark carbon fixation at the chemocline of the meromictic Lake Cadagno (Switzerland) and its relevance to the food web. Aquatic Sciences 63: 91–106.

Carlsson, P. & D. A. Caron, 2001. Seasonal variation of phosphorus limitation of bacterial growth in a small lake. Limnology and Oceanography 46: 108–120.

Edler, L. (ed), 1979. Phytoplankton and Chlorophyll. The Baltic Marine Biologists, 38 pp.

Gasol, J. M., 1994. A framework for the assessment of top-down vs bottom-up control of heterotrophic Nanoflagellate abundance. Marine Ecology–Progress Series 113: 291–300.

Hoch, M. P. & D. L. Kirchman, 1993. Seasonal and Inter-Annual Variability in Bacterial Production and Biomass in a Temperate Estuary. Marine Ecology–Progress Series 98: 283–295.

Jespersen, A.-M. & K. Christoffersen, 1987. Measurements of chlorophyll *a* from phytoplankton, using ethanol as an extraction solvent. Archiv für Hüdrobiologie Hydrobiol 109: 445–454.

Kasprzak, P., F. Gervais, R. Adrian, W. Weiler, R. Radke, I. Jager, S. Riest, U. Siedel, B. Schneider, M. Bohme, R. Eckmann & N. Walz, 2000. Trophic characterization, pelagic food web structure and comparison of two mesotrophic lakes in Brandenburg (Germany). International Review of Hydrobiology 85: 167–189.

Loopmann, A., 1984. Suuremate Eesti järvede morfomeetrilised andmed ja veevahetus. Tallinn, 150 lk. [Morphometrical data and water exchange of larger Estonian lakes. In Estonian].

Nõges, P., 2005. Water and nutrient mass balance of the partly meromictic temperate Lake Verevi. Hydrobiologia 547: 21–31.

Nõges, T. & K. Kangro, 2005. Primary production of phytoplankton in a strongly stratified temperate lake. Hydrobiologia 547: 105–122.

Nõges, T. & I. Solovjova, 2000. The influence of different solvents and extraction regimes on the recovery of chlorophyll *a* from freshwater phytoplankton. Geophysica 36: 161–168.

Ott, I., T. Kõiv, P. Nõges, A. Kisand, A. Järvalt. & E. Kirt, 2005. General description of partly meromictic hypertrophic Lake Verevi, its ecological status, changes during the past eight decades and restoration problems. Hydrobiologia 547: 1–20.

Porter, K. G. & Y. S. Feig, 1980. The use of DAPI for identifying and counting aquatic microflora. Limnology and Oceanography 25: 943–948.

Reinart, A., Arst, H. & D.C. Pierson, 2005. Optical properties and light climate in Lake Verevi. Hydrobiologia 547: 41–49.

Steeman-Nielsen, E., 1952. The use of radioactive carbon (^{14}C) for measuring primary production in the sea. Journal du Conseil permanent international pour l'exploration del la mer 18: 117–140.

Weinbauer, M. G. & M. G. Höfle, 1998. Distribution and life strategies of two bacterial populations in a eutrophic lake. Applied and Environmental Microbiology 64: 3776–3783.

Wicks, R. J. & R. D. Robarts, 1987. The extraction and purification of DNA labelled with [methyl-3H]thymidine in aquatic bacterial production studies. Journal of Plankton Research 9: 1159–1166.

Hydrobiologia (2005) 547:91–103
I. Ott & T. Kõiv (eds), Lake Verevi, Estonia – A Highly Stratified Hypertrophic Lake
DOI 10.1007/s10750-005-4151-0

Long-term changes and seasonal development of phytoplankton in a strongly stratified, hypertrophic lake

Kersti Kangro*, Reet Laugaste, Peeter Nõges & Ingmar Ott
Institute of Zoology and Botany, Estonian Agricultural University, Võrtsjärv Limnological Station, 61101, Rannu, Tartu County, Estonia
*(*Author for correspondence: E-mail: kiti@ut.ee)*

Key words: phytoplankton, long-term changes, vertical distribution, seasonal dynamics, *Planktothrix agardhii*

Abstract

Changes in the phytoplankton community of the hypertrophic, sharply stratified Lake Verevi have been studied over eight decades. Due to irregular discharge of urban wastewater, the trophic state of the lake has changed from moderately eutrophic to hypertrophic. We found that the trophic state in summer increased in the 1980s and remained at a hypertrophic level since then. *Planktothrix agardhii* was recorded first in the 1950s and became the dominant species in the 1980s, forming biomass maxima under the ice and in the metalimnion during the vegetation period. In summer 1989, *P. agardhii* contributed almost 100% of the phytoplankton biomass. Generally, the highest biomass values occurred in the metalimnion. In spring, when *P. agardhii* was less numerous, diatoms and cryptophytes prevailed. In springs 2000 and 2001 different diatoms dominated – *Synedra acus* var. *angustissima* (18.6 g m^{-3}) and *Cyclostephanos dubius* (9.2 g m^{-3}), respectively. In recent years, the spring overturn has been absent. In the conditions of strong thermal stratification sharp vertical gradients of light and nutrients caused a large number of vertically narrow niches in the water column. During a typical summer stage, the epilimnion, dominated by small flagellated chrysophytes, is nearly mesotrophic, and water transparency may reach 4 m. The lower part of the water column is hypertrophic with different species of cryptophytes and euglenophytes. A characteristic feature is the higher diversity of Chlorococcales. Often, species could form their peaks of biomass in very narrow layers, e.g. in August 2001 *Ceratium hirundinella* (18.6 g m^{-3}) was found at a depth of 5 m (the lower part of the metalimnion with hypoxic conditions), *Cryptomonas* spp. (56 g m^{-3}) at 6 m (with traces of oxygen and a relatively high content of dissolved organic matter) and euglenophytes (0.6 g m^{-3}) at 7 m and deeper (without oxygen and a high content of dissolved organic matter).

Introduction

Being a classical object of limnology, stratified lakes still attract the attention of many researches. These lakes differ greatly from non-stratified shallow lakes, where the water column is constantly mixed. Due to thermal stratification, the upper water layer is isolated from the lower part and also from the sediments (Scheffer, 1998), which causes differences in biological and chemical parameters. The gradients of light, temperature, oxygen and inorganic substances combine and cause a variety of microhabitats (Davey & Heany, 1989; Reynolds, 1992; Gasol et al., 1991; Nõges & Nõges, 1998). The stratification processes are important for phytoplankton, providing advantages to some species and influencing the community structure, which tends to be more complex than in shallow lakes. The situation where the light needed by phytoplankton for photosynthesis is available in the epilimnion while the mineral nutrient pool is located mainly in the hypolimnion has been called the paradox of stratification (Mann, 1991; Klausmeier & Litchman, 2001).

Vertical distribution of phytoplankton affects the distribution and functioning of other components of the food web. The maximum activity of plankton can be found in the lower layers of the water column (Wetzel, 1983) both in oligotrophic and hypertrophic lakes. Nutrients can become available in the epilimnion during short periods of deeper mixing, which allows the coexistence of a great variety of species as well as occasionally high biomasses of phytoplankton in different layers. In the metalimnion, steep environmental gradients causing higher nisches diversity can be found. In addition species from lower and upper layers are present there, which makes the metalimnetic community more diverse compared to other layers. The distribution of the species depends on the nutrient amount, nutrient ratios, turnover speed, sedimentation rate, temperature, water density and viscosity, light attenuation, species mobility, grazers, as well as on internal loading of the lake. Lake Verevi is a special case, where the phytoplankton has been greatly affected by the lack of vernal circulation and by the rapid formation of stratification in the past years.

The aim of this paper is to analyze the changes in the phytoplankton over eight decades. This long period allows to follow the development of the lake from a natural moderately eutrophic to a hypertrophic state.

Material and methods

Lake description

L. Verevi is located in the town of Elva (6400 inhabitants) in S-E Estonia. The lake has an elongated shape with a deeper and broader part at the southern end. The southern and eastern shores are sandy, sloping towards the lake; the other shores are flat, muddy, or peaty framed by a swampy bank or reed belt. Both the sandy and swampy areas are covered mostly by pine forest. A road with heavy traffic passes the lake from the east. The area of the lake is 12.6 ha; maximum depth is 11 m, the mean depth is 3.6 m. The watershed area is 1.1 km^{-2} including the lake area. Small ditches and bottom springs in the narrow northern part form the bulk of the inflowing water; the outflow is via a larger ditch from the western side. The water exchange rate is 0.63 times per year (Loopmann, 1984). Mostly the water of the surface layers is exchanged as water from the deeper layers can flow out only during a short vernal and a longer fall turnover. No vernal turnover occurred in 2000 and 2001. The ice-free period lasts on the average from April to November. The lake is sheltered, which further enhances stratification. The temperature gradient in the metalimnion may exceed 10 °C m^{-1} (Nõges & Nõges 1998) and is accompanied by steep gradients in dissolved oxygen content, nutrients, and biota. Below 6 m the water is usually anoxic during the summer stagnation, winter anoxia may occupy the entire water column. This caused several fish kills in eighties (Kangur, 1991). The lake has been polluted by irregular discharge of urban wastewaters from oxidation ponds probably since the 1970s.

Historical plankton records

The first data about the lake were collected in the 1920s by Riikoja (1930) and since then the lake has been investigated repeatedly and more thoroughly in the 1980s and in 2000 and 2001.

Samples for phytoplankton analysis were collected at the deepest point of the lake located in the broader part near the western shore. Most samples were taken from the surface layer (0.5 m) and at depths of 4–5 and 7–9 m by the Ruttner or van Dorn sampler. In the 1920s and 1950s, only qualitative samples were taken by a 85 μm net. Since the 1980s quantitative samples were taken as well (Fig. 1). After settling from 500 ml, phytoplankton was counted by a light microscope at 400× magnification.

In 2000 water was taken from eight layers (2 in the epi-, 4 in the meta- and 2 in the hypolimnion) on 17 occasions from April to October and in 2001 from April to August on nine occasions. The absolute sampling depths differed from case to case depending on the temperature and oxygen profiles. The metalimnion was defined as the layer in which the temperature gradient

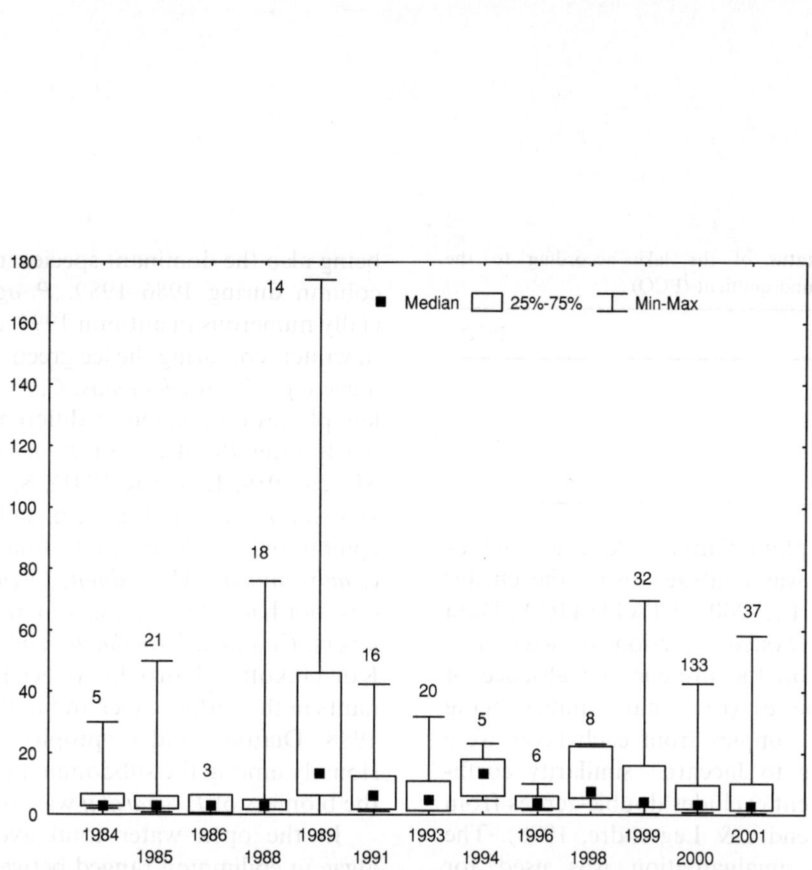

Figure 1. Phytoplankton biomass (g m^{-3}) in Lake Verevi: minimum, maximum and median values in different years. The number above the maximum value represents the number of different samples gathered in that year.

was ≥1.5 °C m^{-1} (Nõges & Nõges, 1998). At that time the studies focused on the formation and loss of stratification and for this reason sampling was more frequent in spring and autumn. More precise investigations were conducted in August 2001 when samples were taken from 0 to 7 m with 1 m interval.

Samples were taken by a water pump (Masterflex N 7533–60) with a capacity 2 l/min. The pump was equipped with a flexible tube ending with a T-shaped nozzle in order to prevent the mixing of water layers by the sucking stream (for more details see Zingel & Ott, 2000). Phytoplankton samples were preserved with Lugol's solution and kept in dark at 4 °C until counting.

Since 1998 phytoplankton samples were counted by an inverted microscope at 400× magnification using the Utermöhl (1958) technique. At least 600 counting units (cells, filaments, colonies) were counted. The mean volume of each species was measured in all samples by approximating the shape of species to the closest simple geometric form (Wetzel & Likens, 1991). In filamentous cyanobacteria the lengths of at least 50 trichomes were measured in each sample, and the mean length was used in biomass calculations.

Modified Nygaard's (1949) phytoplankton compound quotient (PCQ) was used to characterize the ecological status of the lake. PCQ gives quite good estimation to lake tropic condition, although algal groups in formula may contain species with different preferences to tropic conditions. Ott & Laugaste (1996) added to the original formula 2 extra taxons: Cryptophyta to numerator and Chrysophyceae to denominator. Modified index gives more precise estimation about Estonian lakes, because the abundance of Desmidiales in open water and in littoral zone has declined during last decade.

PCQ, modified by Ott & Laugaste (1996):

$$PCQ = \frac{\text{Cyanophyta}^* + \text{Chlorococcales}^* + \text{Centrales}^* + \text{Euglenophyceae}^* + \text{Cryptophyta}^* + 1}{\text{Desmidiales}^* + \text{Chrysophyceae}^* + 1}$$

where * is the number of different species. The classification of values is in Table 1.

Table 1. Ecological status of the lake according to the phytoplankton compound quotient (PCQ)

Lake status	PCQ
Oligo- or dystrophic	< 2
Mesotrophic	2–5
Eutrophic	5–7
Hypertrophic	> 7

Seasonality of phytoplankton data in different layers from 2000 was analyzed using the cluster analysis (StatSoft, Inc. 2001. STATISTICA. Data analysis software system, version 6. www.statsoft.com). Based on the presence or absence of phytoplankton species (the total number being 208), 17 seasonal samples from each layer were grouped according to Jaccard's similarity coefficient. This coefficient excludes double-zeroes from comparison (Legrendre & Legrendre, 1998). The complete linkage amalgamation was used for clustering the samples. According to Legendre & Legendre (1998), this method is best for delineating clusters with clear discontinuities.

Results

A historical review of the phytoplankton community

The first phytoplankton samples were taken from Lake Verevi in 1928 and 1929. Inter-annual variability was low: the following algae with different ecological demands prevailed: (a) mesotrophic chrysophytes (*Dinobryon* spp.); (b) moderate eutrophic diatoms *Asterionella formosa* Hassal, *Fragilaria crotonensis* Kitton, cyanobacteria *Anabaena lemmermanni* P. Richt. and relatively large green algae (*Pediastrum* spp., *Staurastrum* spp., *Botryococcus braunii* Kütz.); (c) eurytopic dinoflagellates *Ceratium hirundinella* (O. F. Müller) Schrank and *Peridinium* spp.; (d) various green algae. The number of species in summer was 18–26. The lowest (13) as well as the highest number (30) of species was found in October. Generally the same species dominated in 1956 and 1957 (Eesti järved, 1968), but *Planktothrix agardhii* Anagn. & Komarek (syn. *Oscillatoria agardhii* Gom.) already occurred in small numbers.

The relatively wide *P. agardhii* with its form *aequicrassa* dominated in the deeper layers in 1984, being also the dominant species in the whole water column during 1986–1989. *P. agardhii* was especially numerous in autumn 1991 and occurred even in winter, colouring the ice green. Flagellated green algae e.g. *Chlamydomonas*, *Carteria*, and also euglenophytes dominated in different periods, particularly after the decay of *P. agardhii* (April 1985, March 1989; Laugaste 1991). Species of the genus *Dinobryon* were dominant or subdominant in the epilimnion in May and June. Euglenophytes (*Rhabdomonas*, *Menoidium*, *Euglena*) were numerous, but had a low biomass in anoxic near-bottom layers. *Ceratium hirundinella* and *Phacotus coccifer* Korschikoff (Chlorophyta) occurred as co dominants in the surface layer and in the metalimnion in 1988. Diatoms and cryptophytes occurred occasionally among the subdominants, especially when the biomass of *P. agardhii* was low.

In the open water total average biomass of algae in epilimnion ranged between 1.8–174 g m^{-3} (Fig. 2a), between 1.77–50 g m^{-3} in metalimnion and between 0.6–30 g m^{-3} in hypolimnion. The highest biomass (724 g m^{-3}) was found in October 1985 in the epilimnion near the eastern shore (this value is not used in calculations) due to physical accumulation.

The average biomass was especially high in 1989 due to high values in summer (174 g m^{-3}). The maximum biomasses exceeded 10 g m^{-3} every year except in 1986 (Fig. 1). Lower biomass (<6 g m^{-3}) occurred in the epilimnion at the beginning of the 1980s, but in winter 1988 the biomass was significantly higher in comparison with the other seasons of the same year. The average biomass of *P. agardhii* in 1989 compared to the previous and following years was also higher. Unfortunately, samples were not taken from the metalimnion at that time, but, generally, biomasses in the metalimnion were higher than in the epilimnion (Fig. 2a, b).

In the hypolimnion average biomass was lower (Fig. 2c) than in the epi- and metalimnion, except in 1989, when the hypolimnion showed a maximum. In winter 1988, the average biomass was low as well.

The species composition in the 1990s was rather similar to that in the 1980s. In spring *Synedra* spp. were numerous among the diatoms. The average total biomass in autumn during the 1990s was lower compared to the previous decade.

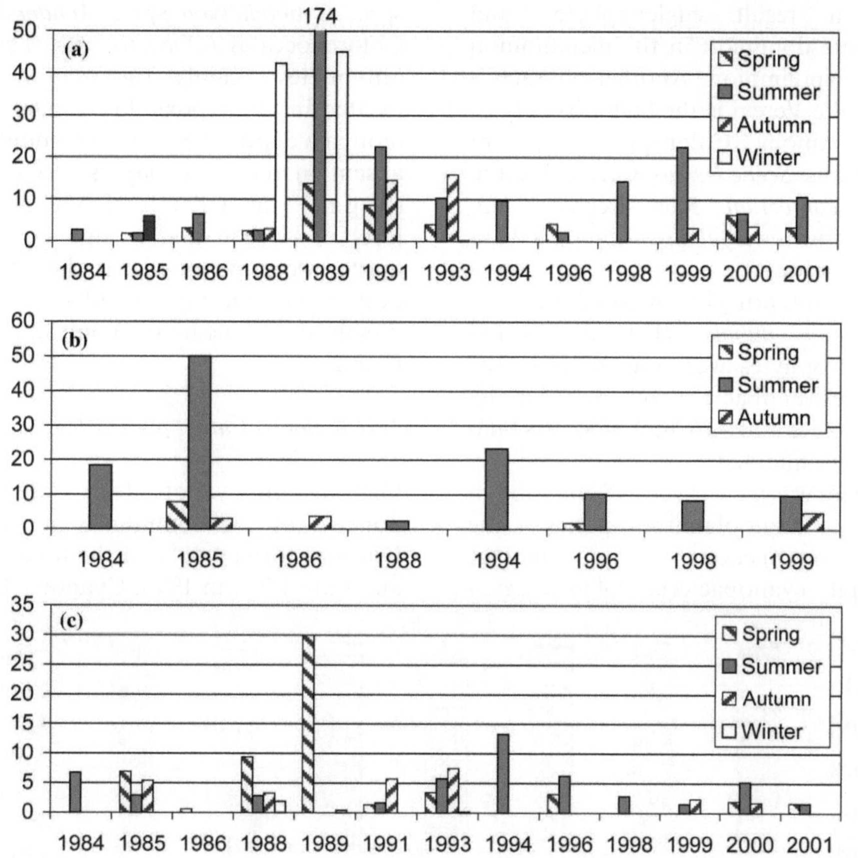

Figure 2. Total average biomass (g m^{-3}) of phytoplankton in Lake Verevi in the epilimnion (a), in the metalimnion (b) and in the hypolimnion (c).

Phytoplankton in 2000 and 2001

In spring (April and May) 2000 the phytoplankton from the mixed layer consisted of small flagellates (mainly Chrysophyta) and diatoms (the dominant species being *Synedra acus* var. *angustissima* Grun.). The biomass of diatoms reached its maximum (18.6 g m^{-3}) in the metalimnion at the beginning of May. In addition to the diatoms, cryptophytes and cyanobacteria were important components of the phytoplankton. In springtime also the biomass of cryptophytes reached its peak (14.8 g m^{-3}) in the metalimnion. Small-celled colonies of cyanobacteria (*Aphanocapsa* spp., *Aphanothece* spp.) dominated in the epilimnion while filiform *Limnothrix redekei* (van Goor) Meffert (syn. *Oscillatoria redekei*) and *Planktothrix agardhii* were important components of the metalimnetic community.

In summer consisted the diatom coenosis of characteristic spring species, and a few of them, like *Asterionella formosa* and *Nitzschia* spp. were present throughout the whole vegetation period. *Acanthoceras zachariasii* (Brun) Simonsen and *Rhisozolenia longiseta* Zacharias occurred only in summer and early autumn. *P. agardhii* reached its peak (37.7 g m^{-3}) in July in metalimnion, the filiform diazotrophic cyanobacterium *Aphanizomenon klebahnii* (Elenk.) Pechar & Kalina being codominant. Together with these two cyanobacteria the dinoflagellate *Ceratium hirundinella* reached its peak (13.8 g m^{-3}).

In autumn one could observe colonial species of cyanobacteria – *Cyanodictyon* spp., *Aphanocapsa* spp., *Aphanothece* spp. in the epilimnion. Lake Verevi is characterized by high diversity of euglenides in the metalimnion, especially in autumn, when the depth of the metalimnion increases. As a result euglenophytes and *P. agardhii* became dominant in the metalimnion similarly to the hypolimnion. Another characteristic feature of Lake Verevi is the high diversity of Chlorococcales. Among Chlorophyta different species of the genus *Scenedesmus* were common. *Monoraphidium contortum* Kom.-Legn., *Chlamydomonas* spp. and *Scourfeldia cordiformis* Takeda were also numerous.

In 2001, the diatom spring bloom was dominated by *Cyclostephanos dubius* (Fricke) Round (9.2 g m^{-3}). Cryptophytes were numerous in the metalimnion while cyanobacteria were rare. Among chlorophytes (max. 2 g m^{-3}) Chlorococcales and *Chlorogonium* sp. dominated.

In August, there was a clear vertical zonation in the distribution of phytoplankton species – the upper 4 m were homogenous with a variety of different colonial cyanobacteria (*Aphanocapsa* spp., *Cyanodictyon* spp., *Aphanothece* spp.) and Chlorococcales (*Oocystis* sp.). The maximum of dinophytes, mainly *Ceratium hirundinella*, was located in the hypoxic lower part of the metalimnion at a depth of 5 m. Cryptomonads, which were absent in the upper layers, occurred abundantly (56 g m^{-3}) at a depth of 6 m, where only trace level of oxygen and a high content of organic matter were measured. At the depth of 7 m and deeper (without oxygen and a very high content of dissolved organic matter) only euglenophytes were found.

Proportions of algal groups between 1984 and 2001

Diatoms and cyanobacteria were important components of the community in spring (Fig. 3a). Diatoms formed 40% of the total biomass in 2001 and only 1.7% in 1996. Cyanobacteria formed the

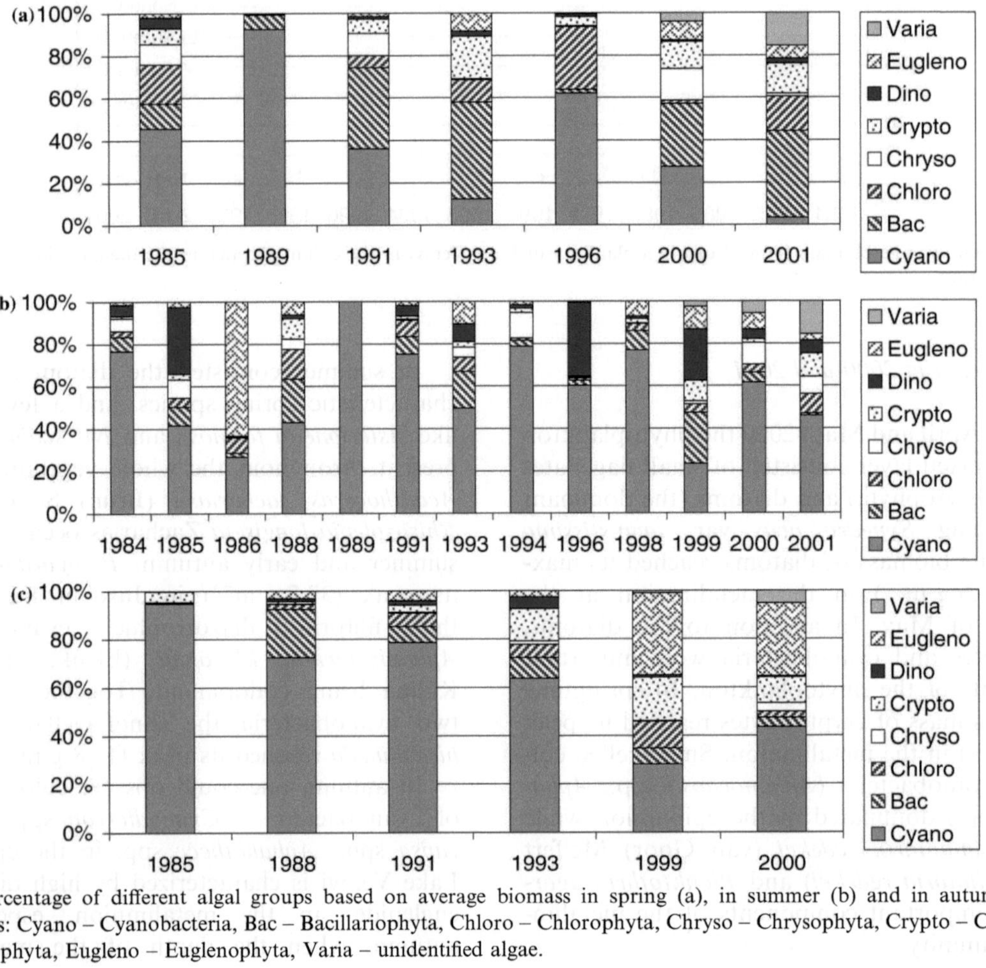

Figure 3. Percentage of different algal groups based on average biomass in spring (a), in summer (b) and in autumn (c). The abbreviations: Cyano – Cyanobacteria, Bac – Bacillariophyta, Chloro – Chlorophyta, Chryso – Chrysophyta, Crypto – Cryptophyta, Dino – Dinophyta, Eugleno – Euglenophyta, Varia – unidentified algae.

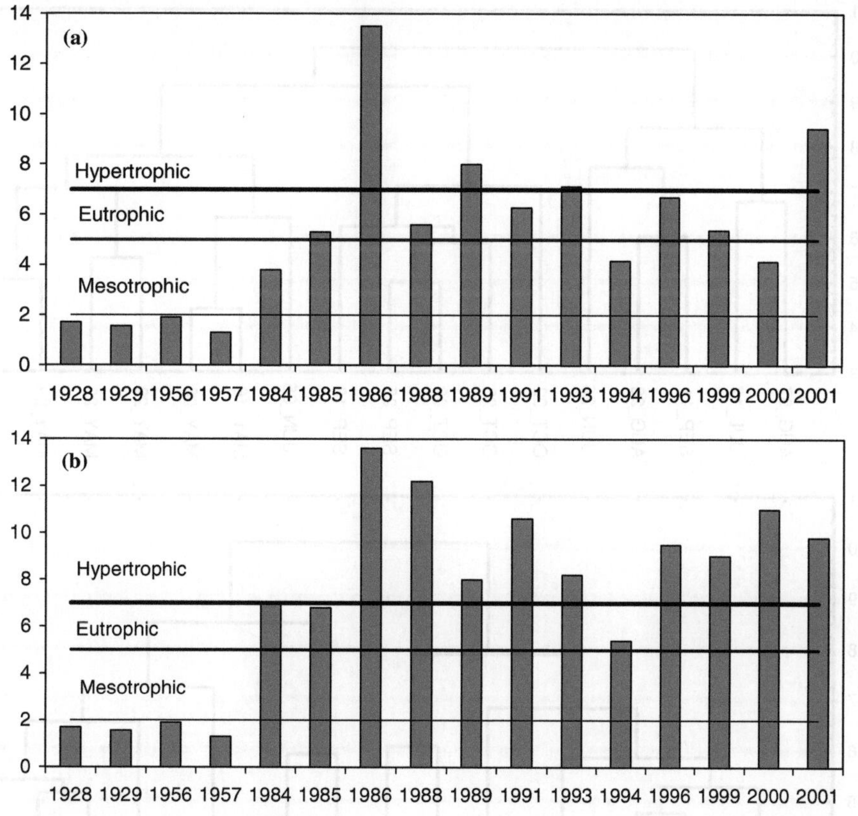

Figure 4. Phytoplankton compound quotient in summer in the epilimnion (a) and in the entire water column (b).

largest part in 1989 (93%) and the smallest in 2001 (3%). Cryptophytes and chlorophytes were numerous occasionally (1993 and 1996, respectively).

In summer, cyanobacteria formed an important part of the community (Fig. 3b). Their proportion ranged from 23% in 1999 to 100% in 1989. Dinophytes made up to 35% of the biomass in 1996. Sometimes also euglenophytes were important – 63% in 1986. The average percentage of other groups was always lower than 15%. Cyanobacteria were also numerous in autumn, being the main part of coenosis (Fig. 3c).

Phytoplankton compound quotient

Modified Nygaard's (1949) compound quotient (Ott, 1987; Ott & Laugaste, 1996) provides information about the ecological status of the lake on the basis of species composition (Fig. 4a). As a rule, the epilimnetic values were lower in summer in comparison with the entire water column (Fig. 4b). These values, calculated on the basis of summer data, showed a remarkable increase in the whole water column comparing to the 1950s. Since 1986, the quotient exceeded the hypertrophic limit (7) and remained on a very high level throughout the entire investigated period, except in 1994.

Cluster analysis (data from the year 2000)

In the epilimnion three periods could be clearly distinguished (Fig. 5a). The samples taken in the summertime differentiated from the samples taken in spring and autumn.

The samples from the metalimnion could be divided into two clusters-from April until the beginning of June and from the end of June until the October (Fig. 5b). Inside the branch of spring samples was also one sample taken in the end of October.

The samples taken from the hypolimnion could be divided into two clusters: from April to July and from August to October (Fig. 5c). In the first

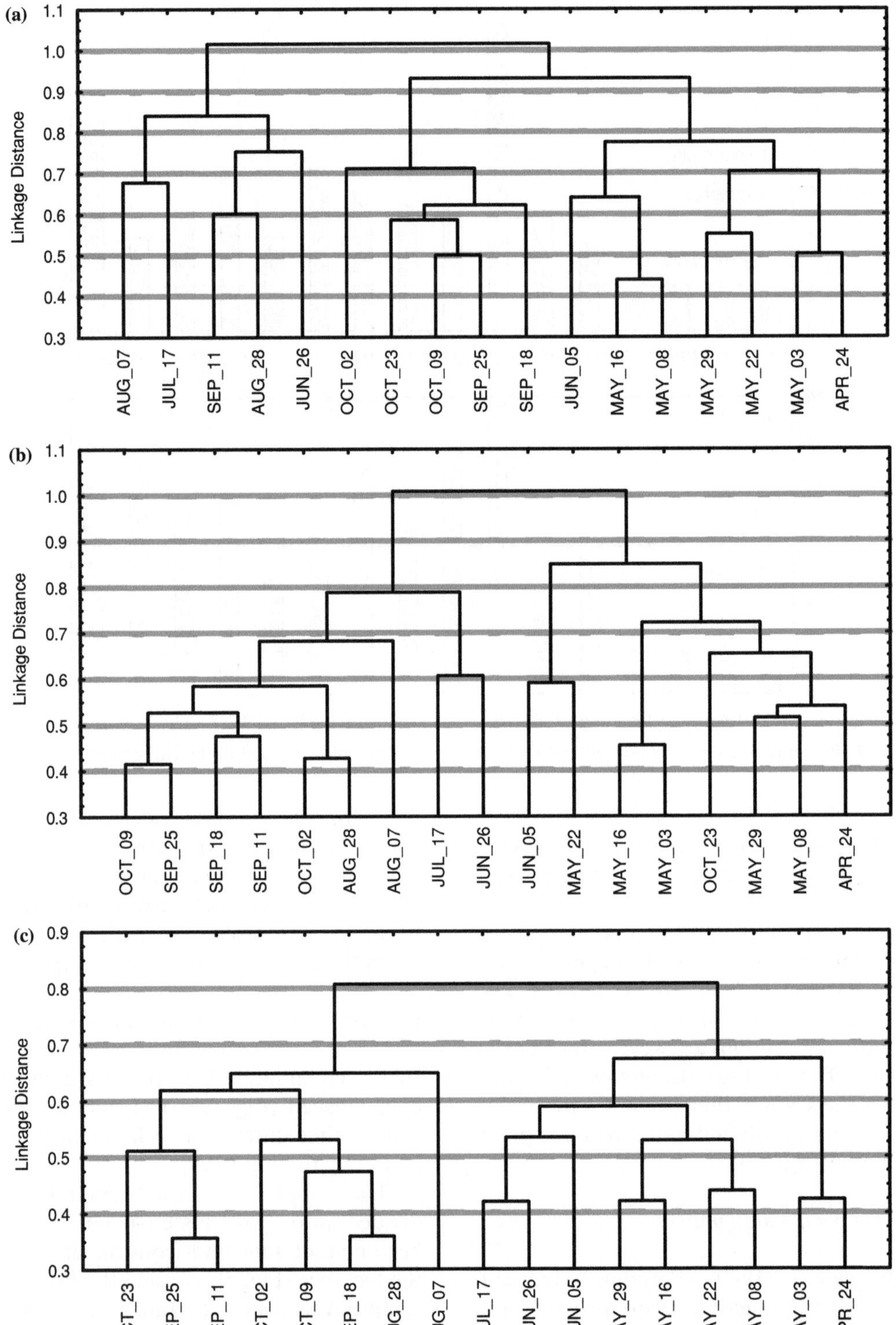

Figure 5. Grouping of samples based on phytoplankton species composition in 2000 in the epilimnion (a), in the metalimnion (b) and in the hypolimnion (c).

cluster samples from early spring differentiated from the samples taken in later springtime. In the second cluster, the samples from the beginning of August differentiated from samples taken later. The linkage distance between the different clusters was lower in the hypolimnion compared to the epi- and metalimnion.

Discussion

Species composition

Three periods characterized by different ecological status can be distinguished on the basis of phytoplankton species composition. The ecological status of the lake was close to natural (moderately eutrophic) in the 1920s and the 1950s. During the 1980s and the 1990s, the ecosystem was disturbed by anthropogenic eutrophication and indicators of hypertrophic conditions prevailed. The past years reveal fluctuations of species composition and abundance in both seasonal and vertical aspects. The phytoplankton of stratified hypertrophic lakes is generally characterized by uneven vertical distribution (Laugaste, 1991; Adler et al., 2000).

The main reason for the great changes of the phytoplankton community has been the irregular nutrient influx originating from the oxidation ponds filled with urban wastewaters (Ott et al., 2005; Nõges, 2005). Therefore, the situation is different from other lakes in Estonia, which generally follow the dynamics of agricultural pollution from diffuse and point sources. Lake Verevi has high nutrient reserves in the sediments, and internal loading is the main reason why there are still high nutrient concentration, the presence of communities characteristic of hypertrophic conditions and high phytoplankton biomass. The changes between the 1950s and the 1980s caused by human impact were favoured by the small lake volume and a low water exchange rate making the lake susceptible to eutrophication. The seasonal dynamics of phytoplankton is rather similar from year to year in lakes with a moderate trophic state (Hutchinson, 1967), but accelerated eutrophication makes the biomass of cyanobacteria and seasonal patterns less predictable. The increasing importance of cyanobacteria, especially *P. agardhii*, was also observed in Lake Verevi in the 1980s. The sharp gradients of light and nutrients (not shown), formed already in early spring in 2000 and 2001 due to strong thermal stratification, resulted in a large number of vertically separated niches in the water column. This led to a specialization and stratification of biota, which presumably alleviated inter-specific competition.

The formation of gradients is crucial for different phytoplankton species, which could take advantage of rapid changes and afterwards persist for the whole vegetation period in the community. This might explain difference between communities in years 2000 and 2001.

Cluster analysis (data from 2000)

The grouping of seasonal samples gives a possibility to find similar periods and periods of greatest changes in the community. In spring and autumn, the epilimnetic communities were rather similar consisting mainly of diatoms, complemented by small numbers of cyanobacteria and chrysophytes. The summer stage differed clearly due to abundance of nitrogen-fixing cyanobacteria and dinophytes. At the end of June there was a clear water period characterized by low phytoplankton biomass and species diversity. This was presumably caused by the cladocerans grazing on phytoplankton (Kübar et al., 2005). Another possible reason might be the decline of the diatom peak, which led to loss of nutrients from the epilimnion through sedimentation. July was characterized by a higher biomass of cyanobacteria (mainly *Aphanizomenon klebahnii*) and by the biomass maximum of *Ceratium hirundinella*. The species diversity was the highest after the decline of *A. klebahnii*. The presence of N_2-fixing cyanobacteria with heterocysts in the epilimnion could be due to temporal nitrogen limitation. The period from late June to September differed from other periods by a higher temperature (up to 22.4 °C) (Nõges & Kangro, 2005). The species with better buoyancy get an advantage in warm stratified conditions.

In the metalimnion samples from early spring differed from the rest of the vegetation period by a rapid growth of diatoms. Diatoms were also in the samples in the end of October, which caused the similarity to spring phytoplankton community. Species diversity was consistently higher in the metalimnion compared to the epi- or hypolimnion, except at the end of August. The high diversity could be caused by the multitude of vertically separated niches in the strongly stratified metalimnion. Light can penetrate the deeper parts of the metalimnion and, at the same time, nutrients diffuse from the hypolimnion into the metalimnion (Stauffer, 1987). Sedimentation of living cells and detritus are important factors affecting the composition of the phytoplankton community. The sharp drop in temperature and the increasing viscosity and water density diminish the sedimentation rate of sinking particles (Kufel & Kalinowska, 1997). The particulate matter accumulates in the metalimnion where high rates of remineralization support algal growth (Miracle & Alfonso, 1993).

Euglenophytes became more abundant after the decline of the metalimnion in autumn, when the thermocline was located in nearly anaerobic conditions. Erosion and deepening of the metalimnion towards autumn was reflected practically by all indicators. Conditions in the metalimnion in autumn were similar to those in the hypolimnion: low oxygen content, high values of ammonium, total phosphorus, silicon, sulphides and soluble reactive phosphorus (Kõiv & Kangro, 2005).

The higher stability of the hypolimnion, which caused the smaller variability of the community structure, was the main reason for the shorter linkage distance between clusters compared to that in the upper layers.

Autecology and demecology of phytoplankton

Predictably, stratified lakes have a high phytoplankton biomass near the chemocline if their trophic status is moderate (Moll & Stoermer, 1982; Wetzel, 1983; Lindholm, 1992; Pedrós-Alió & Guerrero, 1993). Moll & Stoermer (1982) explained it by a higher amount of algae in the epilimnion by shading and negatively affecting the photosynthesis of the algae growing in the deeper layers. Thus increasing eutrophication leads to disappearence of deep phytoplankton maximum in stratified waters.

Lake Verevi is hypertrophic (the phytoplankton biomass exceeds 10 g m^{-3}, for other parameters see Ott et al., the present issue). According to Lindholm et al. (1985) the deep maximum can develop especially in stratified and wind-sheltered small lakes, where stratification is quite permanent. It means, that in meromictic lakes are good conditions for the development of more deeply located phytoplankton. Meromictic lakes are characterized by anoxic conditions and the presence of H_2S in high concentration. In fact, several species that form metalimnetic maxima in Lake Verevi like *L. redekei, P. agardhii* and *Cryptomonas* sp. have been earlier described as indicators of H_2S (Huber-Pestalozzi, 1938, 1941; Nicklisch et al., 1991).

According to Krienitz et al. (1996) *P. agardhii* dominates in lakes with a high N:P ratio, which is often caused by discharge of allochthonous organic matter. Mass growth of *P. agardhii* and *L. redekei* is often observed in eutrophic lakes (Berger, 1984; Henning & Kohl, 1981). These species can occupy two alternative habitats: usually they have been observed in shallow lakes where mixing and frequent sediment resuspension create high turbidity (Berger, 1984; Riddols, 1985; Moed et al., 1988; Rücker et al., 1997; Scheffer et al., 1997), but they can grow also in stratified lakes (Meffert, 1989; Reynolds & Bellingher, 1992; Rojo & Alvarez-Cobelas, 1994; Olli, 1996), where they usually inhabit the metalimnion. According to Klein & Chorus (1991), metalimnetic maxima are caused by *Planktothrix rubescens* and *P. agardhii* var *isothrix* while the typical *P. agardhii* can dominate only in shallow polymictic lakes. Laugaste (1991) pointed out, that under a light microscope is difficult to distinguish close species of genus *Planktotrix*. According to Boone et al. (2001) *P. rubescens* and *P. agardhii* differ only in subspecies level.

P. agardhii and *L. redekei* are both favoured in hypertrophic conditions and because of accelerated eutrophication in the 1970s these species expand into many Estonian lakes (Ott et al., 1997). The main reason for their success is the tolerance to low light and high nutrient concentration. According to Lindholm & Meriluoto (1991), some strains of *P. agardhii* may be hepatotoxic and thus be a potential health hazard.

For the first time, *P. agardhii* was discovered in Lake Verevi in the 1950s. Since the 1980s the species has a high biomasses both in the surface layer and in the metalimnion, whereas the differences between the layers have been manifold. The fact that *P. agardhii* occurrs on a massive scale right under the ice is also known from other lakes

(Lindholm, 1992). It could be explained by a similar light climate under the ice and in the lower metalimnetic layers.

In August 2001, cryptophytes were located in thin layer around 6 m. They were completely absent in the upper and lower parts of the water column. The presence of cryptophytes in the metalimnion and the development of deep biomass maxima have been observed for several times in stratified lakes (Gasol et al., 1993; Gervais, 1997). The efficient growth of cryptophytes could probably be explained by the absence of *P. agardhii* (better light climate) or grazers. The maximum of cryptophytes was located at the border of the euphotic layer, in almost anaerobic conditions. Cryptophytes are tolerant to the anaerobic environment as they can migrate to the metalimnion, being in the upper layers in the daytime and in the lower part of the metalimnion at night (Gasol et al., 1991). According to Lindholm et al. (1985), the vertical maximum may be very thin (only a few centimeters). Due to the small number of samples from the water column, earlier maxima of cryptophytes could have been missed.

According to Gervais (1997), several species of cryptophytes (*C. rostratiformis, C. phaseolus, C. undulata*) do not tolerate oxygen contents over 1 mg l^{-1}. The cryptophytes flora in Lake Verevi is variable, and they are found both in deeper and in oxygenated surface layers. The cryptophytes are typically present in low numbers during most of the year, but commonly their increase is associated with the decline of some major dominating species (Stewart & Wetzel, 1986). Being tolerant to low light intensities, cryptophytes tend to develop rather in the metalimnion than in epilimnion. The reason for this may be the higher phosphorous availability compared to upper water column or due to reduced predation near the sulphide-rich waters (Gasol et al., 1993).

The flora of euglenophytes in Lake Verevi is rather rich and inhabits mainly the lower metalimnion or the hypolimnion. Only the genus *Trachelomonas* prefers the upper parts of the water column. According to Lee (1999), euglenophytes are usually numerous in ponds that are polluted by organic material. Wetzel (1983) observed a dense population of euglenophytes in the aerobic-anaerobic interface zone. Nygaard (1977) explains that euglenophytes prefer oxygen deficiency or some positively or negatively correlated factor with oxygen concentration.

The high trophic state is not always reflected by a steadily high phytoplankton biomass in the epilimnion. In fact, high water transparency and a moderate amount of phytoplankton in the upper layers in summer could be accompanied by a biomass maximum in the lower metalimnion while the hypolimnion is anaerobic (Nõges & Nõges, 1998). Lack of nutrients in the epilimnion forces the community to change from diatoms to the species adapted to high-light and low-nutrient conditions (Sommer et al., 1986). Small flagellates and different chlorococcales are successful in these conditions. *Dinobryon* is known to be successful at phosphorus-limited conditions due to its phagotrophy (Nygaard & Tobiesen, 1993; Granéli et al., 1999). The presence of *Acanthoceras zachariasii* and *Rhizosolenia longiseta* in Lake Verevi might indicate improvement in the water quality of the lake because these species tend to disappear with eutrophication (Kümmerlin, 1998).

Compared to the 1980s, the condition of the lake has improved, and the mass occurrences of *P. aghardhii* have declined. This can be attributed to warmer springs, when the lake is warming up and stratifying so quickly that the upper layers will lack nutrients for sufficient phytoplankton growth. The danger of *P. agardhii* bloom still remains, because of the large pool of phosphorous will be released from sediments during turnover, occurrence of which depends on weather conditions.

Acknowledgements

This work was supported by the core grant of the Estonian Ministry of Education No. 0370208s98 "The influence of thermal stratification on biological matter circulation in lakes" and by grants of the Estonian Science Foundation Nos. 3579, 3689, and 4835. We would also like to thank Dr. Enn Veldi for revising the English.

References

Adler, M., F. Gervais & U. Siedel, 2000. Phytoplankton composition in the chemocline of mesotrophic lakes. Archiv für Hydrobiologie – Advances in Limnology 55: 513–530.

Berger, C., 1984. Consistent blooming of *Oscillatoria agardhii* Gom. in shallow hypertrophic lake. Verhandlung Internationale Vereinigung für theoretische und angewande Limnologie 22: 910–916.

Boone, D. R., R. W. Castenholz & G. M. Garrity, (eds) 2001. Bergeýs Manual of Systematic Bacteriology. The Archaea and the Deeply Branching and Phototrophic Bacteria. Springer, New York, I: 1–721.

Davey, M. C. & S. I. Heany, 1989. The control of sub-surface maxima of diatoms in a stratified lake by physical, chemical and biological factors. Journal of Plankton Research 11: 1185–1189.

Gasol, J. M., J. Garcia-Cantizano, R. Massana, F. Peters, R. Guerrero & C. Pedros-Alio, 1991. Diel changes in the microstratification of the metalimnetic community in Lake Cisó. Hydrobiologia 211: 227–240.

Gasol, J. M., J. Garcia-Cantizano, R. Massana, F. Peters, R. Guerrero & C. Pedros-Alio, 1993. Physiological ecology of a metalimnetic *Cryptomonas* population: relationship to light, sulfide and nutrients. Journal of Plankton Research 15: 255–275.

Gervais, F., 1997. Diel vertical migration of *Cryptomonas* and *Chromatium* in the deep chlorophyll maximum of an eutrophic lake. Journal of Plankton Research 19: 533–550.

Granéli, E., P. Carlsson & C. Legrand, 1999. The role of C, N and P in dissolved and particulate organic matter as a nutrient source for phytoplankton growth, including toxic species. Aquatic Ecology 33: 17–27.

Henning, M. & J.-G. Kohl, 1981. Toxic blue-green algae water blooms found in some lakes in the German Democratic Republic. Internationale Revue der gesamten Hydrobiologie 66: 553–561.

Huber-Pestalozzi, G., 1938. Das Phytoplankton des Süβwassers. Systematik und Biologie. Die Binnengewässer, Bd XVI, 1. Teil, Allgemeiner Teil. Blaualgen. Bakterien. Pilze, E. Schweizerbartische Verlagsbuchhandlung, Stuttgart: 236 pp.

Huber-Pestalozzi, G., 1941. Das Phytoplankton des Süβwassers: Chrysophyten, Farbulose Flagellaten, Heterokonten. E. Schweizerbartische Verlagsbuchhandlung, Stuttgart: 365 pp.

Hutchinson, G. E., 1967. A treatise on limnology. Vol. 2: Introduction to lake biology and the limnoplankton. New York, John Wiley & Sons: 1115 pp.

Kangur, A., 1991. Kalad. In Timm, H. (eds), State of Lake Verevi. Hydrobiological researches XVII. Tartu:' pp. 114–119 [The fishes in Lake Verevi. In Estonian].

Klausmeier, Ch. A. & E. Litchman, 2001. Algal games: The vertical distribution of phytoplankton in poorly mixed water columnsLimnology and Oceanography, 46: 1998–2007.

Klein, G. & I. Chorus, 1991. Nutrient balances and phytoplankton dynamics in Schlachtensee during oligotrophication. Verhandlung Internationale Vereinigung für theoretische und angewande Limnologie 24: 873–878.

Krienitz, K., P. Kasprzak & R. Koschel, 1996. Long-term study on the influence of eutrophication, restoration and biomanipulation on the structure and development of phytoplankton communities in Feldberger Haussee (Baltic Lake District), Germany. Hydrobiologia 330: 89–110.

Kufel, L. & K. Kalinowska, 1997. Metalimnetic gradients and the vertical distribution of phosphorus in a eutrophic lake. Archiv für Hydrobiologie 140: 309–320.

Kõiv, T. & K. Kangro, 2005. Resource ratios and phytoplankton species composition in a strongly stratified lake. Hydrobiologia 547: 123–135.

Kübar, K., H. Agasild, T. Virro & I. Ott, 2005. Vertical distribution of zooplankton in a strongly stratified hypertrophic lake. Hydrobiologia 547: 151–162.

Kümmerlin, R. E., 1998. Taxonomical response of the phytoplankton community of Upper Lake Constance (Bodensee-Obersee) to eutrophication and re-oligotrophication. Archiv für Hydrobiologie Special Issues in Advanced Limnology 53: 109–117.

Laugaste, R., 1991. Fütoplankton. In Timm, H. (eds), State of Lake Verevi. Hydrobiological Researches XVII. Tartu: pp. 69–89 [Phytoplankton. In Estonian, summary in English].

Lee, R. E., 1999. Phycology. Cambridge University Press, UK, 613 pp.

Legendre, P., & L. Legendre, 1998. Numerical Ecology. Elsevier Science B. V., The Netherlands: 853 pp.

Lindholm, T., 1992. Ecological role of depth maxima of phytoplankton. Archiv für Hydrobiologie Beiheft, Ergebnisse der Limnologie 35: 33–45.

Lindholm, T. & J. A. O. Meriluoto, 1991. Recurrent depth maxima of the hepatotoxic cyanobacterium Oscillatoria agardhii. Canadian Journal of Fisheries & Aquatic Sciences 48: 1629–1634.

Lindholm, T., K. Weppling & H. S. Jensen, 1985. Stratification and primary production in a small brackish lake studied by close-interval siphon sampling. Verhandlung Internationale Vereinigung für theoretische und angewande Limnologie 22: 2190–2194.

Loopmann, A., 1984. Suuremate Eesti järvede morfomeetrilised andmed ja veevahetus. Tallinn: 150 pp. [Morphometrical data and water exchange of larger Estonian lakes. In Estonian].

Mann, K. H., 1991. Organisms and Ecosystem. In Barnes, R. S. K. & K. H. Mann (eds.), Fundamentals of Aqatic Ecology. EdinburghBlackwell Science, London: 3–26.

Meffert, M.-E., 1989. Planktic unsheated filaments (Cyanophyceae) with polar and central gas vacuoles 2. Biology, population dynamics and biotopes of Limnothrix redekei (van Goor) Meffert. Archiv für Hydrobiologie 116: 257–282.

Miracle, M. R. & M. T. Alfonso, 1993. Rotifer vertical distributions in a meromictic basin of Lake Banyoles (Spain). Hydrobiologia 255–256: 371–380.

Moed, J. R., H. L. Hoogveld & H. De Haan, 1988. A study of factors regulating the succession of cyanobacteria in Lake Tjeukemeer, The Netherlands. Verhandlung Internationale Vereinigung für theoretische und angewande Limnologie 23: 1894–1897.

Moll, R. A. & E. F. Stoermer, 1982. A hypothesis relating trophic status and subsurface chlorophyll maxima of lakes. Archiv für Hydrobiologie 94: 425–440.

Eesti järved, 1968. Tallinn, "Valgus":532 pp. [Estonian lakes. In Estonian].

Nicklisch, A., B. Roloff & A. Ratsch, 1991. Competition experiments with two planktic blue-green algae (Oscillator-

iaceae). Verhandlung Internationale Vereinigung für theoretische und angewande Limnologie 24: 889–892.
Nõges P. & Nõges T. 1998. Stratification of Estonian small lakes studied during hydrooptical expeditions in 1995–97. Proceedings of Estonian Academy of Sciences. Biology. Ecology 47: 268–281.
Nõges, T. & K. Kangro, 2005. Primary production of phytoplankton in a strongly stratified temperate lake. Hydrobiologia 547: 105–122.
Nõges, P., 2005. Water and nutrient mass balance of temperate partly meromictic Lake Verevi. Hydrobiologia 547: 21–31.
Nygaard, G., 1949. Hydrobiological studies on some Danish ponds and lakes II. The quotient hypothesis on some new or little known phytoplankton organisms. Det Kongelige Danske Videnskabernes Selskab 7: 293 pp.
Nygaard, G., 1977. Vertical and seasonal distribution of some motile freshwater plankton algae in relation to some environmental factors. Archiv für Hydrobiologie Supplements 51 Algological Studies 18: 67–76.
Nygaard, K. & A. Tobiesen, 1993. Bacterivory in algae: A survival strategy during nutrient limitation. Limnology and Oceanography 38: 273–279.
Olli, K., 1996. Mass occurrences of cyanobacteria in Estonian waters. Phycologia 36 (6th suppl): 156–159.
Ott, I., 1987. Mnogoletnie izmenenija letnego fitoplanktona i ih svjazi s ekologičeskimi faktorami v Estonskih ozerah. Dissertation. Manuscript at Tartu University, 203 pp. [Long-term changes of summer phytoplankton in Estonian lakes and their relations with ecological factors. In Russian].
Ott, I., T. Kõiv, P. Nõges, A. Kisand, A. Järvalt, & E. Kirt, 2005. General description of Lake Verevi, its ecological status, changes during the past eight decades and restoration problems. Hydrobiologia 547: 1–20.
Ott, I. & R. Laugaste, 1996. Fütoplanktoni koondindeks (FKI), üldistus Eesti järvede kohta. Eesti Keskkonnaministeeriumi infoleht 3: 7–8[The Phytoplankton Compound Quotient (PCQ), generalisation about Estonian small lakes. In Estonian].
Ott, I., S. Lokk, A. Mäemets, & R. Laugaste. 1997. Plankton changes in Estonian small lakes in 1951–1993. Proceedings of Estonian Academy of Sciences. Biology. Ecology 46 (1/2): 58–79.
Pedrós-Alió, C. & R. Guerrero, 1993. Microbial ecology of Lake Cisó. In Jones, J.G. (eds), Plenum PressNew York: 155–209.
Reynolds, C. S., 1992. Dynamics, selection and composition of phytoplankton in relation to vertical structure in lakes. Archiv für Hydrobiologie Beiheft Ergebnisse der Limnologie 35: 13–31.
Reynolds, C. S. & E. G. Bellingher, 1992. Patterns of abundance and dominance of the phytoplankton of Rostherne Mere, England: evidence from an 18-year data set. Aquatic Science 54: 10–36.
Riddols, A., 1985. Aspects of nitrogen fixation in Lough Neagh. 1. Acetylene reduction and the frequency of Aphanizomenon flos-aquae heterocysts. 2. Competition between Aphanizomenon flos-aquae, Oscillatoria redekei and Oscillatoria agardhii. Freshwater Biology 15: 299–306.
Riikoja, H., 1930. Zur Morphometrie eineiger Seen im Estland. 15: 116–201.
Rojo, C. & M. Alvarez-Cobelas, 1994. Population dynamics of *Limnothrix redekei, Oscillatoria lancaeformis, Planktothrix agardhii.* Hydrobiologia 275–276: 165–171.
Rücker, J., C. Wiedner & P. Zippel, 1997. Factors controlling the dominance of *Planktothrix agardhii* and *Limnothrix redekei* in eutrophic shallow lakes. Hydrobiologia 342–343: 107–115.
Scheffer, M., 1998. Ecology of shallow lakes, In M. B. Usher (ed.) Chapman & Hall, London, Weinheim, New York: 357 pp.
Scheffer, M., S. Rinaldi, A. Gragnani, L. R. Mur & E. H. van Nes, 1997. On the dominance of filamentous cyanobacteria in shallow, turbid lake. Ecology 78: 272–282.
Sommer, U., Z. M. Gliwicz, W. Lampert & A. Duncan, 1986. The PEG-model of seasonal succession of planktonic events in freshwater. Archiv für Hydrobiologie 106: 433–471.
Stauffer, R. E., 1987. Vertical nutrient transport and its effects on epilimnetic phosphorus in four calcareous lakes. Hydrobiologia 154: 87–102.
Stewart, A. J. & R. G. Wetzel, 1986. Cryptophytes and other microflagellates as couplers in planktonic community dynamics. Archiv für Hydrobiologie 106: 1–19.
Utermöhl, H., 1958. Zur Vervollkomnung der quantitativen Phytoplankton Methodik. Mitteilungen internationale Vereingung für theoretische und angewandte Limnologie 9: 1–38.
Wetzel R. G., 1983. Limnology. Saunders College Publishing, Philadelphia, London, Toronto: 767 pp.
Wetzel R. G. & Likens G. E., 1991. Limnological analyses. Springer-Verlag, New York, Inc: 391 pp.
Zingel, P. & I. Ott, 2000. Vertical distribution of planktonic ciliates in strongly stratified temperate lake. Hydrobiologia 435: 19–26.

Hydrobiologia (2005) 547:105–122
I. Ott & T. Kõiv (eds), Lake Verevi, Estonia – A Highly Stratified Hypertrophic Lake
DOI 10.1007/s10750-005-4152-z

Primary production of phytoplankton in a strongly stratified temperate lake

Tiina Nõges* & Kersti Kangro
Institute of Agricultural and Environmental Sciences, Estonian Agricultural University, Centre for Limnology, 61101, Rannu, Tartu County, Estonia
*(*Author for correspondence: E-mail: tnoges@zbi.ee)*

Key words: primary production, sharply stratified lake, interannual, seasonal, and diurnal dynamics, metalimnetic chlorophyll maximum, ecological status

Abstract

Lake Verevi (12.6 ha, maximum depth 11.0 m, mean depth 3.6 m) is a strongly eutrophic and stratified lake. *Planktothrix agardhii* is the most characteristic phytoplankton species in summer and autumn, while photosynthesizing sulphur bacteria can occur massively in the metalimnion. Primary production (PP) and chlorophyll *a* concentration (Chl *a*) were seasonally studied in 1991, 1993, 2000, and 2001. Vertical distribution of PP was rather complex, having usually two peaks, one at or near the surface (0–1 m), and another deeper (at 3–7 m) in the metalimnion. The values of dark fixation of CO_2 in the metalimnion were in most cases higher than those in the upper water layer. Considering the average daily PP 896 mg C m^{-2} and yearly PP 162 mg C m^{-2}, Secchi depth 2.34 m, and epilimnetic concentrations of chlorophyll *a* (19.6 mg m^{-3}), total nitrogen and total phosphorus (TP, 52 mg m^{-3}) in 2000, L. Verevi is a eutrophic lake of a 'good' status. Considering the total amounts of nutrients stored in the hypolimnion, the average potential concentrations in the whole water column could achieve 1885 mg m^{-3} of TN and 170 mg m^{-3} of TP reflecting hypertrophic conditions and a 'bad' status. Improvement of the epilimnetic water quality from the 1990s to the 2000s may have resulted from incomplete spring mixing and might not reflect the real improvement. A decreased nutrient concentration in the epilimnion has supported the establishment of a 'clear epilimnion state' allowing light to penetrate into the nutrient-rich metalimnion and sustaining a high production of cyanobacteria and phototrophic sulphur bacteria.

Introduction

Primary production (PP) of photoautotrophic organisms, first of all phytoplankton, is a process that depends on and influences many biotic and abiotic mechanisms representing a major synthesis of organic matter in aquatic systems. It initiates food chains and forms a basis of the ecological pyramid. Uncertainties in PP will influence all other calculations of productivity, for example zooplankton and fish. PP also forms the most relevant basis for the classification of lakes into different trophic categories (Hakason & Boulion, 2002).

Photosynthesis is often low at the lake surface, due to inhibition by ultraviolet light, with a peak at some depth below. This peak is set at light saturation (I_k); its size and position depend on the light-saturated rate of photosynthesis per unit of biomass (P_{max}) and the total biomass (B, usually expressed as chlorophyll *a*) of the photosynthetic cells present. I_k is often low for cyanobacteria and higher for green algae. The P_{max} value tends to increase with temperature, almost doubling for every 10 °C. Photosynthesis-depth curves for stratified and non-stratified lakes are similar if primary production is expressed as per unit biomass of phytoplankton (Moss, 1998).

The rate of photosynthesis is mostly limited by the energy source (light), inorganic substrate (CO_2, H_2S for sulphur bacteria) and inorganic nutrients

(N, P, Si). In stratified lakes, light is sufficiently available in the epilimnion while the deep hypolimnion is often rich in nutrients (N, P). If surface light intensity is adequate, most phytoplankton species inhabit the fraction of the water column with the optimal light conditions. Because wind mixes the uppermost water layer, the phytoplankton is usually exposed to light intensity equal to the average light intensity of the mixing zone. During the period of stratification, the epilimnetic phytoplankton is usually nutrient-limited while low-light-adapted species (e.g. genera *Limnothrix* and *Planktothrix*) can develop pronounced maxima in the metalimnion as long as the light intensity does not drop below the compensation point for those species. Photosynthetic bacteria can develop below the phytoplankton zone in lakes with an anaerobic hypolimnion and a buildup of H_2S, as long as light penetrates into the H_2S zone (for more detailed discussion see Lampert & Sommer, 1997).

Lake Verevi (12.6 ha, maximum depth 11.0 m, mean depth 3.6 m) is a strongly eutrophic and stratified lake situated in Elva, a small summer-resort town (population 7000) in South Estonia. The situation of the lake has been observed since the 1950s, based on occasional sampling as well as some periods of seasonal measurements including water chemistry, species composition, and biomass of hydrobionts. The results of previous investigations were published in a book in Estonian (Timm, 1991). In the 1980s, *Planktothrix agardhii* (Gomont) Anagn. & Kom. was reported to be the most characteristic phytoplankton species in L. Verevi in summer and autumn. Phytoplankton biomass in summer were about 20 gWW m^{-3} and could exceed 50 gWW m^{-3} in autumn, and also under the transparent ice in February (Laugaste, 1991). In 1995–1997 an extremely pronounced seasonally changing biological stratification was found in Lake Verevi. In June, the surface layer was dominated by two filamentous cyanophytes, *Aphanizomenon flos-aquae* and *Planktothrix agardhii* together with small flagellates. At a depth of 5–6 m *Cryptomonas* sp. was a codominant to *P. agardhii.* A mass occurrence of photosynthesizing purple sulphur bacterium *Thiopedia rosea*, an obligate anaerobe, gave evidence of oxygen depletion at 5–6 m. In August, the epilimnion was inhabited by small flagellates, *A. flos-aquae* formed a dense layer in the upper part of the metalimnion, while a maximum of *P. agardhii* together with *Thiopedia rosea* was located deeper in the anoxic layers. In September, when the turbulent mixing extended to a depth of 5 m, both cyanophyte species occurred in the epilimnion. At that time green sulphur bacteria *Pelodictyon luteolum* peaked in the anoxic layer (Nõges & Nõges, 1998a).

The present paper deals with vertical distribution and diurnal, seasonal and interannual dynamics of primary production and related indices in Lake Verevi with the aim to:

1. quantify the annual rate of primary productivity; study the patterns of vertical distribution of primary production and their relationship to the physical stratification of the water column as well as to the amount of phototrophic plankton, light conditions, and nutrient regime;
2. study the seasonal dynamics of primary production, the interannual variation and relationship with the meteorological conditions and physico-chemical stratification;
3. evaluate the present ecological status of the lake.

Material and methods

Primary production in L. Verevi was measured during the period of 1991–2001 by ^{14}C-technique (Steeman-Nielsen, 1952) and *in situ* exposition. Frequency of the measurements varied on a large scale (Table 1). In 1991, 1993 and 1995–1997 measurements of particulated PP were made with $NaH^{14}CO_3$ (Izotop, St. Petersburg), final activity of 0.08 μCi ml^{-1}. The metalimnion was defined as the layer in which the temperature gradient was $\geq$1.5 °C m^{-1} (Nõges & Nõges, 1998a). The water from the epi- and metalimnion was poured into 24 ml glass scintillation vials. When handling the anaerobic water from the metalimnion, special attention was paid to maintaining the anaerobic and low light conditions – the vials were kept in a dark box, the tube from the water sampler was placed on the bottom of the vial and the water was allowed to overflow slowly from the vial for some seconds. The vials were incubated for 2 h at midday (usually from 11 a.m. to 1 p.m.) in the lake, at the same depths where the water was taken from

Table 1. Dates, Secchi depths, and the depths and methods of primary production measurements in L. Verevi

Date	Secchi (m)	PP fraction measured	Exposition depths (m)						
17.04.1991	0.5	Particulated	0	0.1	0.3	0.5	1.0	1.5	
22.05.1991	1.8	Particulated	0	0.5	0.9	1.8	3.6	5.4	
26.06.1991	1.1	Particulated	0	0.3	0.6	1.1	2.2	3.3	
17.07.1991	0.6	Particulated	0	0.1	0.3	0.6	1.1	1.7	
14.08.1991	1.2	Particulated	0	0.3	0.6	1.2	2.4	3.6	
18.09.1991	0.8	Particulated	0	0.2	0.4	0.8	1.5	2.3	
23.10.1991	0.9	Particulated	0	0.2	0.5	0.9	1.8	2.7	
21.04.1993	0.9	Particulated	0	0.2	0.5	0.9	1.8	2.7	
12.05.1993	1.8	Particulated	0	0.5	0.9	1.8	3.6	5.4	
16.06.1993	1.6	Particulated	0	0.4	0.8	1.6	3.2	4.8	
20.07.1993	2.0	Particulated	0	0.5	1.0	2.0	4.0	6.0	
17.08.1993	2.1	Particulated	0	0.5	1.1	2.1	4.2	6.3	
23.09.1993	1.3	Particulated	0	0.3	0.6	1.3	2.5	3.8	
19.10.1993	1.1	Particulated	0	0.3	0.6	1.1	2.2	3.3	
17.11.1993	2.2	Particulated	0	0.6	1.1	2.2	4.4	6.6	
3.05.1995	2.0	Particulated	0	0.5	1.0	2.0	4.0	6.0	
9.06.1995	2.0	Particulated	0	0.5	1.0	2.0	4.0	6.0	
1.08.1995	3.0	Particulated	0	0.8	1.5	3.0	6.0	9.0	
4.06.1996	3.0	Particulated	0	0.8	1.5	3.0	6.0	9.0	
19.08.1996	3.5	Particulated	0	0.9	1.8	3.5	7.0	10.5	
17.09.1996	2.4	Particulated	0	0.6	1.2	2.4	4.8	7.2	
6.08.1997	2.6	Particulated	0	0.7	1.3	2.6	5.2	7.8	
28.07.1999		Total		0.5	2.0	3.0	3.5	4.0	4.5
7.09.1999		Total		0.5	1.5	3.0	3.5	4.0	4.5
24.04.2000	1.4	Total	0	0.5	1.5	2.0	2.5	3.5	
3.05.2000	2.0	Total	0	0.5	3.0	3.5	4.0	4.5	
8.05.2000	2.0	Total	0	0.5	2.0	3.0	3.5	4.0	4.5
16.05.2000	2.0	Total	0	0.5	2.0	3.5	4.0	4.5	5.0
22.05.2000	1.8	Total	0	0.5	1.5	2.5	3.0	4.0	4.5
29.05.2000	2.0	Total	0	0.5	1.5	3.0	3.5	4.0	4.5
5.06.2000	2.7	Total	0	0.5	2.0	3.5	4.0	4.5	5.5
26.06.2000	3.1	Total	0	0.5	1.5	3.0	4.0	4.5	5.5
17.07.2000	2.6	Total	0	0.5	2.0	4.0	4.5	5.0	5.5
7.08.2000	1.5	Total	0	0.5	2.0	3.5	4.0	5.0	6.0
28.08.2000	2.1	Total	0	0.5	3.0	5.0	5.5	6.0	6.5
11.09.2000	2.5	Total	0	0.5	2.5	5.0	5.5	6.0	6.5
18.09.2000	2.3	Total	0	0.5	3.0	5.5	6.0	6.5	7.0
25.09.2000	2.7	Total	0	0.5	3.0	6.0	6.5	7.0	7.5
2.10.2000	3.1	Total	0	0.5	4.5	6.5	7.0	7.5	8.0
9.10.2000	3.0	Total	0	0.5	4.5	6.5	7.0	7.5	8.0
23.10.2000	3.0	Total	0	0.5	2.0	4.0	5.0	6.0	7.0
19.04.2001	1.2	Total	0	0.5	1.0	2.5	3.5	4.0	
23.04.2001	1.2	Total	0	0.5	1.0	1.5	2.0	2.5	4.0
26.04.2001	1.6	Total	0	0.5	1.0	1.5	2.0	2.5	4.0
30.04.2001	2.5	Total	0	0.5	1.0	1.5	2.0	2.5	3.5
3.05.2001	2.5	Total	0	0.5	1.0	2.0	2.5	3.0	3.5

Continued on p. 108

Table 1. (Continued)

Date	Secchi (m)	PP fraction measured	Exposition depths (m)								
7.05.2001	2.3	Total	0	0.5	1.0	2.0	2.5	3.0	3.5		
10.05.2001	3.3	Total	0	0.5	1.5	2.0	3.0	3.5	4.0		
24.05.2001	2.8	Total	0	0.5	1.5	2.0	3.0	4.0	4.5		
5.06.2001	4.0	Total	0	0.5	2.0	3.0	4.0	4.5	5.0	5.5	6.0
2.08.2001	2.5	Total	0	0.5	1.0	1.5	2.0	3.0	4.0	5.0	
3.08.2001	2.5	Total	0	0.5	1.0	1.5	2.0	3.0	4.0	5.0	

(Table 1). After incubation, the water was filtered through membranes of 0.45 μm pore size (Millipore HA), the filters were treated with concentrated HCl fumes for 5 min to remove excessive inorganic ^{14}C and air-dried. 100 μl of water from each bottle was mixed with 0.5 ml β-phenylethylamine (PEA) for the assessment of total radioactivity. In 1999–2001 total primary production was measured. About 6 ml of water from each sample was poured into a clean glass scintillation vial and acidified (pH < 2) by adding 150 μl of 0.5 N HCl. Inorganic ^{14}C was assumed to be removed during 24 h (Niemi et al., 1983; Hilmer & Bate, 1989; Lignell, 1992). Next, 5 ml subsamples were poured into new plastic vials. The radioactivity was assessed by LSC RackBeta 1211 (Wallac, Finland) using external standardization for DPM calculations. The radioactivity of the filters was measured in a toluene-PPO-POPOP cocktail, and OPTIPHASE (Wallac, Finland) was applied for the assessment of liquid samples. In all cases PP was calculated according to the standard formula (Guidelines, 1994). Non-photosynthetic carbon fixation was measured in dark vials and subtracted from light assimilation. The trapeze integration over depth and time was applied for calculating values per m^2 and per year. Accordingly, daily PP values were calculated using the equation relating daily PP (PPday; mg C m^{-2} day^{-1}) to integral PP at midday (PPint; mg C m^{-2} h^{-1}) and with the length of the light day (DL; h) obtained for Lake Võrtsjärv (Nõges & Nõges, 1998b), which is situated 20 km from L. Verevi:

$$\text{PPday} = \text{PPint}/(0.230 - 890 \times 10^{-5} \times \text{DL});$$
$$R^2 = 0.66, p < 0.01$$

In order to calculate the annual PP, a 200 day vegetation period (April 15–October 31) was assumed.

The concentration of chlorophyll *a* (Chl *a*) was estimated from all the water samples where PP was measured, and also in metalimnetic and hypolimnetic water. In most cases eight vertical samples were gathered. The plankton was filtered on Whatman GF/F filters. In 1991–1995, pigments were extracted by 90% acetone, in 1996–1997 by 96% ethanol, and in 1999–2001 in parallel by 90% acetone and 96% ethanol. Filters were soaked into the solvent either for 4 h at room temperature or for 24 h at +4 °C, after which the extract was vortexed and centrifuged for 10 min at 3000 rev/min. The absorption of the extract was determined in the region of 430–750 nm by a scanning UV–vis spectrophotometer Cecil-3000 (Great Britain). Chl *a* concentration was calculated by the equations of Jeffrey & Humphrey (1975) and Lorenzen (1967). If parallel extraction by acetone and ethanol was applied, the concentration of Chl *a* achieved by maximum recovery was used in further analysis as recommended by Nõges & Solovjova (2000).

In August 1997 and 2001, the diurnal dynamics of PP was studied. *In situ* expositions on 6 August 1997 at 8–10, 12–14, 16–18, on 2 August 2001 at 12–14 and 16–18, and on 3 August 2001 at 12–14 were carried out as described above. At the beginning of each exposition vertical water samples for pigment analysis were taken, and vertical distribution of light (W m^{-2}) was measured using an underwater 4-π PhAR collector (Williams & Jenkinson, 1980). The vertical light attenuation coefficient (K_d) was calculated according to Lambert–Beer's law:

$$I_z = I_0 e^{(-K_d Z)}$$

Hourly and daily sums of total irradiance (MJ m^{-2}) were obtained from Tõravere Meteorological Station located 5 km NE from L. Verevi.

The program STATISTICA (StatSoft Inc., 2001) was used for statistical analysis. As the data were not normally distributed, nonparametric methods were applied.

Results

Physical environment

In Estonian lakes (latitude range 57–59 N) the ice break-up generally takes place in mid-April and the freeze-up in mid-November. In 2000, the ice break-up occurred on 15 April, and in 2001 on 13 April. We visited the lake correspondingly 9 and 3 days after the ice break-up and it was already stratified. Oxygen stratification was established very quickly (Fig. 1), and the lake remained stratified throughout the vegetation period. Measurements of conductivity also pointed to early stratification (Fig. 2). Despite lack of measurements every day during the ice break, we have an opinion, that spring overturn of water column was absent in 2000 and 2001. We term it 'partly meromixis'. May be the correct term for that case is monomixis. However, Wetzel (1983) explains two possibilities for monomixis in cold and warm lakes, and these descriptions are different from our case. Dynamics of conductivity, temperature and oxygen content and during whole vegetation period in 2000 is depicted in Figures 2, 3 and 4. Complete mixing occurred before the ice cover was formed. The exact date of the ice break-up in

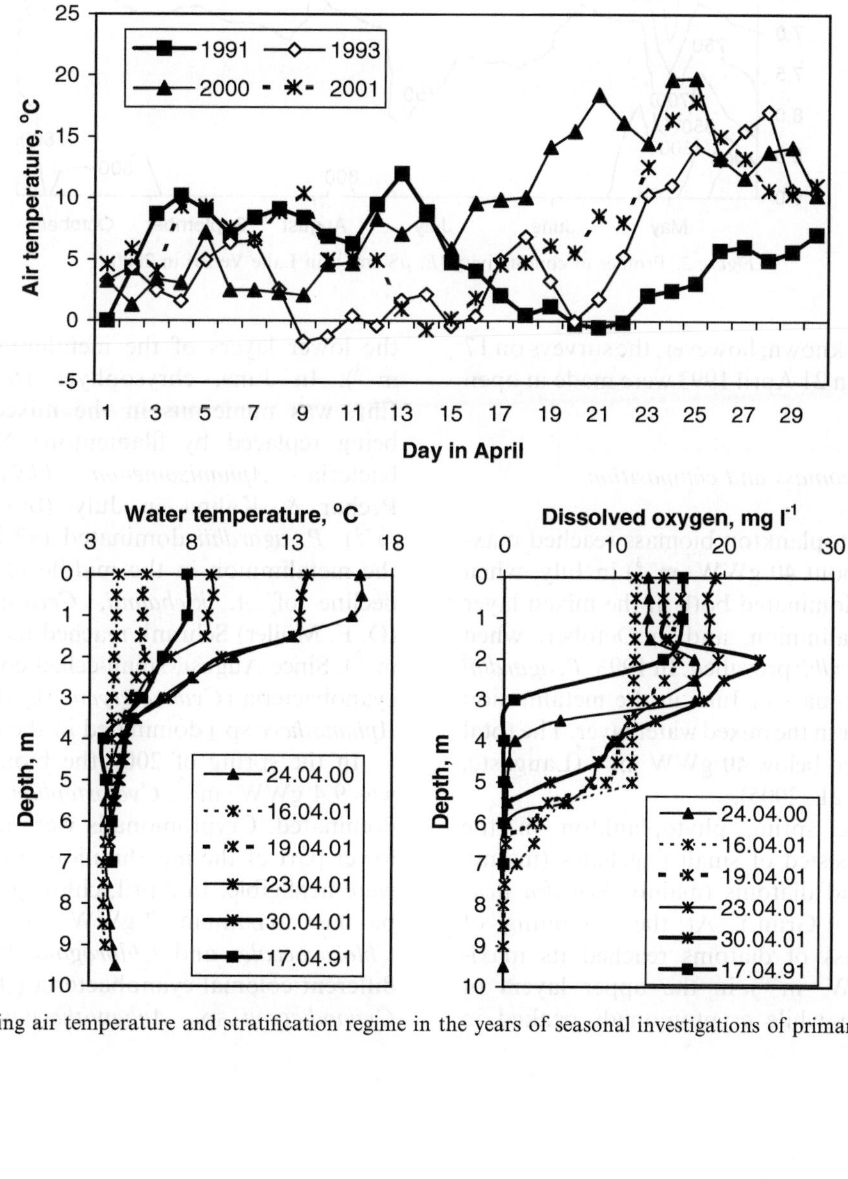

Figure 1. Spring air temperature and stratification regime in the years of seasonal investigations of primary production.

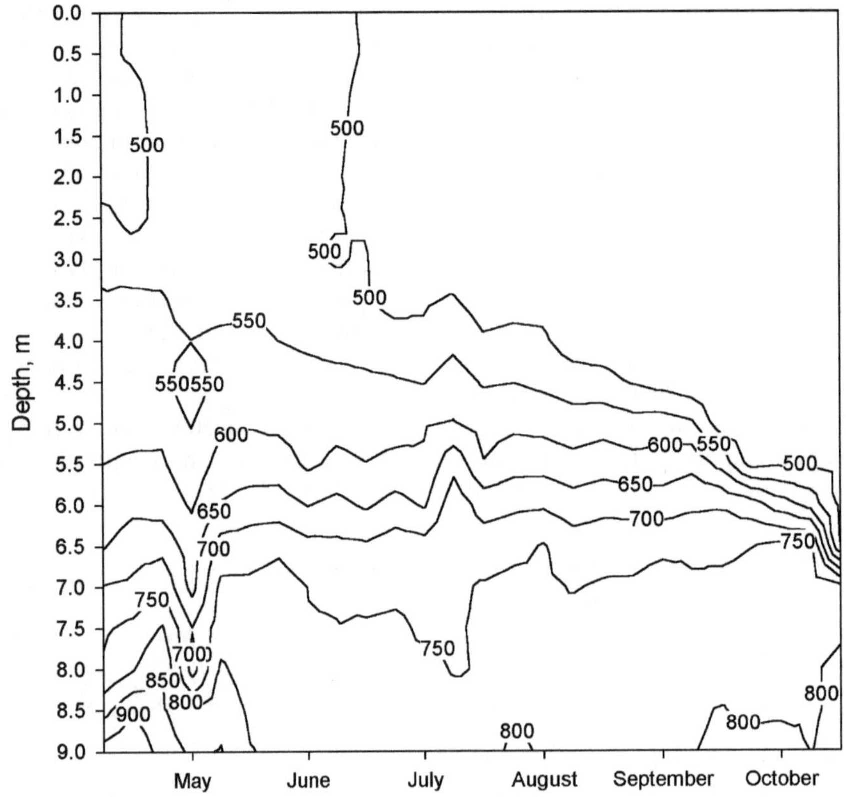

Figure 2. Profiles of conductivity (E, μS cm^{-1}) in Lake Verevi in 2000.

other years is not known; however, the surveys on 17 April 1991 and on 21 April 1993 were made in open water.

Phytoplankton biomass and composition

In 1991, the phytoplankton biomass reached maximum values (about 40 gWW m^{-3}) in July, when *Anabaena* spp. dominated both in the mixed layer and in the metalimnion, and in October, when *Planktothrix agardhii* prevailed. In 1993, *P. agardhii* built up high biomass in June in the metalimnion and in September in the mixed water layer. The total biomass remained below 40 gWW m^{-3} (Laugaste, 1994; Kangro et al., 2005).

In 2000, the spring phytoplankton in the mixed layer consisted of small flagellates (mainly *Chrysophyta*) and diatoms (mainly *Synedra acus* var *angustissima* Grun.). At the beginning of May the biomass of diatoms reached its maximum (18.6 gWW m^{-3}) in the upper layers of the metalimnion while cryptomonads peaked in the lower layers of the metalimnion (14.8 gWW m^{-3}). In June, chrysophyte *Dinobryon sociale* Ehr. was numerous in the mixed surface layer being replaced by filamentous N fixing cyanobacteria *Aphanizomenon klebahnii* (Elenk.) Pechar & Kalina in July (biomass 6.1 gWW m^{-3}). *P. agardhii* dominated (37.7 gWW m^{-3}) in the metalimnion in the middle of July. After the decline of *A. klebahnii*, *Ceratium hirundinella* (O. F. Müller) Schrank reached its peak (4.6 gWW m^{-3}). Since August, small-celled colonial species of cyanobacteria (*Cyanodictyon* sp., *Aphanocapsa* sp., *Aphanothece* sp.) dominated in the mixed layer.

In the spring of 2001 the biomass of diatoms was 9.4 gWW m^{-3}, *Cyclostephanos dubius* Round dominated. Cryptomonads were numerous in the lower part of the metalimnion, and cyanobacteria were negligible. In April, chlorophytes reached its biomass maximum (2 gWW m^{-3}) dominated by *Chlorococcales* and *Chlorogonium* sp. In August different colonial cyanobacteria (*Aphanocapsa* sp., *Cyanodictyon* sp., *Aphanothece* sp., biomass ca

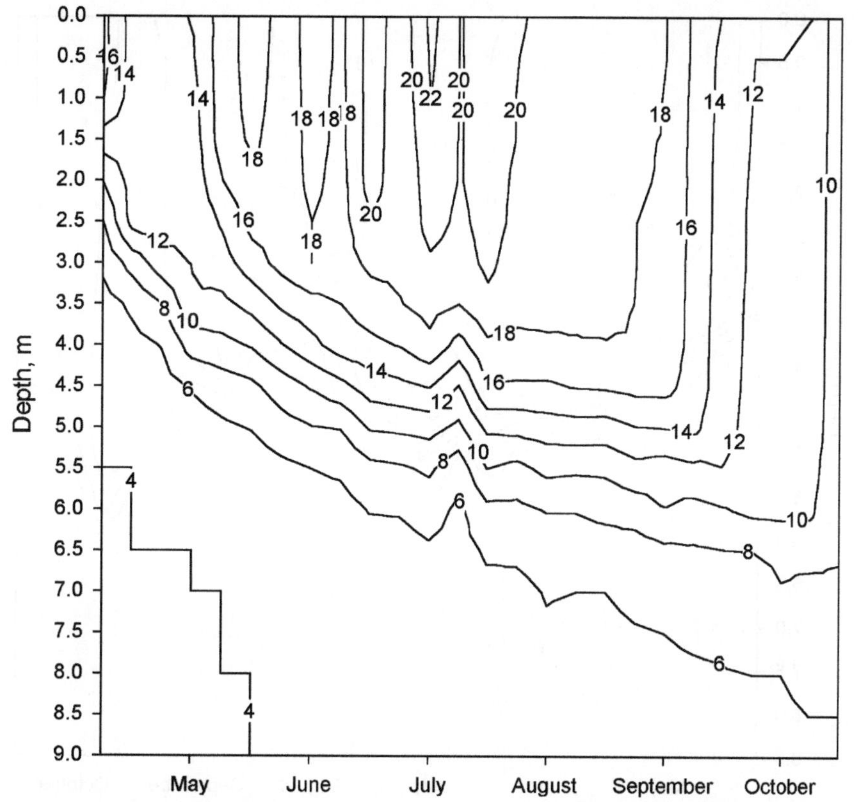

Figure 3. Profiles of water temperature (*T*, °C) in Lake Verevi in 2000.

14 gWW m^{-3}) and Chlorococcales (*Oocystis* sp.) occurred in the homogenous upper 4 m water column. At the depth of 5 m *Dinophyta*, mainly *Ceratium hirundinella* dominated (18.6 gWW m^{-3}), at 6 m was the maximum of cryptomonads (biomass 58.6 gWW m^{-3}), and at 7 m different euglenophytes prevailed.

Vertical distribution of primary production and chlorophyll

In L. Verevi the vertical distribution of PP was rather complex having usually two peaks – one at or near the surface layer (0–1 m) and the other deeper (at 3–7 m) in the metalimnion. As also the chlorophyll concentration expressed the metalimnetic maximum, PP expressed per Chl *a* unit (assimilation number, AN) had the depth distribution with a near-surface maximum value (Fig. 5). The values of dark fixation of CO_2 in the metalimnion were in most cases higher than those in the upper water layer (Fig. 6). The depth of the maximum PP varied from 0 to 7 m, and the depth of maximum AN from 0 to 3.2 m (see Table 2 for average values).

The depth of the maximum PP (Z_{PPmax}) was in good correlation (see Table 3 for all significant correlation coefficients) with the Secchi depth (Z_S) and the depth of the upper boundary of metalimnion (Z_M). The depth of the maximum AN value (Z_{ANmax}) was not significantly correlated either with Secchi depth or with Z_M while the positive correlation with total daily irradiance (Q; MJ m^{-2}) was evident. The Z_{PPmax} and Z_{ANmax} were weakly correlated with each other, and Z_S was weakly correlated with Z_M while the strong positive correlation of day number with Z_M and with the depth of oxygen depletion (Z_H) was evident. The metalimnion was the thinnest in summer (Fig. 7).

Seasonal dynamics and interannual variation of primary production and chlorophyll

In summer of 1991 and 1993 Z_S was relatively low (0.5–2 m) and PP and Chl *a* in the upper mixed layer (PP_{mix} and Chl_{mix}, respectively) were

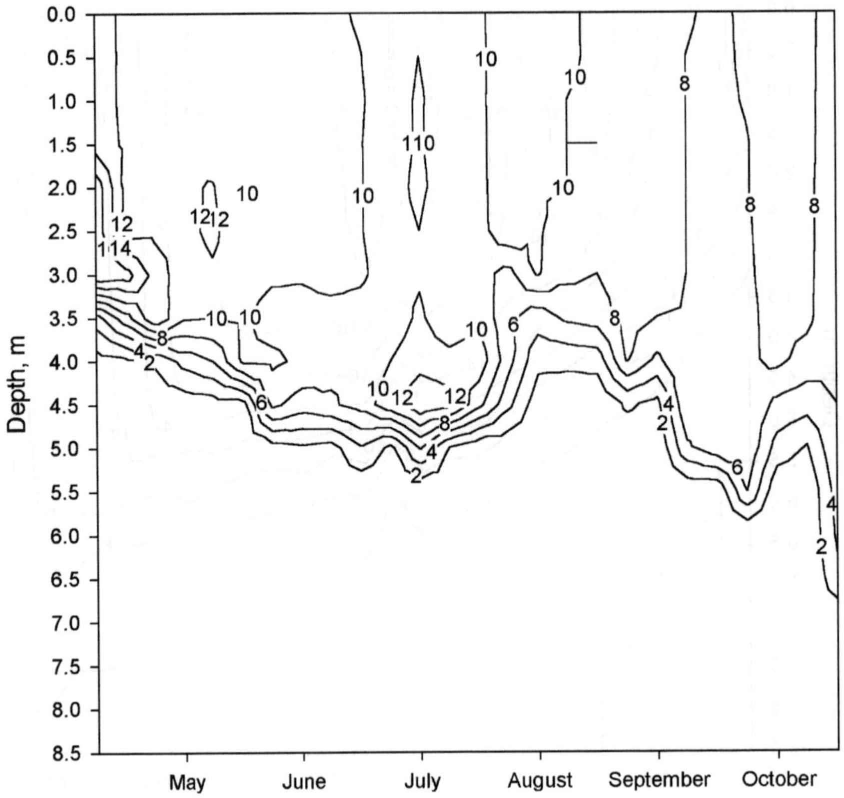

Figure 4. Profiles of dissolved oxygen concentration (O_2, mg l^{-1}) in Lake Verevi in 2000.

quite high (up to 100 mg C m^{-3} h^{-1} and 90 mg m^{-3}, correspondingly). In 2000 and 2001 PP_{mix} and Chl_{mix} were the highest (up to 55 mg C m^{-3} h^{-1} and 55 mg m^{-3}, correspondingly) and Z_S the lowest (ca. 1 m) in spring, and during the whole summer the epilimnetic water was quite clear (Z_S about 2–4 m) and unproductive – PP_{mix} below 20 mg C m^{-3} h^{-1}, Chl_{mix} below 30 mg C m^{-3} (Fig. 9). Seasonal peaks of PP occurred in July (1991), April (1993 and 2001) and October (2000).

PP_{mix} and Chl_{mix} were positively related to each other and with the concentration of total phosphorus in the mixed water layer (TP_{mix}) and inversely related to the ratio of total N and P in the mixed water layer (N/P_{mix}) as shown in Table 3. The assimilation number (AN_{max}) was not significantly correlated either with water temperature or with the concentration of nutrients. A regression model was developed (Fig. 8) enabling to predict the daily PP on the basis of the value of the maximum PP at noon (PP_{max}).

The average yearly PP constituted 220 g C m^{-2} being the highest in 1993 and the lowest in 2000 (Table 2). In 1990s, the average Chl_{mix} and PP_{mix} were higher (43 mg m^{-3} and 34 mg C m^{-3} h^{-1}, correspondingly) and Secchi depth lower (1.3 m) than in 2000 and 2001 – 20 mg m^{-3}, 16 mg C m^{-3} h^{-1} and 2.4 m, correspondingly. The hypolimnetic concentrations of chlorophyll (Chl_H), total nitrogen (TN_H), and total phosphorus (TP_H) were significantly higher in the 2000s (190, 12,000 and 1600 mg m^{-3}, correspondingly) than in the 1990s – respectively 50, 5000 and 500 mg m^{-3} (Fig. 10).

Diurnal dynamics of primary production

On 6 August 1997, the deep chlorophyll maximum was located at a depth of 4 m and was rather stable during the day. Epilimnetic primary production and assimilation numbers were the highest at noon while the vertical maximum was located at the water surface in the morning (8–10) and at a

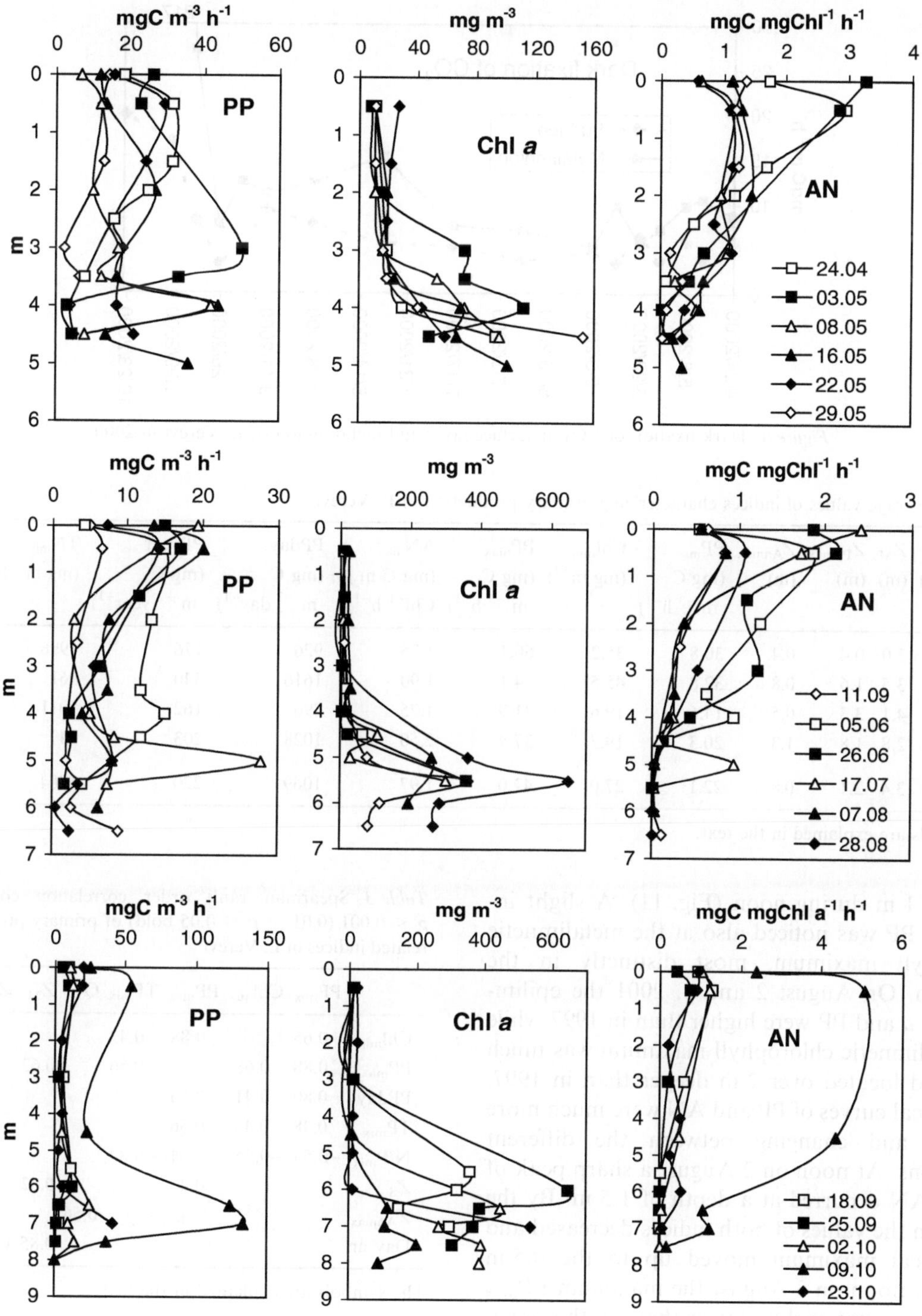

Figure 5. Seasonal depth profiles of primary production (PP, mg C m^{-3} h^{-1}), chlorophyll concentration (Chl *a*, mg m^{-3}) and assimilation number (AN, mg C mg Chl^{-1} h^{-1}) in L. Verevi in 2000.

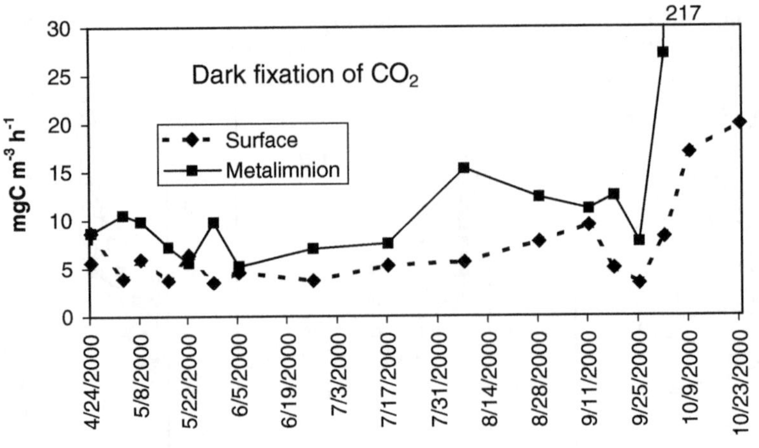

Figure 6. Dark fixation of CO_2 in surface layer and metalimnion of L. Verevi in 2000.

Table 2. Average values of indices characterizing primary production in L. Verevi

Year	Z_S (m)	Z_M (m)	Z_{Ppmax} (m)	Z_{Anmax} (m)	PP_{mix} (mg C m^{-3} h^{-1})	Chl_{mix} (mg m^{-3})	PP_{max} (mg C m^{-3} h^{-1})	AN_{max} (mg C m g Chl^{-1} h^{-1})	PPday (mg C m^{-2} day^{-1})	PPyear (mg C m^{-2} year^{-1})	TN_{mix} (mg m^{-3})	TP_{mix} (mg m^{-3})
1991	1.3	3.0	0.4	0.4	30.8	35.2	60.1	1.75	926	176	1996	202
1993	1.8	3.5	1.6	0.8	32.8	45.5	54.1	1.90	1616	340	961	80
2000	2.3	4.1	3.3	0.5	14.6	19.6	31.9	1.75	896	162	1011	52
2001	2.6	2.8	1.8	1.3	20.3	19.7	37.4	2.50	1028	203	847	38
Aver.	2.0	3.4	2.1	0.8	22.1	27.0	42.0	1.97	1069	220	1204	93

The symbols are explained in the text.

depth of 1 m during noon (Fig. 11). A slight increase of PP was noticed also at the metalimnetic chlorophyll maximum, most distinctly in the afternoon. On August 2 and 3, 2001 the epilimnetic Chl *a* and PP were higher than in 1997 while the metalimnetic chlorophyll maximum was much lower and located over 2 m deeper than in 1997. The vertical curves of PP and AN were much more complex and changing between the different expositions. At noon on 2 August a sharp peak of PP and AN occurred at a depth of 1.5 m. By the afternoon the values of both indices decreased and the vertical maximum moved up to the 0.5 m depth. At noon on 3 August the maximum PP_{max} and AN_{max} were slightly lower than on the previous day, and they occurred at a depth of 2 m (Fig. 11). Light conditions of different expositions varied substantially, though no clear relationship was found between light intensity and primary production (Fig. 12).

Table 3. Spearman rank order correlation coefficients at $p < 0.001$ ($0.01 < p < 0.05$ bold) of primary production and related indices in L. Verevi

	PP_{mix}	Chl_{mix}	PP_{max}	TP_{mix}	Q	Z_M	Z_H	Z_{PPmax}
Chl_{mix}	0.65		0.88	0.43				
PP_{max}	0.88	0.68		0.56		0.62		
PPday	0.80	0.41	0.73					
TP_{mix}	0.48	0.43	0.56					
N/P_{mix}	−0.59	−0.45	−0.64	−0.87				
Z_S			0.4			**0.32**		0.6
Z_{Anmax}					0.56			**0.36**
Day nr						0.85	0.64	

The symbols are explained in the text.

Discussion

According to the Directive 2000/60/EC of the European Parliament, so called Water Framework Directive (WFD), member states of the European

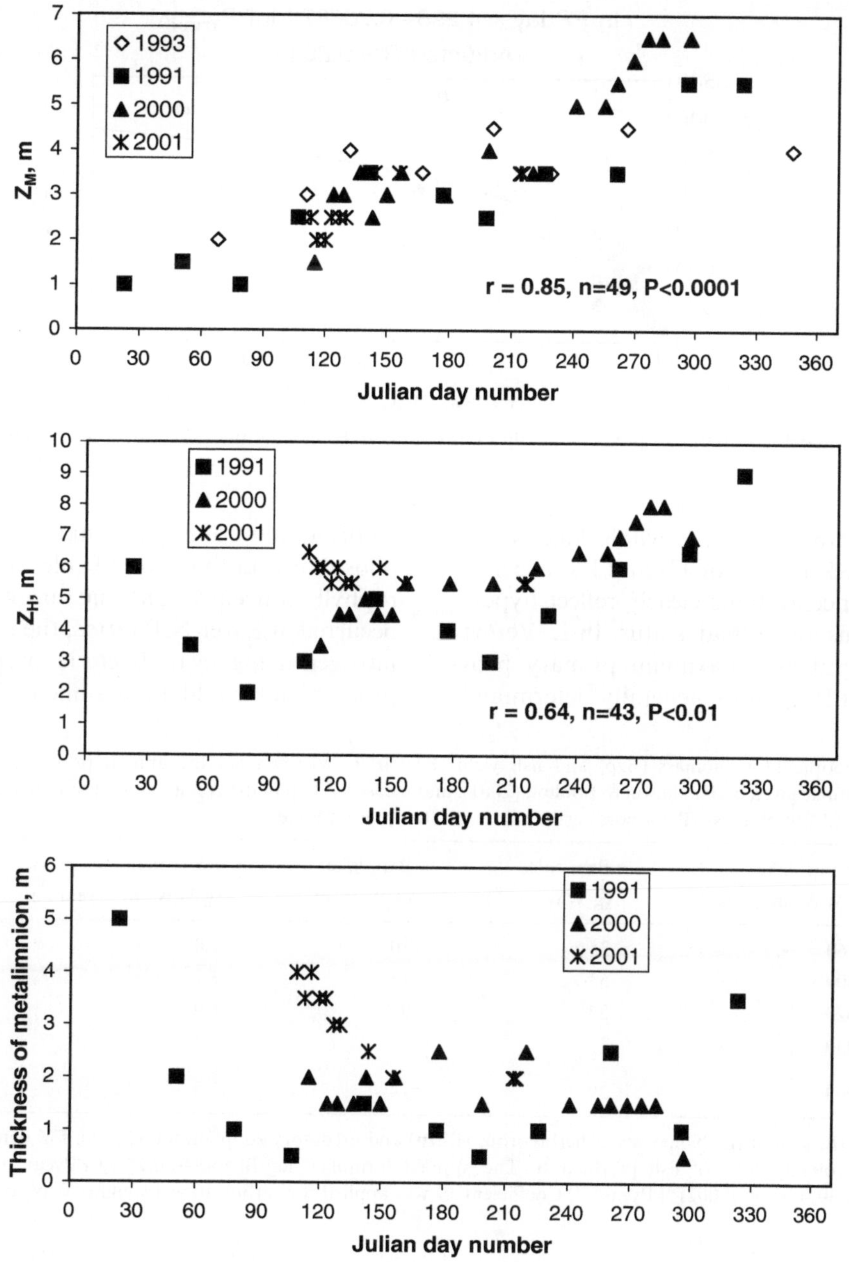

Figure 7. Seasonal dynamics of the depth of metalimnion (Z_M) and hypolimnion (Z_H) and the thickness of metalimnion ($= Z_H - Z_M$) in L. Verevi.

Union should distinguish ecological status of their waters compared with high quality sites. To evaluate the present status of L. Verevi, we chose the year 2000 as the most frequently studied year (Table 1). Considering the average daily PP 896 mg C m^{-2} and the annual PP 162 mg C m^{-2}, Secchi depth 2.34 m, epilimnetic concentrations of chlorophyll *a* (19.6 mg m^{-3}) and total phosphorus (TP, 52 mg m^{-3}), L. Verevi is a eutrophic lake (Table 4) of a 'good' status (Table 5). However, if to consider the total amounts of these nutrients stored in the lake, which, for example, in the year 2000 were on the average 877 kg of TN and 79 kg of TP (P. Nõges, 2005), and the volume of the lake 465,271 m^3, the average

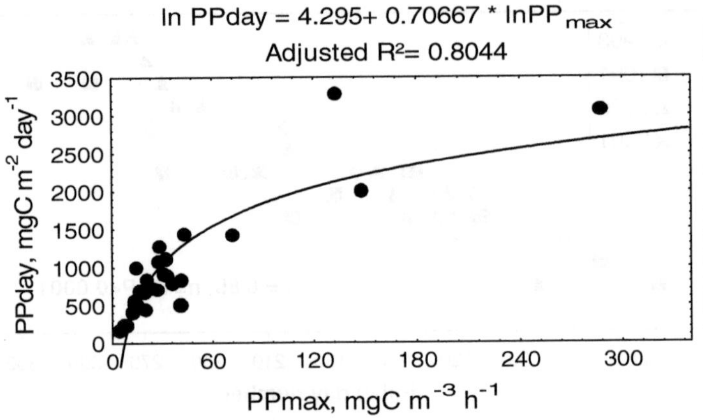

Figure 8. Regression model ($p < 0.01$) for predicting daily PP (Ppday) on the basis of the value of vertically maximum PP in midday (PP_{max}).

potential concentrations in the whole lake water could achieve 1885 mg m^{-3} of TN and 170 mg m^{-3} of TP. These concentrations clearly reflect hypertrophic conditions and a 'bad' status. In L. Verevi epilimnetic integral and maximum primary production (PP_{mix}, PP_{max}) were generally determined by the amount of phytoplankton (chlorophyll) and phosphorus in this layer (Table 3). As higher chlorophyll concentration and primary production occurred at lower N/P ratios, the important role of nitrogen-fixing cyanobacteria in periods of high productivity could be assumed. The sharper the

Table 4. Predicted zooplankton biomass (Bzp) and fish yield (FY) in L. Verevi from the annual primary production (PPyear) according to the formulae given by Hakanson & Boulion (2001), and measured Bzp in the epi- and metalimnion according Kübar et al. (the present issue), and fish biomass (Bf) according Järvalt et al. (the present issue)

Year	Ppyear, lake (g WW m^{-2} $year^{-1}$)	Bzp, calc (g WW m^{-2})	Bzp, lake, (g WW m^{-2}	FY, calc, (g WW m^{-2} $year^{-1}$)	Bf, lake (g WW m^{-2})
1991	1760	34	10	2.0	
1993	3400	53	15	3.7	
2000	1620	32	9	1.9	
2001	2030	37	20	2.3	
Average	2203	39	14	2.5	36

Bzp is calculated as the sum of the biomasses of herbivorous (Bhzp) and predatory zooplankton (Bpzp); FY denotes the total yearly fish catch forming roughly 30% of fish production. The applied formulae: log Bhzp = 0.7*log PPyear − 0.83; log Bpzp = 0.67*log PPyear − 1.39; FY = 0.0021*$PPyear^{0.92}$. Coefficient 10 was applied for carbon to wet weight (WW) conversion in the case of PP.

Table 5. Fixed boundary values for trophic classification acoording to OECD (1982)

Trophic category	Total phosphorus, (mg m^{-3})	Chlorophyll *a* (mg m^{-3})	Secchi depth, (m)	Daily PP (mg C m^{-2})	Annual PP (gC m^{-2})
Oligotrophic	<10	<2.5	>6	<200	<30
Mesotrophic	10–35	2.5–8	6–3	200–700	30–100
Eutrophic	35–100	8–25	3–1.5	700–2000	100–350
Hyper-eutrophic	>100	>25	<1.5	>2000	>350

Boulion (1983) and Wetzel (1983).

stratification, the less phosphorus can penetrate from hypolimnion into the epilimnion, and the better the status of the epilimnion could appear throughout the vegetation period. Consequently, the stratification regime is an important factor which determines the 'apparent status' of such sharply stratified lakes.

If to consider water transparency, Lake Verevi has changed quite substantially from 1990s to 2000s – the average Secchi depth has increased

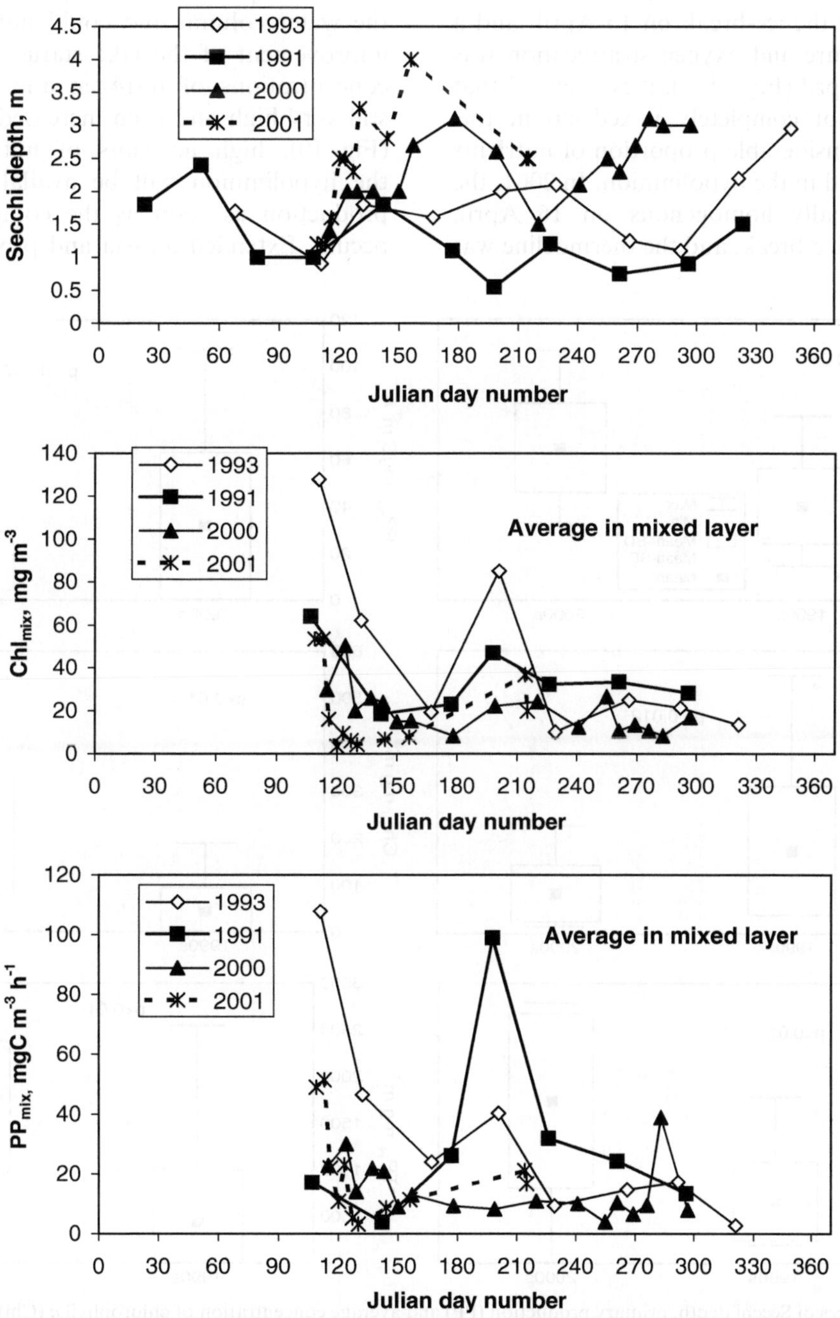

Figure 9. Seasonal dynamics of Secchi depth and mixed layer average concentration of chlorophyll *a* (Chl_{mix}) and primary production (PP_{mix}) in L. Verevi.

from 1.5 to 2.5 m. Also seasonal dynamics of PP and related indices in the 1990s were profoundly different from that in 2000 and 2001 (Fig. 9). This was most probably caused by a rather different stratification regime. It is assumed that in 1991 and 1993 the lake was dimictic with complete spring and autumn circulation. As the spring of 2000 was very warm after the ice-break on 15 April, and a strong temperature and oxygen stratification was formed by 24 April (Figs. 1–4), it is assumed that the lake was not completely mixed up in this spring, and a considerable proportion of nutrients remained trapped in the hypolimnion. In 2001, the lake was thermally homogenous on 16 April, 3 days after the ice-break, and the thermocline was formed one week later (Fig. 1). However, the anoxic bottom layer was present already on 16 April indicating the possibility of incomplete mixing in this spring as well. So, increased water transparency and the improvement of the epilimnetic water quality from the 1990s to the 2000s (Fig. 10) may be a consequence of incomplete spring mixing of the water column and could not reflect the real improvement of the lake status. As near-bottom concentrations of nitrogen and phosphorus are still very high and even increased from the 1990s (Fig. 10), high amounts of nutrients stored in the hypolimnion will be available for euphotic production as soon as the complete circulation occurs. Extended anoxia and production of toxic

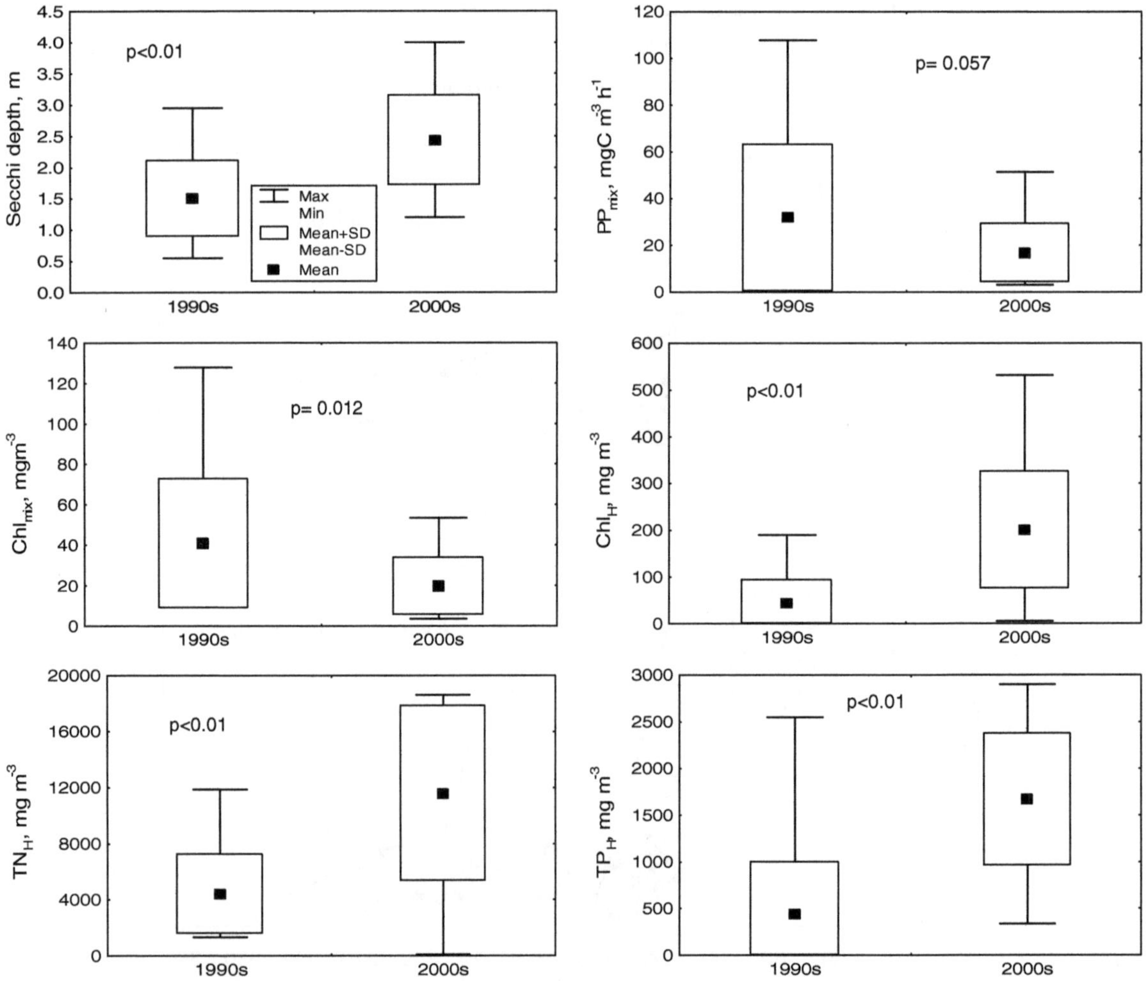

Figure 10. Differences of Secchi depth, primary production (PP) and average concentration of chlorophyll *a* (Chl), total nitrogen (TN) and total phosphorus (TP) in mixed layer ($_{mix}$) and hypolimnion ($_{H}$) between studied 1990s (1991and 1993) and 2000s (2000 and 2001) in L. Verevi. P values determined by Tukey's post hoc test ANOVA.

compounds like H_2S in the hypolimnion makes the life conditions unbearable for most of eucaryotic organisms. The mixing up of these compounds into the water column in autumn increases the risk of winter anoxia and fish-kill. On the other hand, the nutrient budgets calculated for 1991 and 1993 are negative and show that on a yearly basis L. Verevi releases more nitrogen and phosphorus than receives from the inflows, and the total amount of these nutrients stored in the water column has diminished by 1.2 times (P. Nõges, 2005). It is possible that the decreased nutrient concentration has supported the establishment of a 'clear' epilimnion allowing light to penetrate into the nutrient-rich metalimnion and sustaining a high production of cyanobacteria and phototrophic sulphur bacteria (Nõges & Solovjova, 2005). In the years when the lake was meromictic, clearer water supported lowering of the upper boundary of metalimnion allowing light for primary production to penetrate deeper. Metalimnion and hypolimnion started deeper and their ability to trap nutrients was assumingly more pronounced in years when lake was meromictic (Fig. 7). In all years starting depths of both meta- and hypolimnion moved deeper during the vegetation period (Table 3, Fig. 7), which supported the more effective trapping of nutrients in late summer and early autumn. The comparison of most different years, 1991 and 2000 showed that in 1991 average Z_S was 1.3 m and the average Z_{PPmax} was only 0.4 m (most intensive PP in upper epilimnion) while in 2000 the corresponding values were substantially higher (2.3 and 3.3 m) showing that the zone of most intensive PP was quite close to the upper border of metalimnion (average Z_M in 2000 was 4.1 m). The depth of the maximum assimilation number, however, was in 1991 and

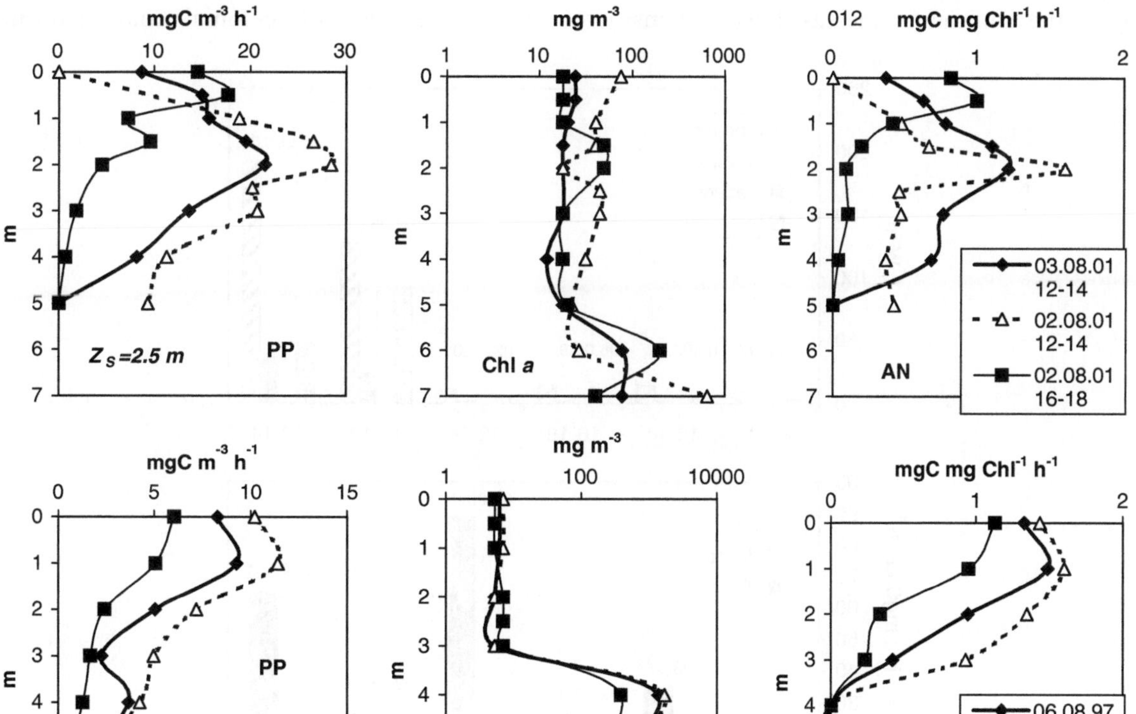

Figure 11. Diurnal depth profiles of primary production (PP), chlorophyll concentration (Chl *a*) and assimilation number (AN) in L. Verevi in August 1997 and 2001.

2000 quite similar (0.4 and 0.5 m, correspondingly), showing that the production efficiency of the phytoplankton biomass (chlorophyll) unit was always highest in upper epilimnion. In course of the vegetation period both the upper and lower borders of the metalimnion deepened and metalimnion got also thinner, and this could support the growth of phototrophic bacteria in this layer. An extremely dense metalimnetic chlorophyll layer reflecting the presence of phototrophic bacteria (Fig. 5) could also trap nutrients released from the hypolimnion and stabilize the 'clear epilimnion state'. Additionally, metalimnetic chemoautotrophs causing a high dark fixation of CO_2 (Fig. 6) could also trap a portion of the mineral nutrients (Kuznetsov & Dubinina, 1989). The situation could be compared with the occurrence of two stable states – a 'clear macrophyte state' and a 'turbid plankton state' in shallow lakes (Scheffer, 1998). It is also possible that the concentration of dissolved substances in hypolimnetic water will result in the sufficiently increased water density preventing the complete circulation also in autumn and making the lake meromictic. This would even stabilize the summer 'clear epilimnion state'. We could find support to the assumed stabilizing role of the deep chlorophyll maximum (DCM) if comparing diurnal dynamics of PP and chlorophyll in 1997 and 2001. In 1997 when DCM was extremely strong, diurnal vertical profiles were rather stable while in 2001 DCM was weaker by an order of magnitude allowing a much stronger variation of diurnal profiles (Fig. 11).

Basing on the model developed by Hakason & Boulion (2002), the annual PP of L. Verevi could support such production of zooplankton, which could sustain average zooplankton biomass (Bzp) of 32 gWW m^{-2} and an annual fish yield (FY) of 2.5 kg ha^{-1} (Table 6). Real Bzp is about three times lower than this predicted value while the fish biomass of 355 kg ha^{-1} (Järvalt et al., 2005) seems to support even higher FY than predicted from PP values. As phytoplankton of L. Verevi consists mainly of colonial and filamentous

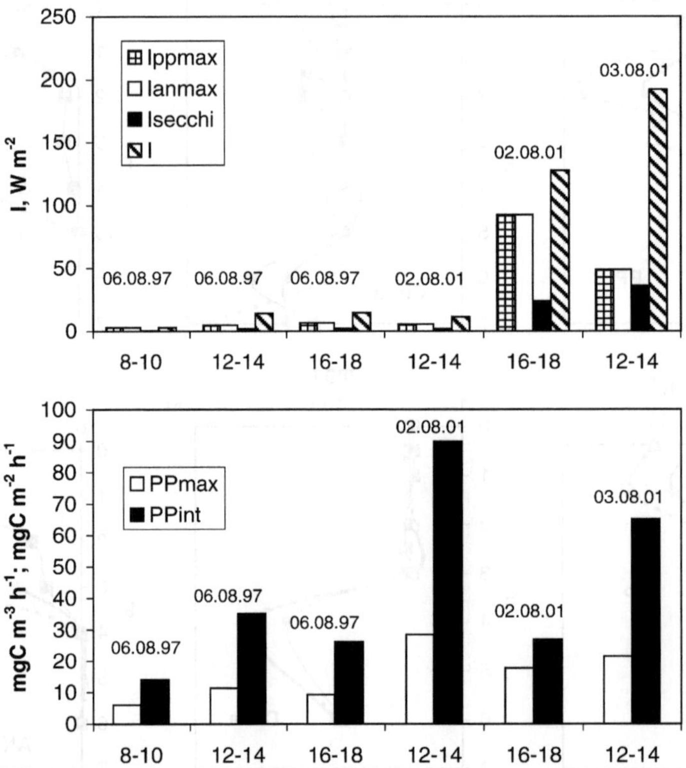

Figure 12. The irradiance entering the water surface (I), reaching to the Secchi depth (Isecchi), and to the depths of maximum primary production (Ippmax) and maximum assimilation number (Ianmax), and maximum and integral primary production (PP_{max}, PP_{int}) during diurnal studies in L. Verevi in August 1997 and 2001.

Table 6. Classification of the water quality of light-coloured Estonian lakes (Ott, 2001)

Parameter	Class 1 Excellent	Class 2 Good	Class 3 Moderate	Class 4 Poor	Class 5 Bad
Secchi depth (m)	> 3	2–3	1–2	< 1	< 1
Total phosphorus (mg m^{-3})	< 30	30–60	60–80	80–100	> 100
Total nitrogen (mg m^{-3})	< 500	500–700	700–1000	1000–1300	> 1300
Chlorophyll *a* (mg m^{-3})	< 10	10–20	20–40	40–50	> 50

cyanobacteria, the feeding conditions for zooplankton are not favourable sustaining a comparatively low biomass. A rather high fish biomass and the dominance of planktivores, as roach and young perch, form also the basis for a strong top-down control on zooplankton.

Conclusions

1. Considering the annual primary production, Secchi depth and epilimnetic concentrations of chlorophyll *a* and total phosphorus, L. Verevi is a eutrophic lake of a 'good' status. The hypolimnion consists high amounts of nutrients, which, if spread over the whole water column, could make the lake strongly hypertrophic and qualify its water as 'bad'.
2. An extremely dense metalimnetic chlorophyll layer reflects the presence of photoautotrophs like sulphur bacteria, cyanobacteria and algae. This layer creates and stabilizes the 'clear epilimnion state' by trapping the nutrients released from the hypolimnion, and does not allow primary production to reach a high level.
3. The yearly primary production of L. Verevi could support three times higher zooplankton biomasses than at present in the lake. Poor feeding conditions due to the dominance of large algae, as well as the unsuitable living conditions in the extended hypolimnion and intensive fish pressure, could explain the situation.

Acknowledgements

This work was supported by the core Grants No 0370208s98 and 0362480s03 of Estonian Ministry of Education, and by Grants No 3579, 4080 and 4835 of the Estonian Science Foundation. We would like thank all the members of the research group (I. Ott, K. Ott, A. Rakko, D. Sarik, T. Kõiv, P. Nõges, E. Lill, H. Tammert, H. Künnap, V. Kisand, H. Starast, A. Lindpere, I. Solovjova, R. Laugaste etc.) who participated in the project. We would also like to thank Dr Enn Veldi, who revised the English, and Anne Jõeveer and Ingrid Niklus of Tõravere Station of the Estonian Institute of Meteorology and Hydrology for climatic data.

References

Boulion, V. V., 1983. Pervichnaya produktsiya planktona vnutrennyikh vodoyoemov. Leningrad, Nauka, 149 pp. [Primary production of inland water bodies. In Russian].

Guidelines for the Baltic monitoring programme for the third stage, 1984. The Baltic Marine Biologists. Publ. 1, 2nd edn.

Hakanson, L. & V. Boulion, 2002. The lake foodweb – modeling predation and abiotic/biotic interactions. Backhus Publishers, Leiden, 344 pp.

Hilmer, T. & G. C. Bate, 1992. Filter types, Filtration and Post-Filtration Treatment in Phytoplankton Production Studies. Applied and Environmental Microbiology 33: 1225–1228.

Järvalt, A., T. Krause & A. Palm, 2005. Diel migration and spatial distribution of fish in a small stratified lake. Hydrobiologia 547: 197–203.

Kangro, K., R. Laugaste, P. Nõges & I. Ott, 2005. Long-term changes and special features of seasonal development of phytoplankton in a strongly stratified, hypertrophic lake. Hydrobiologia 547: 91–103.

Kuznetsov, S. I. & G. A. Dubinina, 1989. Metody izucheniya vodnykh mikroorganizmov. Moskva, Nauka. [Methods of the study of water microorganisms. In Russian].

Kübar, K., H. Agasild, T. Virro & I. Ott, 2005. Vertical distribution of zooplankton in a strongly stratified hypertrophic lake. Hydrobiologia 547: 151–162.

Laugaste, R., 1991. Fütoplankton. In Timm, H. (ed.), State of Lake Verevi. Hydrobiological researches XVII,Tartu: 69–90. [Phytoplankton. In Estonian].

Laugaste, R., 1994. Verevi järve seisund, biogeensete ainete päritolu ja tervistamise abinõud. In Järvekülg, A. (ed.), Eesti jõgede ja järvede seisund ning kaitse. Teaduste Akadeemia Kirjastus, Tallinn, 47–64. [The state, the origin of nutrients and the measures necessary for recovering Lake Verevi. In Estonian].

Lampert, W. & U. Sommer, 1997. Limnoecology: The Ecology of Lakes and Streams. Oxford University Press, New York & Oxford, 382 pp.

Lignell, R., 1992. Problems in filtration fractionation of ^{14}C primary productivity samples. Limnology and Oceanography 37: 172–178.

Moss, B. 1998. Ecology of Fresh Waters. Man and Medium, Past to Future. 3rd edn. Blackwell Science Ltd., 557 pp.

Niemi, M., J. Kuparinen, A. Uusi-Rauva & K. Korhonen, 1983. Preparation of algal samples for liquid scintillation counting. Hydrobiologia 106: 149–159.

Nõges, P., 2005. Water and nutrient mass balance of temperate partly meromictic Lake Verevi. Hydrobiologia 547: 21–31.

Nõges, P. & T. Nõges, 1998a. Stratification of Estonian lakes studied during hydrooptical expeditions in 1995–97. Proceedings of the Estonian Academy of Sciences. Biology. Ecology 47, 268–281.

Nõges, T. & P. Nõges, 1998b. Primary production of Lake Võrtsjärv. Limnologica 28(1): 29–40.

Nõges, T. & I. Solovjova, 2000. The influence of different solvents and extraction regimes on the recovery of chlorophyll *a* from freshwater phytoplankton. Geophysica 36 (1–2): 161–168.

Nõges, T. & I. Solovjova, 2005. The formation and dynamics of deep bacteriochlorophyll maximum in the temperate and partly meromictic Lake Verevi. Hydrobiologia 547: 73–81.

OECD, 1982. Eutrophication of Waters. Monitoring, Assessment and Control. OECD, Paris, 154 pp.

Ott, I., 2001. Typology and ecological classification of Estonian Lakes. In Bäck, S. & K. Karttunen (eds.) Classification of Ecological Status of Lakes and Rivers. TemaNord 584, Nordic Council of Ministers, Helsinki, 52–63.

StatSoft Inc., 2001. STATISTICA (data analysis software system), version 6. www.statsoft.com.

Steeman-Nielsen, E., 1952. The use of radioactive carbon (^{14}C) for measuring primary production in the sea. Journal du Conseil permanent international pour l'exploration del la mer 18: 117–140.

Scheffer, M., 1998. Ecology of Shallow Lakes. Chapman and Hall, London.

Timm, H. (ed.). Verevi järve seisund. 1991. A Monograph. Tartu: 139. [State of Lake Verevi. In Estonian, English and Russian summary].

Wetzel, R. G., 1983. Limnology. Saunders College Publishing, 767 pp.

Williams, P. J., le, B. & N. W. Jenkinson, 1980. A simple and inexpensive 4-π light collector and two designs for a light meter for light attenuation studies. Freshwater Biology 10: 491–496.

Hydrobiologia (2005) 547:123–135
I. Ott & T. Kõiv (eds), Lake Verevi, Estonia – A Highly Stratified Hypertrophic Lake
DOI 10.1007/s10750-005-4154-x

Resource ratios and phytoplankton species composition in a strongly stratified lake

Toomas Kõiv* & Kersti Kangro
Institute of Zoology and Botany, Võrtsjärv Limnological Station, Estonian Agricultural University, 61101 Rannu, Tartu County, Estonia
*(*Author for correspondence: E-mail: tom@zbi.ee)*

Key words: resource ratios, phytoplankton, stratification, seasonal dynamics

Abstract

The epilimnetic phytoplankton and its relations to nutrient content in Lake Verevi through the whole vegetation period in 2000 were studied. Lake Verevi (surface 12.6 ha, mean depth 3.6 m, maximum depth 11 m) is a hypertrophic hard-water lake, where the so-called spring meromixis occurs due to an extremely warm spring. Most dissolved nutrients in the epilimnion were low already in spring, and their concentrations were quite stable during the study period. The concentration of total silicon was very low in spring but increased rapidly in summer. Total phosphorus followed the pattern for stratified eutrophic lakes, and total nitrogen was quite high. The stoichiometric N:P ratio fluctuated between 25 and 81. The dynamics of phytoplankton biomass with a spring peak from April to May and a late summer peak from July to August is typical of Estonian eutrophic lakes. Green algae and chrysophytes occurred in the phytoplankton throughout the vegetation period. The spring peak was dominated by diatoms (*Synedra ulna* and *Synedra acus* var. *angustissima*) and the summer peak was caused by *Aphanizomenon klebahnii* and *Ceratium hirundinella*. The study showed that in physically stratified systems, the total concentration of limiting resources and plain physical factors (light and temperature) may be more important in the determination of phytoplankton dominants than different resource ratios. A combination of light and temperature optimum, along with nutrient utilization and transport capacity, effectively segregates phytoplankton species and can be used for the explanation of seasonal succession pattern.

Introduction

Our knowledge about the abiotic and biotic factors regulating the species composition of natural phytoplankton communities is still very sketchy. The mechanisms explaining the species dominance patterns in lakes during the vegetation period vary greatly and sometimes even show discrepancies. Among numerous hypotheses, the use of resource ratios, as predictor of microalgal species composition, is the most applicable in limnological practice. The resource–ratio hypothesis is the backbone in Tilman's mechanistic theory of competition (Tilman, 1977, 1982), which uses simple graphical models to predict the steady state outcome of competition for two potentially limiting resources from single species physiological traits. Tilman's theory has been thoroughly tested by chemostat experiments in the laboratory with algae from unialgal strains and with natural phytoplankton communities (Sommer, 1989, 1993; Smith & Bennet, 1999). However, these arguments did not sound convincive to all limnologists, and some skepticism exists in the literature concerning the potential applicability of this equilibrium theory to real-world ecological systems (Reynolds,

1996, 1999). Although there is sample evidence to support a strong effect of the resource ratios on phytoplankton structure and functioning, some additional factors are clearly important, and these moderating factors may override other factors in some systems. For example, stratification, buoyancy regulation, light regime and grazing could potentially affect the phytoplankton biomass and species composition (Smith & Bennet, 1999).

The aim of the study was to describe the phytoplankton species composition and succession in the epilimnion of a strongly stratified lake, and clarify how the ratios of growth limiting resources or single abiotical factors determine biomasses of species in a community.

Materials and methods

Lake Verevi is a small and relatively deep lake situated in town Elva in South Estonia. Table 1 shows the main morphometric and limnological characteristics of this dimictic, strongly stratified hard-water lake. Additional morphometrical, hydrological and limnological data are presented by Ott et al. and Nõges & Kangro, in the present issue. Human activities in the watershed cause considerable input of domestic pollutants and Lake Verevi has changed from moderately eutrophic at the beginning of 20th century to a hypertrophic lake in the 1980s. The highly eutrophic phase persists still today.

The lake was sampled from April to October 2000 at the deepest point located in the broader part near the western shore. Physical data (transparency, temperature) was recorded weekly. The temperature profiles were measured vertically using the multiprobe Aqua-Check Water Analyzer (USA). Underwater light intensity was not measured. The availability of the light resource (light index) was calculated indirectly from the Secchi-disc readings according to Sommer (1993):

$$LI = 2(SD/Z_{mix}) \times (D/24)$$

where LI is the light index, SD is Secchi depth (*m*), Z_{mix} is mixing depth, and *D* is day length (*h*). Mixing depth was defined as upper water layer in which the temperature gradient was < 1.5 °C m^{-1}.

Biological and chemical data were collected by a pre-planned programme (for details see Nõges & Solovjova, 2005) that covers densely mixing periods in spring and autumn and has longer intervals in the stagnation period. Water samples were taken at two depths, at 0.5 m (Layer 1) and in the middle of the epilimnion (Layer 2). The latter sampling depth differed from case to case depending on the evolution of the stratification. The phytoplankton and water samples were gathered using a special vacuum probe. A Masterflex pump (model N 7533-60) with an easy-load pump head (model 7518-12) was used for pumping water to the surface. A hose with inner Ø 8 mm was placed vertically into the water. The lower tip of the hose was closed and the water was sucked through a horizontal tube in order to obtain water from horizontal layers. The capacity of the complex device was approximately 2 l min^{-1}. For more details see Zingel & Ott (2000).

Chemical analyses were performed using the methods described by Grasshoff et al. (1982). Dissolved reactive phosphorus (SRP) was measured by the molybdate blue method using ascorbic acid as reductant. Nitrate and nitrite were determined by reduction with a cadmium column. In order to determine total nitrogen (TN) and total phosphorus (TP), organic compounds were mineralized into nitrite and phosphate, using persulphate. Standard photometric analysis was applied to complete each named estimation (Grasshoff et al., 1982). The silicon content was

Table 1. Morphological and limnological characteristics (epilimnion in summer during 1998–2000) of Lake Verevi, South Estonia

Morphometrical parameters		Limnological parameters min...max	
Surface area (ha)	12.6	Secchi depth (m)	0.7 ... 4.2
Volume, 10^6 m^3	453.6	Tot P, μg l^{-1}	22 ... 153
Maximum depth (m)	11.0	Tot N, μg l^{-1}	670 ... 2450
Mean depth (m)	3.6	Chlorophyll *a*, μg l^{-1}	4 ... 110
Water exchange times per year	0.63	Silicon, mg l^{-1}	100 ... 8

determined by the silicomolybdic blue method (Hansen & Koroleff, 1999).

The phytoplankton samples were preserved and fixed with acidified Lugol's solution. The samples were counted by an inverted microscope at 400× magnification using the Utermöhl (1958) technique. The algae were counted along the transect the chamber diameter until reaching the number of 600 counting units (cells, filaments, colonies). The mean volume of each species was estimated in all samples by approximating the shape of species to the nearest simple geometric solid. At least 30 trichomes of filamentous cyanobacteria were measured in each sample and the mean length was used when calculating the biomass.

Because normal distribution could not be assumed, the analysis of data was restricted to non-parametric methods. Spearman's rank correlation (Siegel & Castellan, 1988) coefficient was used for cross-correlation analysis between the resource ratios, chemical data and species biomass. For data presentation and statistical evaluation, software packages SigmaPlot 8.0 and Statistica 6.0 (StatSoft, Inc. 2001) were applied.

Results

The year 2000 was anomalous in a number of ways in Lake Verevi. The ice broke up in 15 April and due to sudden heating, permanent thermal stratification was established few days later. On April 24, the permanent thermocline was located at a depth of 2 m. Such a short overturn time was apparently insufficient for complete mixing, and the so-called spring meromixis (Arvola & Rask, 1984) occurred (Nõges & Kangro, 2005). Though May was considerably cooler than April, the thermocline had lowered quickly to the common stable depth at 4.5–5 m in this month.

The light climate in Lake Verevi varies remarkably seasonally. The light index reflects the high light intensity and long day length relatively well, while the maximum was reached at the beginning of June (Fig. 1), when high water transparency (2.7–3.1 m) and long day length combined favourably. High values in April were mainly caused by the low mixing depth; thickness of epilimnion was less than 1.5 m at that moment.

After the onset of stratification, the epilimnetic waters were isolated from the deeper nutrient-rich layers. The most dissolved nutrients were reduced to summer levels already by April 24, and the concentrations remained stable during the rest of the study period. The dissolved phosphorus concentrations were below 0.05 μM, NO_3^- and NH_4^+, respectively were below 0.03 μM and 0.5 μM. The concentrations of NO_2^- was always negligible. Such low concentrations are very common because dissolved nutrients constitute the residual which is left over in the water after consumption by organisms. Higher taxon dominance patterns along the nutrient ratio gradients are rather similar, irrespective of which nutrient fraction was measured (Sommer, 1999). This similarity arises from by the positive correlation between different fractions of the same nutrient element (Vollenweider & Kerekes, 1982). Therefore, we calculated the resource ratios using total nutrient concentrations.

TP concentration decreased instantly at the end of the May and remained low (0.7–1.1 μM) until autumn peak (Fig. 1). The very low silicon concentration (<10 μM) in spring increased rapidly during the summer. Despite the fluctuations and low TP, the high (>60 μM) TN concentrations (Fig. 1) show a potential lack of N-limitation.

Having three limiting essential resources (Si, TP and light), three meaningful pairwise combinations (Si:TP, Si:LI, TP:LI) were calculated (Fig. 2). Although no N-limitation was observed, some ratios (TN:TP, Si:TN, TN:LI) with total nitrogen were calculated because of fluctuating behaviour of the TN concentration and occurrence of nitrogen fixation (Tõnno et al., 2005).

The dynamics of phytoplankton biomass (Fig. 3) in Lake Verevi with the spring peak from April to May and the late summer peak from July to August is typical to Estonian eutrophic lakes. For temperate lakes, the common clear-water phase in early June was reflected mainly in water transparency and not so much in the total phytoplankton biomass. The latter was low prior to mid-July.

The phytoplankton assemblage of Lake Verevi in 2000 was composed of more than 100 taxa. The greatest number of taxa occurred in August (42). During the spring bloom (until 03.05, Figs. 4 and 5) the phytoplankton was mainly composed of

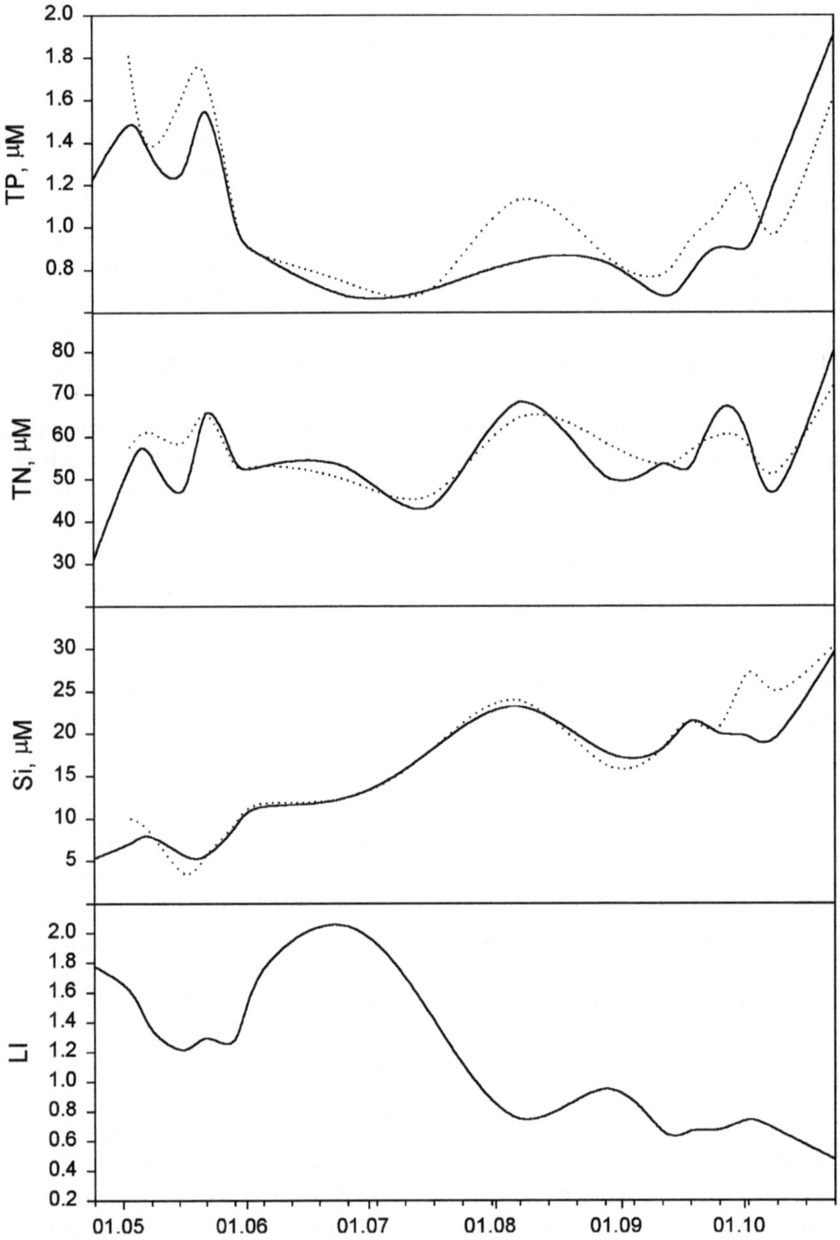

Figure 1. Changes in limiting resources in epilimnion. Lake Verevi in April–October 2000. (—): Layer 1; (. . .): Layer 2. TP – total phosphorus, TN – total nitrogen, Si – silicon, LI – the light index according to Sommer (1993).

diatoms (*Synedra acus* var *angustissima* Grun. and *S. ulna* (Nitzsch) Her.) and chrysophytes (*Uroglena* sp.). Also some blue–green algae (*Aphanocapsa* sp., *Anabaena* sp., *Limnothrix redekei* (Van Goor) Meff.) appeared during the peak period. In the clear-water phase (with water transparency 2.6(3 m), the diatoms disappeared and small chrysophytes (*Uroglena* sp., *Dinobryon sociale*) and cryptophytes (*Chroomonas* sp.) constituted the main component of the phytoplankton biomass. Immediately after the clear-water phase, there was a short peak of the small green algae (*Chlamydomonas* sp., *Korshikoviella limnetica* (Lemm.) Silva). At the end of July and during the first half of August, the cyanobacteria *Aphanizomenon klebahnii* (Elenk.) Pechar &

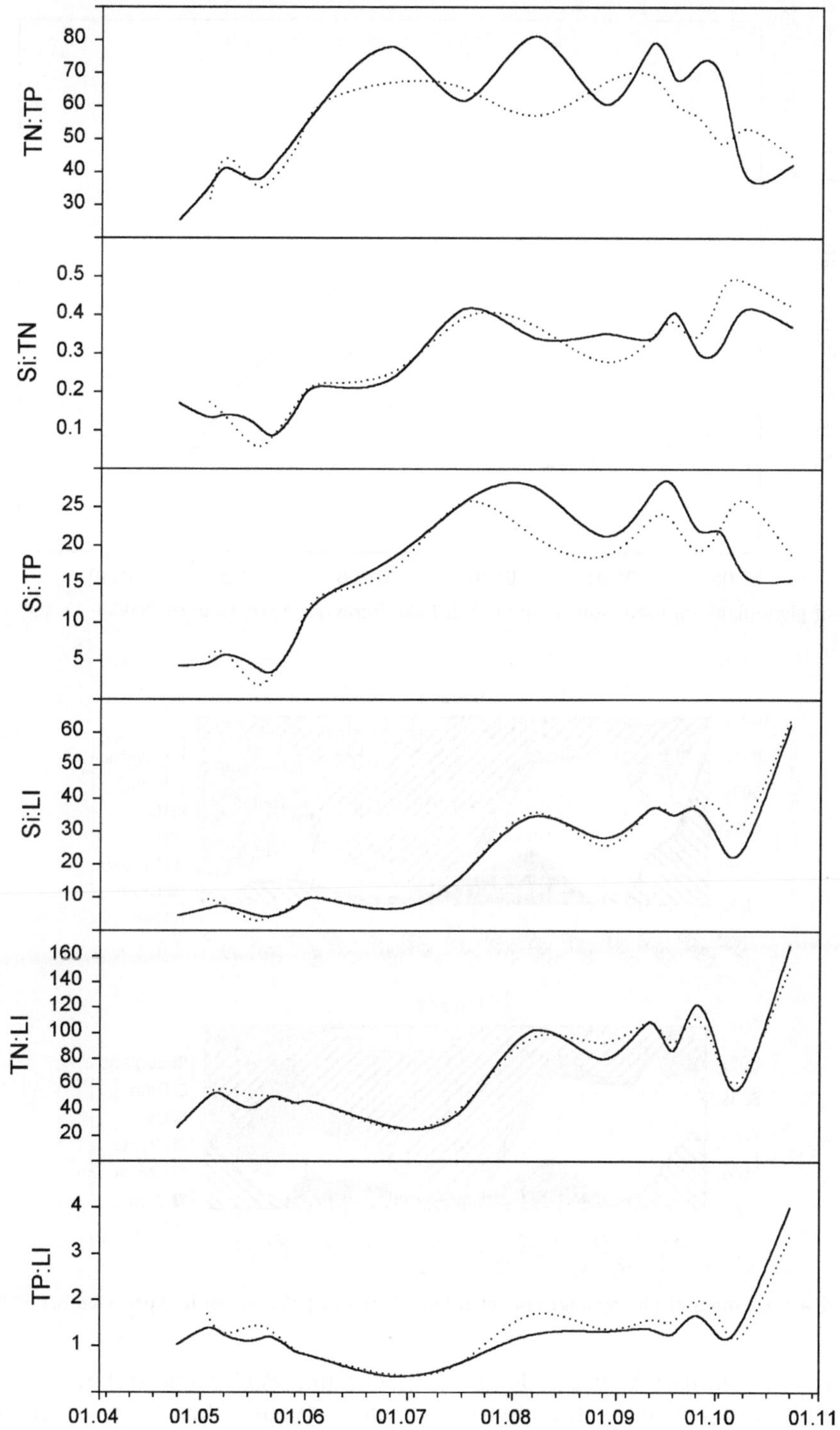

Figure 2. Dynamics of stoichiometric resource ratios in Lake Verevi in April–October 2000. (—): Layer 1; (. . .): Layer 2. Abbreviations as in Fig. 1.

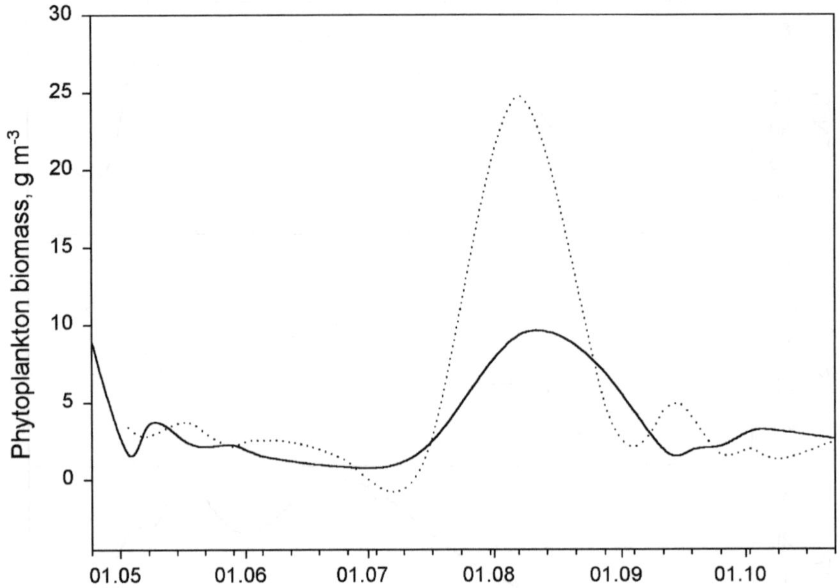

Figure 3. Dynamics of phytoplankton total biomass (g m^{-3}) in Lake Verevi in April–October 2000. (—): Layer 1; (...): Layer 2.

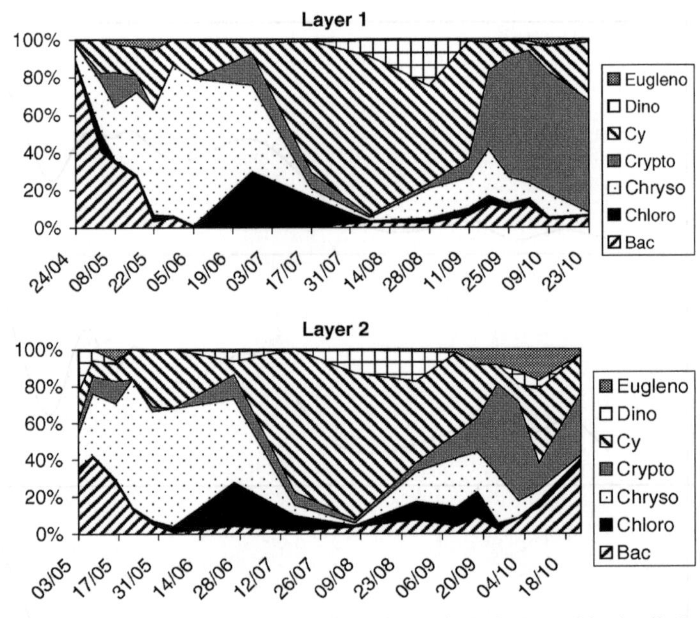

Figure 4. Dynamics of phytoplankton-related biovolumes in Lake Verevi in April–October 2000.

Kalina and *Planktothrix agardhii* Anagn. & Kom. acted as an important components of the phytoplankton. The period of maximum dominance *A. klebahnii* coincided with the period of the seasonal biomass maximum with the total biomass 24.8 g m^{-3}. After the decline of *A. klebahnii*, dinophytes (*Ceratium hirundinella* (O.F. Müller) Schrank, *Peridinium* sp.) were rather numerous. Since August, there were again colonial taxa of cyanobacteria – *Anabaena* sp., *Aphanothece* sp., Cyanodictyon sp. The dominant components of autumnal phytoplankton were cryptophytes (*Cryptomonas* sp., *Chroomonas* sp.). The phytoplankton species composition found in Lake

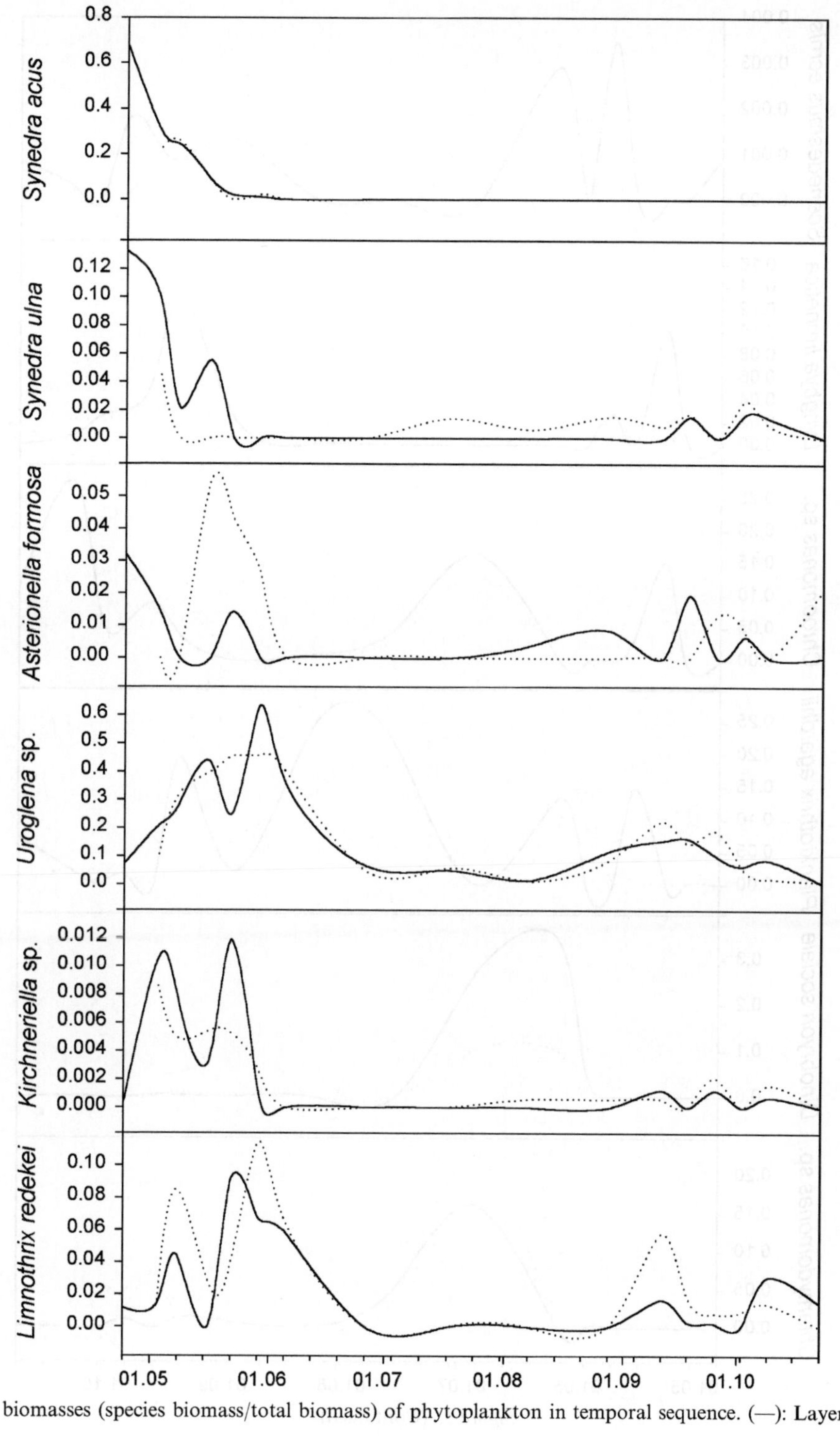

Figure 5. Relative biomasses (species biomass/total biomass) of phytoplankton in temporal sequence. (—): Layer 1; (. . .): Layer 2.

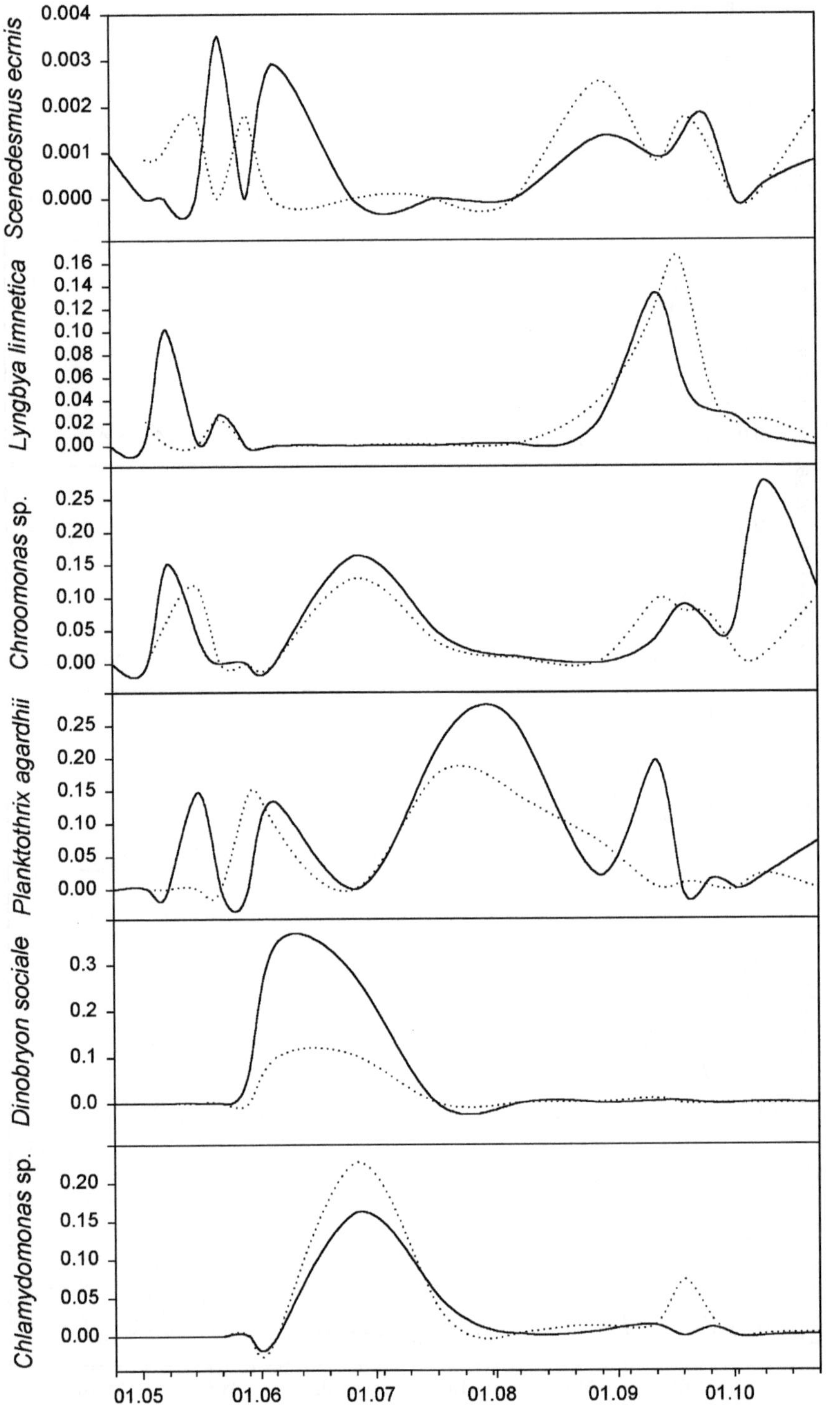

Figure 5. (Continued)

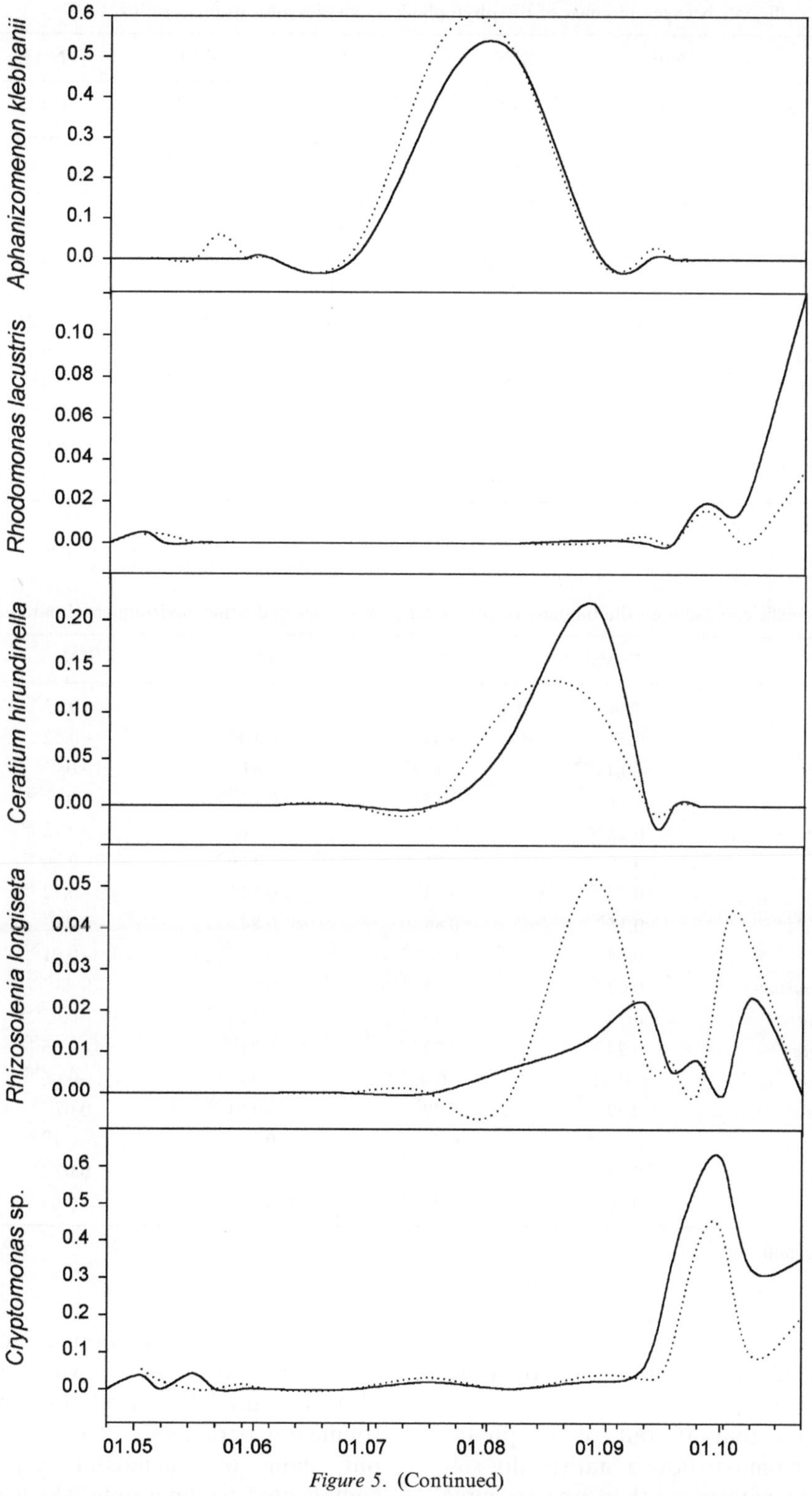

Figure 5. (Continued)

Table 2. Correlation coefficients between biomass of the phytoplankton species and resource ratios

	N:P	Si:N	Si:P	Si:LI	N:LI	P:LI
Aphanizomenon klebhanii	0.01^{NS}	0.66	0.64^{NS}	0.24^{NS}	0.47^{NS}	0.39^{NS}
Chlamydomonas sp.	0.25^{NS}	−0.24^{NS}	−0.17^{NS}	−0.54	−0.63	−0.59
Chroomonas sp.	−0.41	−0.09^{NS}	−0.41	−0.15^{NS}	0.08^{NS}	0.26^{NS}
Cryptomonas sp.	0.01^{NS}	0.35^{NS}	0.03^{NS}	0.42	0.58	0.74
Dinobryon sociale	−0.07^{NS}	−0.82	−0.7	−0.75	−0.82	−0.77
Gymnodinium sp.	−0.36^{NS}	−0.53^{NS}	−0.41^{NS}	−0.72	−0.71	−0.73
Kirchneriella sp.	−0.69	−0.76	−0.81	−0.71	−0.66	−0.24^{NS}
Limnothrix redekei	−0.19^{NS}	−0.54	−0.46	−0.47	−0.45	−0.43
Rhizosolenia longiseta	−0.36^{NS}	−0.01^{NS}	−0.55^{NS}	−0.35^{NS}	−0.7	−0.2^{NS}
Rhodomonas lacustris	−0.24^{NS}	0.21^{NS}	−0.29^{NS}	0.33^{NS}	0.46^{NS}	0.64
Synedra acus	−0.76	−0.52	−0.63	−0.53	−0.75	−0.47^{NS}
Synedra ulna	−0.65	−0.79	−0.84	−0.71	−0.51^{NS}	−0.57
Uroglena sp.	−0.35	−0.69	−0.57	−0.49	−0.3^{NS}	−0.22^{NS}

$p < 0.05$; ^{NS} = not significant.

Table 3. Correlation coefficients between the biomass of phytoplankton species and some environmental parameters

	TEMP	LI	TP	TN	Si
Asterionella formosa	0.54	0.34	0.04^{NS}	−0.23^{NS}	−0.2^{NS}
Chlamydomonas sp.	0.74	0.63	−0.47^{NS}	−0.52	−0.58
Chroomonas sp.	−0.19^{NS}	−0.16^{NS}	0.41	0.08^{NS}	−0.11^{NS}
Cryptomonas sp.	−0.67	−0.57	0.17^{NS}	0.35^{NS}	0.54
Dinobryon divergens	0.23^{NS}	0.61	−0.06^{NS}	−0.12^{NS}	−0.03^{NS}
Dinobryon sociale	0.68	0.71	−0.09^{NS}	−0.26^{NS}	−0.86
Gymnodinium sp.	0.27^{NS}	0.83	0.14^{NS}	−0.12^{NS}	−0.58^{NS}
Kirchneriella sp.	0.15^{NS}	0.84	0.84	0.47^{NS}	−0.71
Limnothrix redekei	0.64	0.33^{NS}	0.18^{NS}	−0.01^{NS}	−0.58
Monoraphidium contortum	0.02^{NS}	0.38^{NS}	0.64	0.48^{NS}	−0.54^{NS}
Planktothrix agardhii	0.74	0.45^{NS}	−0.24^{NS}	−0.07^{NS}	−0.18^{NS}
Rhizosolenia longiseta	0.24^{NS}	0.72	0.31^{NS}	−0.06^{NS}	−0.27^{NS}
Rhodomonas lacustris	−0.64	−0.43^{NS}	0.39^{NS}	−0.06^{NS}	0.8
Scenedesmus ecornis	0.59	0.39^{NS}	−0.01^{NS}	0.07^{NS}	−0.45
Synedra acus	0.07^{NS}	0.77	0.28^{NS}	−0.1^{NS}	−0.65
Synedra ulna	0.34^{NS}	0.87	0.64	0.07^{NS}	−0.66
Uroglena sp.	0.16^{NS}	0.25^{NS}	0.26^{NS}	−0.05^{NS}	−0.73

$p < 0.05$; ^{NS} = not significant.

Verevi in 2000 differed considerably from what had been reported in previous investigations (Kangro et al., 2005).

Table 2 displays that 10 out of 13 phytoplankton species biomass show a statistically significant correlation with more than one resource ratio. This relationship does not necessarily mean that all of these correlations are causally decisive. Correlations of a species with several ratios may also be caused by correlations with single environmental parameters (Table 3). Therefore without data of chemostat experiments, it is complicated to determine which resource ratio is causally important and influential.

Discussion

Due to strong stratification and negligible external input of nutrients (except silicon) the resource ratios (Fig. 2) were highly variable in Lake Verevi. Apparently the periods of nutrient limitations *in situ* are too short to attain a competitive equilibrium like in chemostat experiments where final outcome became apparent after 3–6 weeks (Sommer, 1999). However, they may be long enough to lead to a significant increase in the relative abundance of those taxa, that would eventually prevail.

In all seasons, the stoichiometric N:P ratio stayed within the range of 25(81 (Fig. 2), which by many authors (Bulgakov & Levich, 1999; Smith & Bennet, 1999) indicates the prevalence of green algae. The latter always occurred in the water column throughout the whole vegetation period but the biomasses were low caused by the prevalent small-sized Chlorococcales (*Scenedesmus, Monoraphidium*). Although chlorococcal species play an important role at high N:P ratios, there is a correlation between the amount of easily degradable allochthonous organic matter content and the species composition of Chlorococcales and Volvocales in Estonian lakes (Ott & Laugaste, 1998). Lake Verevi has been polluted by urban runoff, but the values of biochemical and chemical oxygen demand in the epilimnion are moderate for eutrophic lakes (Ott et al., the present issue). Despite the high N:P ratio, the biomasses of the heterocystous filamentous *Aphanizomenon klebahnii* and filamentous *Planktothrix agardhii* peaked during the first half of August. The existing literature (Lindholm & Eriksson, 1990; Sommer, 1999; Oliver & Ganf, 2000, etc.) indicates that both the N_2-fixing cyanophytes and the bloom-forming algae usually dominate in the lakes with relatively low TN:TP ratios. The transition zone of green algal–cyanobacterial coexistence lies around stoichiometric N:P ratios of 15:1 (Sommer, 1999). Reynolds (1987) argues that non-cyanobacterial species tend to be predominate at TN:TP ratios higher than 29:1 by mass. N_2–fixing cyanobacteria tend to be predominate at ratios below 14:1 and non-fixing cyanobacteria dominate at intermediate ratios. These two species may take advantage of strong differences in the N:P ratios of eplimnetic and hypolimnetic waters (Nõges & Solovjova, 2005). Recruitment of populations from the sediments occurs in several bloom-forming cyanobacteria, including *Aphanizomenon* and occasionally they are large enough to influence their population dynamics (Perakis et al., 1996). However, in general, they rely on obtaining on-going supplies of phosphorus. At least, the differences in dominants in the epilimnion may result from differences in the moving speed. Faster moving species, like *Aphanizomenon*, are known to aggregate in the narrow layers for brief periods in the case of suitable photic and thermal characteristics. Therefore, they are less able to maintain discrete layers with respect to depth (Reynolds, 1984), like *Planktothrix* – a typical resident of metalimnetic layers in Lake Verevi (Kangro et al., 2005).

In addition to green algae, chrysophytes (mainly *Uroglena* and *Dinobryon*) were always present in the phytoplankton. Not much is known about optimum resource ratios of chrysophytes. Sandgren (1988) observed higher biomasses of chrysophytes in lakes where the summer mean total TN:TP ratio is between 30 and 60. This range is comparable to our findings. In Lake Verevi, *Uroglena* sp. dominated in the range between 25 and 60 and was replaced by *Dinobryon* and *Chlamydomonas* at higher values of the TN:TP ratio. Reynolds (1984) noted that if the phosphorus supply is critical, a rather different algal assemblage, dominated by colonial chlorophytes or Chrysophyceae, is more typical. Its development is presumably influenced primarily by the outcome of the competition for the available phosphorus, independent of the nitrogen concentration. Both *Uroglena* and *Dinobryon* can utilize bacteria as a substitutable P source (Sandgren, 1988; Urabe et al., 1999) and are facultatively phagotrophic. The shift to *Dinobryon* dominance is well related to the peak of the total number of bacteria at the beginning of June (Tammert et al., 2005). *Chlamydomonas* is successful because its high surface-area-to-volume ratio and small specific biomass might make it a superior nutrient competitor in good light conditions (Reynolds, 1984).

The significant negative correlation between the diatoms (*Synedra acus, Synedra ulna*), silicon, and some resource ratios (with silicon) is surprising but easily explicable. The diatoms bloom started probably under the ice and their sedimentation was apparently accelerated by fast thermal stratification due to an early warm spring and the absence of spring circulation. On the other hand, the concentration of dissolved silicon was observed to fall to 3.5 μM in May and then to rise gradually. This phenomenon can be explained by an intensive inflow during the rains from the beach region, where the formation of new sandbank was in progress. Consequently, silicon concentration may also generally act as a 'time factor', describing the evolution of the stratification and diatom losses. During July, the Si:TP ratio rose over 20, exceeding the transition zone of diatom–non-diatom coexistence (Kilham, 1986; Tilman et al., 1986). The increased Si:TP ratio and mixing depth should have created at least satisfactory conditions for diatoms. On the other hand, during the stratified periods only littoral sediment survivors are able to recolonize the lake. Diatoms without specialized resting stages do not tolerate anaerobic conditions, and many diatoms species lack specialized resting stages, which is especially true for the majority of freshwater planktonic forms. Among freshwater planktonic diatoms, only the *Rhizosolenia* have resting spores (Sommer, 1988). Therefore, autumnal composition of diatoms was rather episodic and variable.

In conclusion, the present study shows that in physically stratified and unstable ecosystems the total concentration of limiting resources and plain physical factors (light, temperature, lake morphometry) may be more important in the determination of phytoplankton dominants than resource ratios. Phytoplankton dynamics in Lake Verevi can be explained mostly by relatively well-known theories. A combination of light and temperature optimum, along with phosphorus utilization and transport capacity, effectively segregates phytoplankton species and can be used to clarify the seasonal succession pattern.

Acknowledgements

The study was supported by the core grants of the Ministry of Education Nos. 0370208s98, 0362482s03 and by grants of the Estonian Science Foundation Nos. 3579 and 4835. We thank Ingmar Ott, Katrin Ott and Helen Tammert for their assistance during the field experiments. Also, we want to express our gratitude to Peeter and Tiina Nõges.

References

Arvola, L. & A. M. Rask, 1984. Relations between phytoplankton and environmental factors in a small, spring-meromictic lake in Southern Finland. Aqua Fennica 14: 129–138.

Bulgakov, N. G. & A. P. Levich, 1999. The nitrogen:phosphorus ratio as a factor regulating phytoplankton community structure. Archiv für Hydrobiologie 146: 3–22.

Grasshoff, K., M. Ehrhardt & K. Kremling (eds.), 1982. Methods of Seawater Analysis. ISBN (Verlag Chemie), 3–527, 25,998–26,008.

Hansen, H. P. & F. Koroleff, 1999. Determination of nutrients. In K. Grasshoff, K. Kremling & M. Ehrhardt (eds.), Methods of Seawater Analysis. WILEY-VCH, Weinheim, New York, Chichester, Brisbane, Singapore, Toronto, 600 pp.

Kilham, S., 1986. Dynamics of Lake Michigan phytoplankton communities in continuous cultures along a Si:P loading gradient. Canadian Journal of Fisheries and Aquatic Sciences 43: 351–360.

Kangro, K., R. Laugaste, P. Nõges & I. Ott, 2005. Long-term changes and seasonal development of phytoplankton in a strongly stratified, hypertrophic lake. Hydrobiologia 547: 91–103.

Lindholm, T. & J. E. Eriksson, 1990. Factors regulating primary productivity in a eutrophic lake with blooms of Microcystis Verhandlung Internationale Vereinigung für theoretische und angewande. Limnologie 24: 657–661.

Nõges, T. & K. Kangro, 2005. Primary production of phytoplankton in a strongly stratified temperate lake. Hydrobiologia 547: 105–122.

Nõges, T. & I. Solovjova, 2005. The formation and dynamics of deep bacteriochlorophyll maximum in the temperate and partly meromictic Lake Verevi. Hydrobiologia 547: 73–81.

Oliver, G. G. & R. L. Ganf, 2000. Freshwater blooms. In Whitton, B. A. & M. Potts (eds.) The Ecology of Cyanobacteria. Their Diversity in Time and Space. Kluwer Academic Publishers, pp 149–194.

Ott, I. & R. Laugaste, 1998. Organic matter and the composition of Volvocales and Chlorococcales in Estonian small lakes. Biologia Bratislava 53/4: 529–535.

Ott, I., T. Kõiv, P. Nõges, A. Kisand, A. Järvalt & E. Kirt, 2005. General description of partly meromictic hypertrophic Lake Verevi, its ecological status, changes during the past eight decades and restoration problems. Hydrobiologia 547: 1–20.

Perakis, S., E. B. Welch & J. M. Jacoby, 1996. Sediment-to-water blue–green algal recruitment in response to alum and environmental factors. Hydrobiologia 318: 165–177.

Reynolds, C. S., 1984. The Ecology of Freshwater Phytoplankton. Cambridge University Press, 384 pp.

Reynolds, C. S., 1987. Community organization in freshwater plankton. In Gee, J. H. R. & P. S. Giller (eds.) Organization of Communities Past and Present. Blackwell Scientific, Oxford, 297–325.

Reynolds, C. S., 1996. Vegetation processes in the pelagic: a model for ecosystem theory. In Kinne, O. (ed.) Exellence in Ecology 9. Ecology Institute,, Oldendorf/Luhe, Germany.

Reynolds, C. S., 1999. Non-determinism to probability, or N:P in the community ecology of phytoplankton. Archiv für Hydrobiologie 146: 23–35.

Sandgren, C. D., 1988. The ecology of Crysophyte flagellates: their growth and perennation strategies as freshwater phytoplankton. In Sandgren, C. D. (ed.) Growth and Reproductive Strategies of Freshwater Phytoplankton. Cambridge University Press,, Cambridge, 9–104.

Siegel, S. & N. J. Castellan, 1988. Nonparametric Statistics for the Behavioral Sciences. (2nd ed., p. 399). McGraw-Hill, New York.

Smith, V. H. & S. J. Bennet, 1999. Nitrogen: phosphorus ratios and phytoplankton community structure in lakes. Archiv für Hydrobiologie, 146: 37–53.

Sommer, U., 1988. Growth and survival strategies of planktonic diatoms. In Sandgren, C. D. (ed.) Growth and Reproductive Strategies of Freshwater Phytoplankton. Cambridge University Press, 227–260.

Sommer, U., 1989.. The role of competition for limiting resources in phytoplancton species replacements. In Sommer, U. (ed.) Plankton Ecology. Springer, 57–106.

Sommer, U., 1993. Phytoplancton competition in Plußsee: A field test of the resource-ratio hypothesis. Limnology and Oceanography 38: 838–845.

Sommer, U., 1999. A comment on the proper use of nutrient ratios in microalgal ecology. Archiv für Hydrobiologie 146: 55–64.

StatSoft, Inc., 2001. STATISTICA (data analysis software system), version 6. www.statsoft.com.

Tammert, H., V. Kisand & T. Nõges, 2005. Bacterioplankton abundance and activity in a small hypertrophic stratified lake. Hydrobiologia 547: 83–90.

Tilman, D., 1977. Resource competition between planctonic algae: an experimental and theoretical approach. Ecology 58: 338–348.

Tilman, D., 1982. Resource Competition and Community Structure. Princeton University Press, Princeton, New Jersey, USA.

Tilman, D., R. Kiesling, R. Sterner, S. Kilham & F. A. Johnson, 1986. Green, blue–green and diatom algae: taxonomic differences in competitive ability for phosphorus, silicon and nitrogen. Archiv für Hydrobiologie 106: 473–485.

Tõnno, I., K. Ott & T. Nõges, 2005. Nitrogen dynamics in steeply stratified temperate Lake Verevi, Estonia. Hydrobiologia 547: 63–71.

Urabe, J., T. K. Gurung & T. Yoshida, 1999. Effects of phosphorus supply on phagotrophy by the mixotrophic alga Uroglena americana (Chrysophyceae). Aquatic Microbial Ecology 18: 77–83.

Utermöhl, H., 1958. Zur Vervollkomnung der quantitativen Phytoplankton Methodik Mitteilungen internationale Vereingung für theoretische und angewandte. Limnologie 9: 1–38.

Vollenweider, R. & J. Kerekes, 1982. Eutrophication of Waters. Monitoring, Assessment and Control. OECD,, Paris.

Zingel, P. & I. Ott, 2000. Vertical distribution of planktonic ciliates in strongly stratified temperate lake. Hydrobiologia 435: 19–26.

Hydrobiologia (2005) 547:137–150
I. Ott & T. Kõiv (eds), Lake Verevi, Estonia – A Highly Stratified Hypertrophic Lake
DOI 10.1007/s10750-005-4155-9

The composition and density of epiphyton on some macrophyte species in the partly meromictic Lake Verevi

Reet Laugaste* & Markku Reunanen
Institute of Zoology and Botany, Estonian Agricultural University, Võrtsjärv Limnological Station, 61101, Rannu, Tartu County, Estonia
*(*Author for correspondence: E-mail: reet@zbi.ee)*

Key words: epiphyton, chlorophyll *a*, biomass, algal groups on different host plants

Abstract

The epiphyton on 22 macrophyte species was studied in the hypertrophic stratified Lake Verevi mainly in the midsummer of 2000 and 2001. Some material from 1998 and 1999 was used as well. Chlorophyll *a* (Chl *a*) level was high: 330–360 μg g dw^{-1} on emergent plants, and an average of 117–200 μg on floating-leaved plants and 820–920 μg g dw^{-1} on submerged plants. Biomass was 15–23, 5–10 and 35–53 mg g dw^{-1}, respectively. The richest in epiphyton were submerged plants with densely growing and fine branchlets such as *Ranunculus*, *Ceratophyllum*, *Myriophyllum*, *Utricularia*, *Potamogeton pectinatus* L. and *P. friesii* Rupr. The share of Chl *a* in biomass was higher in 2001 (2.3%) than in 2000 (1.7%), which can be associated with lower irradiance in summer 2001. Filiform chlorophytes were dominating on most plants; 60% of biomass on submerged, 69% on emergent and 80% on floating-leaved plants; in some cases, the share of filamentous species was 95%. Diatoms formed 29, 12 and 7%, cyanobacteria 8, 16 and 10% of the same ecotopes, respectively. As a rule, the epiphyton was quite sparse on large *Potamogeton* leaves. Cyanobacteria were more abundant on large *Potamogeton* and *Nuphar* leaves, *Elodea*, on stems of *P. natans* L., *Nuphar* and on some emergent plants with a smooth and soft stem surface, as *Butomus* and *Typha*. Diatoms played the most important role on some *Potamogeton* species and in single samples of *Ceratophyllum* and *Ranunculus*. The morphology of plant species appears to be the main factor of epiphyton richness in L. Verevi.

Introduction

Epiphyton plays an important role in the primary production of small lakes even in case they are deep. Its share forms about one-third of total production (Putz, 1997), 31% of phytoplankton in the euphotic zone (Cattaneo et al., 1998), two-thirds of plankton production (Ilmavirta, 1979; cit. Kairesalo, 1984), 23–40% of macrophyte production (Szczepanska, 1970), or most of primary production in the lake (Burkholder & Wetzel, 1989). Abundant *Cladophora* mats among macrophyte stands can form the majority of the biomass of the whole lake, e.g., more than 800 g dw m^{-2} (Marvan et al., 1978). It is obvious that epiphyton is richer in eutrophic lakes, while its share in total primary production of the lake can be larger in oligotrophic ones, as macrophytes reach deeper areas. Light can be the most important factor determining epiphyton abundance and distribution, and the need for light is the most important factor affecting the species composition of epiphyton. Some diatom species prefer heavily shaded biotopes, while others dominate in well-irradiated biotopes (Marvan et al., 1978). According to Havens et al. (1996), the trend of the epiphyton amount is opposite to that of water transparency; however, turbidity does not affect attached algae near the surface layer. Experiments with shading showed that algal abundance reduced about 40%

(Hepinstall & Fuller, 1994). Light is a limiting factor for thick macrophyte beds (Kairesalo, 1984; Cattaneo et al., 1998).

Besides light conditions, several other factors control the abundance and composition of epiphyton. Many papers are dedicated to the role of nutrients; camouflaged by other factors, relationships between nutrients and surrounding water are not clear. Several papers present contradictory results of nutrient enrichment experiments. Most authors assert the relationship with mineral nitrogen, particularly green filiform algae; on the contrary, Cattaneo (1987) notes the relationship of green filiform algae with total P in Canadian lakes. In Florida Everglades, fertilizing caused the disappearance of most filamentous green algae (Vymazal et al., 1994). The N:P ratio can be important in the relationship between algal groups. However, it is not sufficient to be confined to total N:P content ratio; the ratio of the available forms of nutrients, as the mineral forms of N, to suspended particulate P may be totally different (Paterson et al., 2002). Kairesalo (1983, cit. Kairesalo & Uusi-Rauva, 1983) asserts that within an *Equisetum fluviatile* stand the seasonal fluctuation of N:P ratio in epiphytic communities was significantly weaker than that of the surrounding water, suggesting that epiphyton was not solely dependent on the nutrient sources of water. It is proved that the host plant is the source of nutrients for epiphyton. According to Burkholder & Wetzel (1990), epiphyton takes 25–60% of P from its host plant; according to Kairesalo & Uusi-Rauva (1983), all phosphorus released by *Equisetum fluviatile* was fixed by epiphyton. However, the knowledge of the nutrient content of the host plant is not sufficient, since some species excrete nutrients profusely, while others scantily; at the same time, other factors also be involved. According to Sand-Jensen (1990), in Danish lakes of different trophy, Si appears to be an important controlling element of the species composition of epiphytic algae. Organic matter excreted from the host plant also plays an important role causing epiphyton peaks on some submerged plants (Kassim & Al-Saadi, 1995).

The problem is whether the host plant or the growing site is more important in the life of epiphyton. Blindow (1987) in her profound study concludes that the plant species is of more importance; yet, significant differences were found in single epiphyton taxa for the same host plant at different sites. The character of the host plant is undoubtedly of significance; however, this is most obvious in oligotrophic and moderately eutrophic lakes. In highly eutrophic lakes, differences in epiphyton are small (Eminson & Moss, 1980, cit. Kairesalo, 1984). Nevertheless, some differences depend on the plant's morphology and the character of its surface: thick and finely branched submerged plants as *Myriophyllum* or *Ranunculus circinatus* are usually richer in epiphyton than large-leaved pondweeds (*Potamogeton*). Some incrusted *Chara* species have different epiphytes in comparison with non-incrusted species at the same growing site (Blindow, 1987). Stem structure and surface play an important role in the case of emergent plants (Marvan et al., 1978). The toxic influence of the excretions of some emergent plants on algae, particularly cyanobacteria, has been established experimentally: *Myriophyllum* (Gross et al., 2002), *Ceratophyllum* (Iványi et al., 2002) and *Chara* (Wium-Andersen et al., 1982, cit. Blindow, 1987).

Grazing is one of the factors, which can influence the amount and composition of epiphyton and has also been proved by experimental studies (Jernakoff & Nielsen, 1997). Consumers can be scrapers and epiphytic deposit-feeders (snails and amphipods) in thick *Chara* stands, and deposit- and filter-feeders (cladocerans, copepods, chironomid larvae) in sparser stands of *Potamogeton pectinatus* L. (Hart & Lovvorn, 2000). The effect of grazing changes during the vegetation period. In L. Pääjärvi, a maximum of epiphytes was followed by a maximum of grazers (Kairesalo, 1984). According to the same author, gastropods can influence the N:P ratio. Some authors claim that diatoms prove to be more desirable food than green or blue–green algae (cit. Kairesalo, 1984). Thus, many factors can control the amount and composition of epiphyton.

The morphometric characteristics of Lake Verevi (Ott et al., 2005) represent peculiar features of its ecosystem. Different lengths (or total lack) of spring turnover cause shifts in the start of thermal and chemical stratification and nutrient depletion in the lake. In absence of spring turnover, several phenomena unusual for the lake will occur, e.g. changing of ordinary phytoplankton and macrophyte communities during summer stagnation affecting epiphyton. As our study is the first attempt to describe epiphyton in a strongly stratified hypertrophic lake in Estonia, we did not set tasks that are more complicated. Our purpose was (1) to learn the methods of epiphyton study; (2) to find out the chlorophyll content and biomass of epiphyton in the lake; (3) to compare these values for different host plants.

Material and methods

The epiphyton material is quite heterogeneous, sampled in the years 1998–2001 using different methods. In 1998 and 1999, the first attempt was made in this field, and a few plant species were included (Table 1). This paper was completed mainly based on the data from summer 2000 and 2001 as well as some seasonal data (vegetation period 2000, spring and summer 2001).

Emergent plants were picked from dense stands, for seasonal studies possibly from the same location, from the boat or walking, a depth of 0.8–1 m. The upper 50–60 cm was cut and treated as a whole; the epiphyton of different parts of the stems was not studied. The distance between 5 replicates was about 50–100 m. Floating leaves were picked by hand from the boat; submerged plants were picked with a hook; in some cases, the assistance of a diver was used.

Table 1. Material of epiphyton

Taxa (abbreviations)	1998	1999	2000	2001
Emergent plants				
Butomus umbellatus L. (Butom)			Chl, bm	Chl, bm
Carex rostrata Stokes			Chl, bm	Chl, bm
Equisetum fluviatile L. (Equis)		Chl, bm	Chl, bm, s	Chl, bm
Phragmites australis (Cav.) Trin ex Steud. (Phrag)		Chl, bm	Chl, bm, s, r	Chl, bm, s
Sagittaria sagittifolia L. (Sagitt)		Chl	Chl, bm	Chl, bm
Schoenoplectus lacustris (L.) Palla (Schoen)		Chl	Chl, bm	Chl, bm
Typha angustifolia L.	Chl	Chl	Chl, bm, s, r	Chl, bm, s
Floating leaves				
Nuphar lutea (L.) Sm., leaf (Nupl-f)			Chl, bm, r	Chl, bm
Potamogeton natans L., leaf (Pnatl-f)	Chl		Chl, bm	Chl, bm
Submerged plants				
Ceratophyllum demersum L.			Chl, bm, r	Chl, bm
Chara spp.			Chl, bm, r	Chl, bm
Cladophora glomerata (L.) Kütz. (Clado)				Chl, bm
Elodea canadensis L.			Chl, bm	Chl, bm
Fontinalis antipyretica DC. (Fontin)			Chl, bm	Chl, bm
Myriophyllum verticillatum L. (Myrio)				Chl, bm
Nuphar lutea, stem (Nupst)		Chl	Chl, bm, r	Chl, bm
Potamogeton compressus L. (Pcomp)		Chl		Chl, bm
P. friesii Rupr. (Pfries)				Chl, bm
P. lucens L. (Plucen)	Chl	Chl	Chl, bm	
P. natans, stem	Chl	Chl		
P. natans, stem + leaf (Pnatan)			Chl, bm	Chl, bm
P. pectinatus L. (Ppect)			Chl, bm	Chl, bm
P. perfoliatus L. (Ppert)	Chl	Chl	Chl, bm	Chl, bm
Ranunculus circinatus Sibth. (Ranun)			Chl, bm, r	Chl, bm
Utricularia vulgaris L. (Utricul)				Chl, bm

Chl – chlorophyll *a*, pheopigments; bm – biomass; s – seasonal material; r – replicates.

In 1998 and 1999, epiphyton was removed by rubbing with fingers: plant pieces were placed on a cuvette, 100 ml distilled water was added and after pouring the mixture into the jar, further 100 ml was added for rinsing plant parts. The suspension was shaken and divided into two portions, one for pigment analysis and the other fixed with Lugol solution for counting. The plant pieces were measured and their surface area was calculated. The seasonal material of 2000 was also treated in the same way. In 2000, the epiphyton was also removed by shaking and the two methods of removal were compared. There was no reliable difference between the two methods; however, the Chl *a* values were in average 1.2 times higher when removal was done by shaking 2 min in 100 ml distilled water compared with rubbing. Thus, in 2001, the removing only by shaking was used. In 2000 and 2001, the summer material was treated using two different methods: the measured plants were dried at 105 °C for 24 h and weighed: the epiphyton values were calculated per area and weight unit of the host plant. For the epiphyton of most submerged plants, only weighing of the host plant was feasible. Method related data deserve special research and are not discussed here in detail. For the determination of the concentration of Chl *a* and pheopigments, part of a sample was concentrated on Whatman glass fibre filters (GF/C). Pigments were extracted with 90% acetone and/or 96% ethanol and analysed spectrophotometrically (Guidelines for the Baltic Monitoring Programme for the Third Stage, 1984). The equations of Jeffrey & Humphrey (1975) were applied for calculation. The biomass was counted according to the Utermöhl method. Most diatoms were identified in water preparations by counting when possible; in 1999, some identifications of the diatoms of *Phragmites* and *Equisetum* epiphyton were made on microscope slides after heating.

The program STATISTICA 6.0 was used for statistical and cluster analysis. The complete linkage agglomeration was used for clustering the samples.

Results

Chl a content

The epiphyton Chl *a* values are shown in Figures 1–3, content per area unit of the host plant in Figure 1 and per weight on the other figures. In 1999, the poorest in Chl *a* were the leaves of three *Potamogeton* species, leaves of *Nuphar* and stems of *Shoenoplectus* and *Typha*. The richest were stems of *P. natans*, *Nuphar* and *Sagittaria*. In 2000, among emergent plants *Equisetum* and *Phragmites* had more epiphyton, while *Potamogeton* leaves were not measured. The difference between stems and leaves of *Nuphar* and *P. natans* is noteworthy. *Sagittaria* was picked from the formerly polluted region of the lake, the southwestern corner;

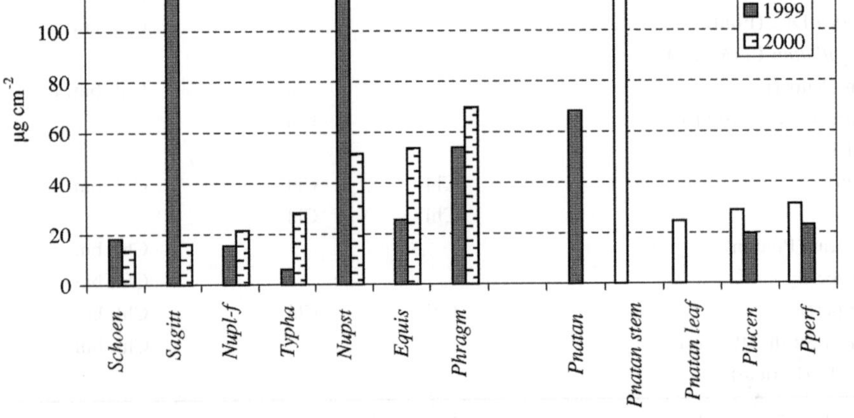

Figure 1. Chl *a* in summer epiphyton in years 1998–2000, per area of host plant. Abbreviations as in Table 1.

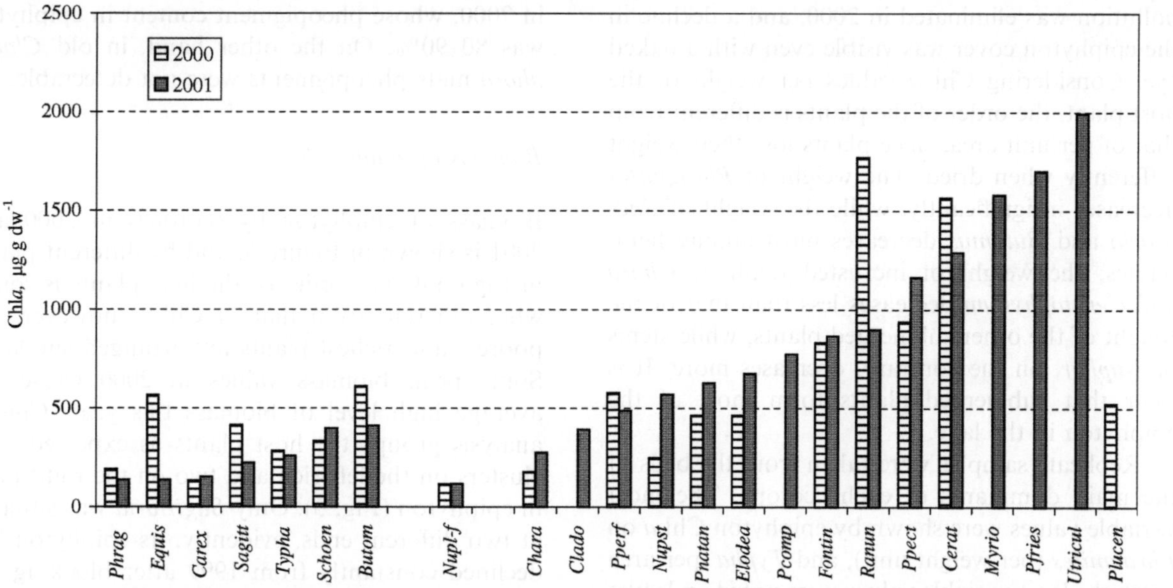

Figure 2. Chl *a* in summer epiphyton in years 2000 and 2001, per weight unit of host plant. Plants are grouped by ecotopes: emergent, floating leaves and submerged. Abbreviations as in Table 1.

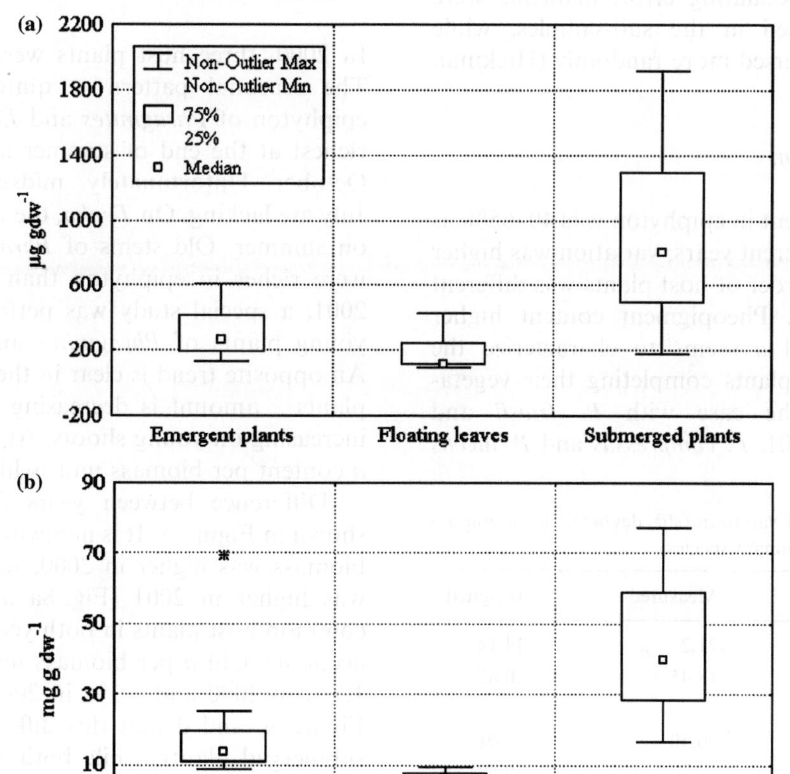

Figure 3. Chl *a* content (a) and biomass (b) of epiphyton in different ecotopes, data of 2000 and 2001.

pollution was eliminated in 2000, and a decline in the epiphyton cover was visible even with a naked eye. Considering Chl *a* values per weight of the host plant, the order of the plants is different from that of per unit area, since plants lose their weight differently when dried. The weight of *Phragmites* decreases insignificantly, while the weight of *Sagittaria* and *Butomus* decreases most among helophytes; the weight of incrusted thalli of *Chara* and *Ceratophyllum* decreases less than that of the weight of the other submerged plants, while stems of *Nuphar*, on the contrary, decreases more. It is clear that submerged plants form most of the epiphyton in the lake.

Replicate samples were taken from the beds of the main dominants of each ecotope. The most variable values were shown by epiphyton Chl *a* on *Phragmites* (per weight unit), and *Typha* (per area unit); the least variable values were noted on leaves and stems of *Nuphar* (Table 2). Presence of filiform algae in the population tended to increase counting and sub-counting error: filaments were not easily dispersed in the sub-samples, while diatoms were dispersed more randomly (Hickman & Clarer, 1973).

Pheopigment content

Pheopigment content in epiphyton was 40–65% as an average for different years; variation was higher than 120%. The order of host plants was different in different years. Pheopigment content higher than 50% of Chl *a* seems to characterize the epiphyton of old plants completing their vegetation; this was the case with *P. friesii* and *R. circinatus* in 2001, *P. compressus* and *P. lucens* in 2000, whose pheopigment content in epiphyton was 80–90%. On the other hand, in old *Cladophora* mats pheopigments were not detectable.

Table 2. Coefficients of variation (std. dev/avg, %) of epiphyton Chl *a* on some dominant species

Taxa	Measured	Weighed
Nuphar, leaf	21.2	14.14
Nuphar, stem	18.45	20.82
Chara sp.		27.84
Typha sp.	70.50	27.01
Ranunculus sp.		30.62
P. natans		35.83
Ceratophyllum sp.		44.68
Phragmites sp.	28.04	66.34

Biomass of epiphyton

Biomass of epiphyton by ecotopes in 2000 and 2001 is shown in Figure 3, and by different plants in Figure 4. The order of the host plants is somewhat different from that of Chl *a*: however, the poorest and richest plants are arranged similarly. Some peak biomass values in 2000 cause the average high level of biomass this year. Cluster analysis groups the host plants as expected: two clusters on the left rich and two on the right poor in epiphyton (Fig. 5). Only *Sagittaria* was situated at two different ends, evidently, its epiphyton has declined constantly from 1999 after blocking the polluting inflow at its growing site.

Seasonal dynamics

In 2000, three host plants were studied (Fig. 6). The seasonal pattern is quite similar for the epiphyton of *Phragmites* and *Equisetum*, which is richest at the end of summer and in September–October. Unfortunately, midsummer data from July are lacking. On *Typha*, the epiphyton peak fell on summer. Old stems of *Phragmites* and *Typha* were richer in epiphyton than young shoots. In 2001, a special study was performed on old and young plants of *Phragmites* and *Typha* (Fig. 7). An opposite trend is clear in the case of both host plants – amount is decreasing on old stems and increasing on young shoots. At the same time Chl *a* content per biomass unit is highly variable.

Difference between years 2000 and 2001 is shown in Figure 8. It is noteworthy that epiphyton biomass was higher in 2000, while Chl *a* content was higher in 2001 (Fig. 8a and b, where only common host plants in both years were taken into account). Chl *a* per biomass unit was on average 1.7% in 2000 and 2.3% in 2001. It follows from Figure 8c and d that this difference is caused by submerged plants, while both epiphyton biomass and Chl *a* content were higher on emergent plants and lower on submerged plants in 2000. The average and median values for the two years did not reveal statistically significant differences.

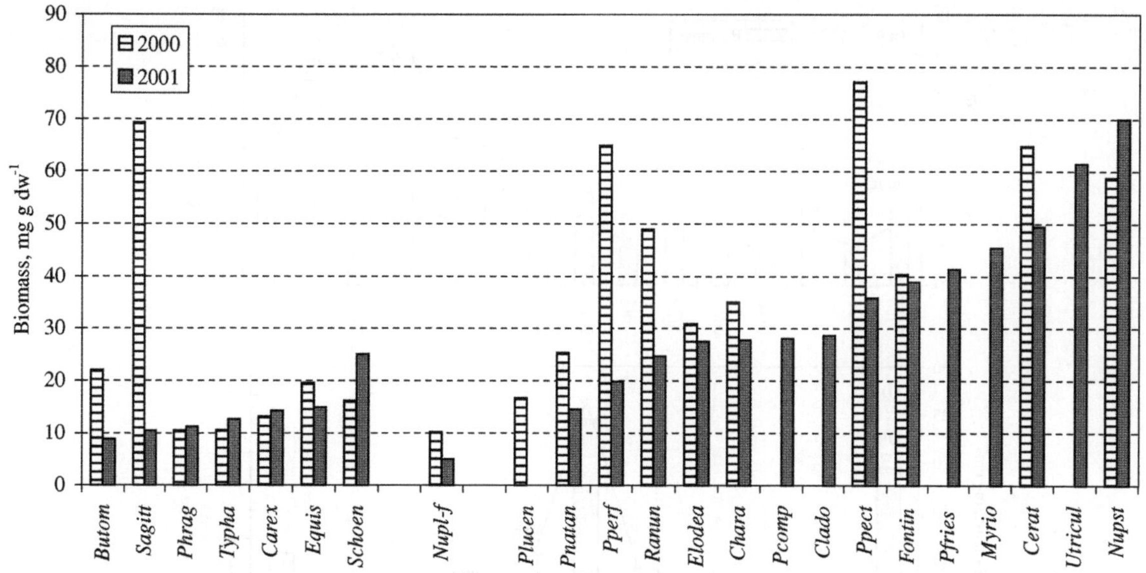

Figure 4. Epiphyton biomass in years 2000 and 2001. Abbreviations as in Table 1.

Relationship of algal groups

In the years 1999–2001, some quite stable features in algal groups are noted in the epiphyton of different host plants (Table 3). As a rule, epiphyton was quite sparse on large *Potamogeton* leaves (*P. lucens*, *P. natans*, *P. perfoliatus*). The most frequent dominants were green filiform algae almost on all plants; relatively sparse were filaments on *Chara, Butomus* and on leaves of *Potamogeton*, except for *P. friesii*, which was covered with a very thick coat of filaments. Blue–greens prevailed on large leaves of *Potamogeton* and *Nuphar*, on *Elodea*, on stems of *P. natans* and *Nuphar* and

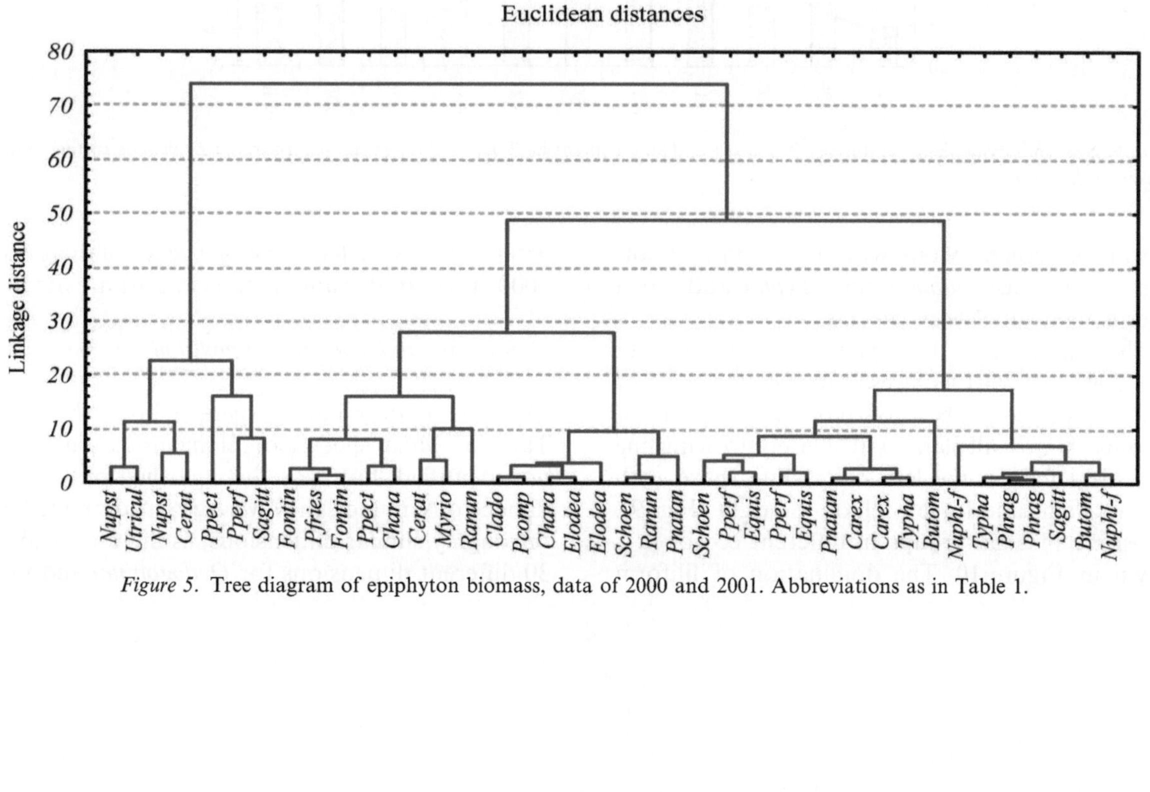

Figure 5. Tree diagram of epiphyton biomass, data of 2000 and 2001. Abbreviations as in Table 1.

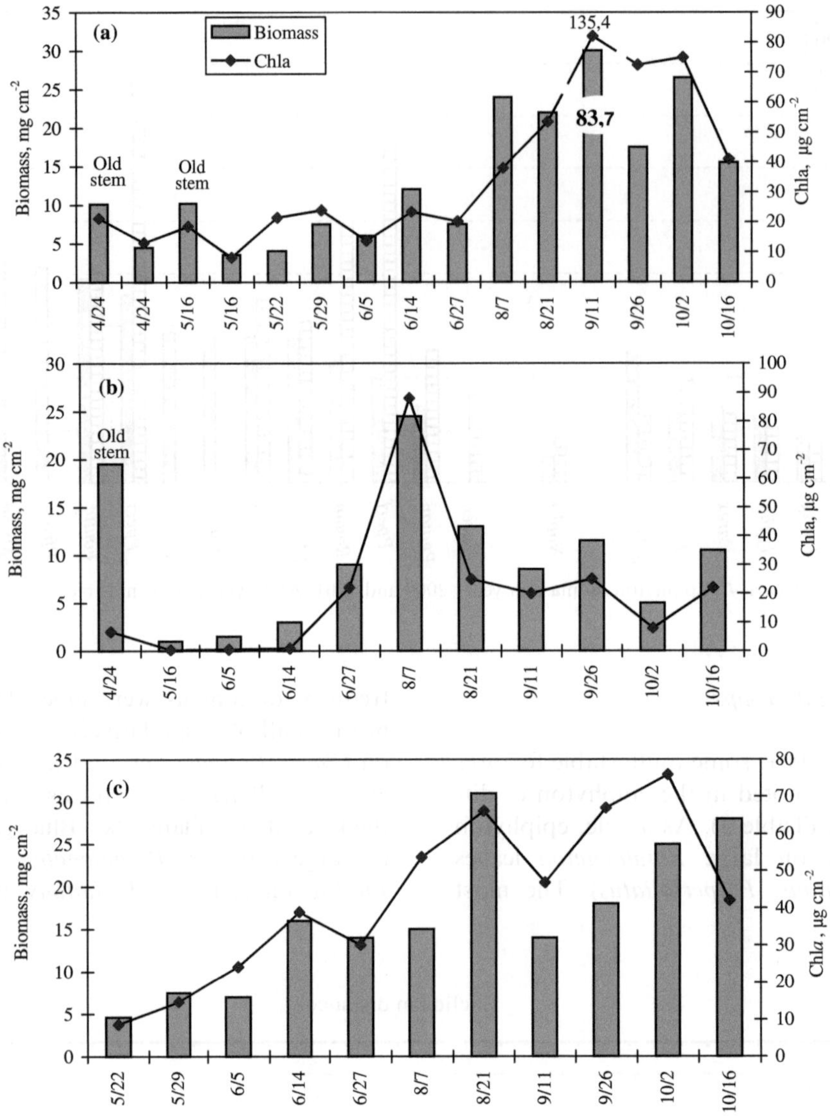

Figure 6. Seasonal dynamics of epiphyton biomass and Chl *a* content in 2000: (a) *Phragmites australis*, (b) *Typha angustifolia* and (c) *Equisetum fluviatile*.

on some emergent plants with a smooth and soft stem surface, as *Butomus* and *Typha* and, to a lesser extent, on *Schoenoplectus*.

The clustering of host plants (epiphyton data from 2001 and 2001 with replicates) by the share of algal groups was the best based on blue–greens and diatoms. Almost all plants with densely growing fine branchlets fell on the left side, while those with smooth stems and leaves, fell on the right (Fig. 9). The share of algal groups in different ecotopes is shown in Figure 10. The domination of filiform green algae as well as the larger share of diatoms in 2000 and that of cyanobacteria in 2001 are obvious.

Dominants and species composition

The main dominating genera are shown in Table 3. Most species of filamentous algae were not identified, but were distinguished only by cell dimensions; in fact, they were grouped on the basis of roughly similar dimensions, as there were about 30 different dimensions for *Oedogonium* and more

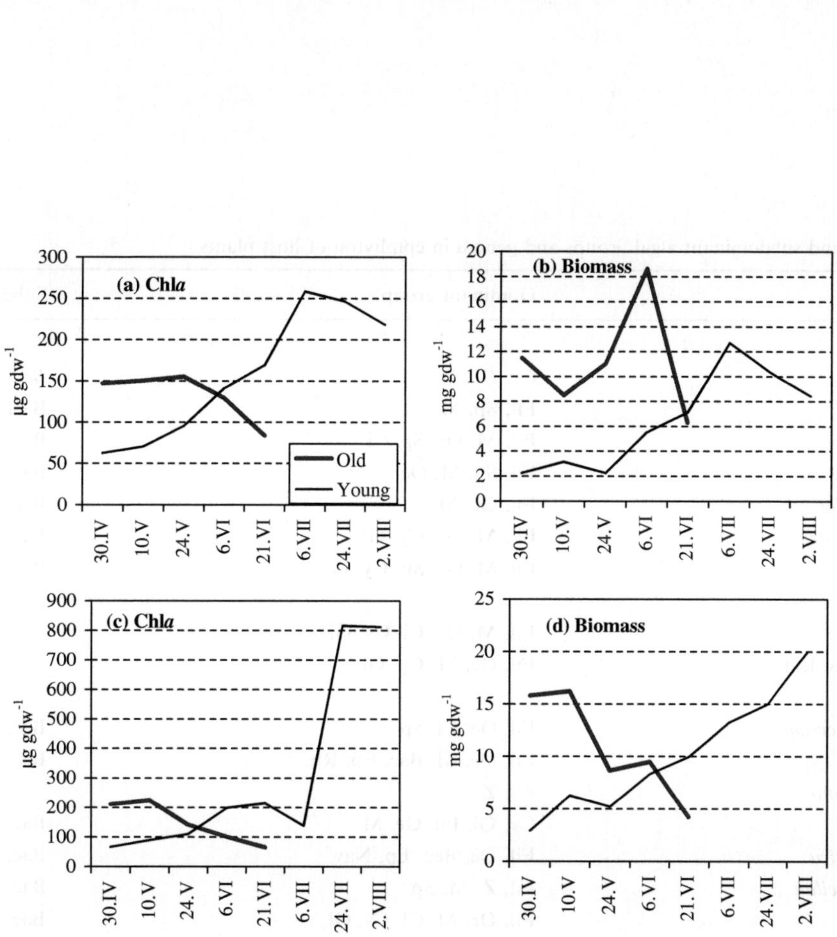

Figure 7. Seasonal dynamics of epiphyton on old and young stems in 2000. (a, b) *Typha angustifolia* and (c, d) *Phragmites australis*.

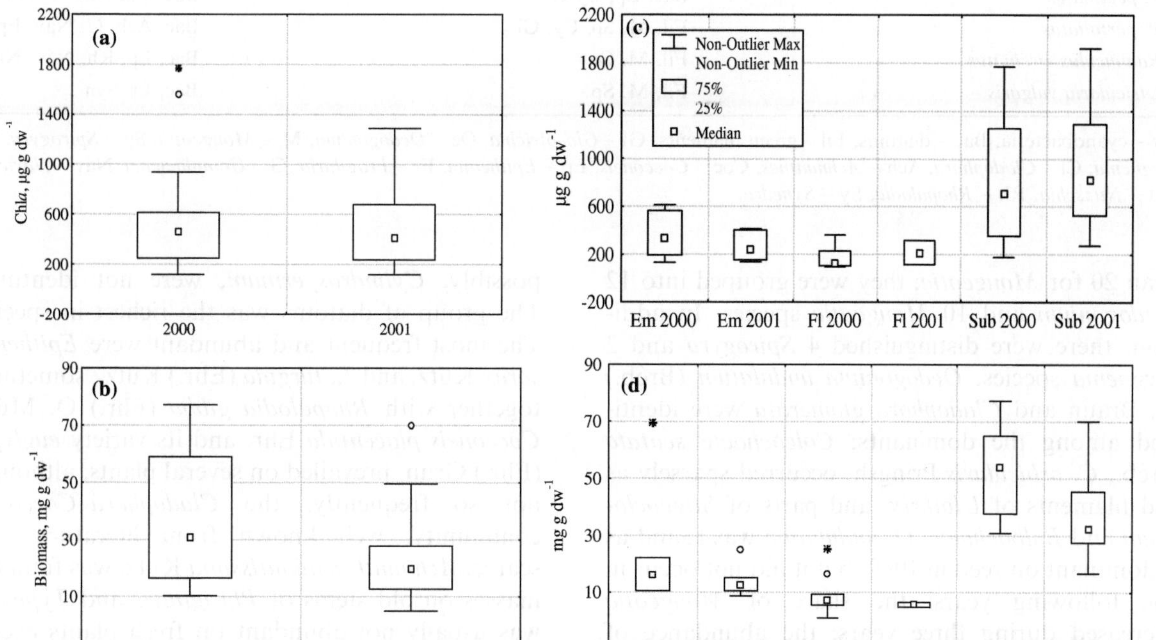

Figure 8. Chl *a* content in years 2000 and 2001, all ecotopes together (a), and in different ecotopes (c); biomass in 2 years, all ecotopes together (b), and in different ecotopes (d). Abbreviations: Em – emergent, Fl – floating-leaved, Sub – submerged plants.

Table 3. Dominant and subdominant algal groups and genera in epiphyton of host plants

Taxa	Dominant group	Subdominant group
Emergent plants		
Butomus umbellatus	*Cy, Gl*	*Bac, Fr*
Carex rostrata	Fil, Sp, M	Bac, Ep
Equisetum fluviatile	Fil, M, Oe, Sp, Cl	Bac, Ep, G, Coc, Nav
Phragmites australis	Fil, Sp, M, Oe	Bac, Nit, Nav, Ep, Ach
Sagittaria sagittifolia	Fil, Oe, M	Bac, Ep
Schoenoplectus lacustris	Fil, M, Oe, Cy, Gl	Bac, Fr, Ep, G
Typha angustifolia	Fil, M. Oe, Sp, Cy, Gl	Bac, Coc, Ep, Nav, Nit
Floating leaves		
Nuphar lutea, leaf	Fil, M, Oe, Cl, Cy, Gl	
Potamogeton natans, leaf	Fil, Oe, M, Cy, Gl	
Submerged plants		
Ceratophyllum demersum	Fil, Oe, Cl, Sp	Bac, Ep, Coc, Nav
Chara sp.	Fil, Oe, Cl, Bac, Ep, Rh	Bac, Ach, Coc
Cladophora glomerata	Fil, Z	
Elodea canadensis	Cy, Gl, Fil, Oe, M	Bac, Ep, Rh, Coc
Fontinalis antipyretica	Fil, Oe, Bac, Ep, Nav	Bac, Coc, Nit
Myriophyllum verticillat.	Fil, Z, M, Sp	Bac, Ach, Syn
Nuphar lutea, stem	Fil, Oe, M, Cl, Cy, Gl	Bac, Ep
Potamogeton compressus	Bac, Coc	Bac, Nit, Ach
P. friesii	Fil, M	Bac, Rh, Ep, Coc
P. lucens	Cy, Gl	
P. natans, stem + leaf	Cy, Gl, Fil, Oe, M, Z, Sp	Bac, Ep
P. pectinatus	Bac, Ep, Cy Gl	Bac, Rh, Nav, Coc, Fr
P. perfoliatus	Fil, M, Sp, Cy, Gl	Bac, Ach, G, Nav, Ep
Ranunculus circinatus	Fil, M, Oe	Bac, Ep, Rh, Nav, Nit
Utricularia vulgaris	Fil, M, Sp	Bac, Fr, Syn

Cy – cyanobacteria, Bac – diatoms, Fil – green filaments, Gl – *Gloeotrichia*, Oe – *Oedogonium*, M – *Mougeotia*, Sp – *Spirogyra*, Z – *Zygnema*, Cl – *Cladophora*, Ach – *Achnanthes*, Coc – *Cocconeis*, Ep – *Epithemia*, Fr – *Fragilaria*, G – *Gomphonema*, Nav – *Navicula*, Nit – *Nitzschia*, Rh – *Rhopalodia*, Sy – *Synedra*.

than 20 for *Mougeotia*; they were grouped into 12 *Oedogonium* and 10 *Mougeotia* species. In addition, there were distinguished 4 *Spirogyra* and 2 *Zygnema* species. *Oedogonium undulatum* (Bréb.) A. Braun and *Cladophora glomerata* were identified among the dominants; *Coleochaete scutata* Bréb., *C. orbicularis* Pringsh. occurred sparsely as did filaments of *Ulothrix*, and parts of *Stigeoclonium* and *Bulbochaete*. *O. undulatum* was found as a dominant on reed in 1999, but it did not occur in the following years; the share of *Mougeotia* increased during three years; the abundance of *Spirogyra* and *Zygnema* were much higher in 2001. *Gloeotrichia pisum* (Ag.) Thur. dominated among blue–greens; some fragments of *Calothrix* and, possibly, *Cylindrospermum*, were not identified. The group of diatoms was the richest in species. The most frequent and abundant were *Epithemia sorex* Kütz. and *E. turgida* (Ehr.) Kütz., sometimes together with *Rhopalodia gibba* (Ehr.) O. Müll.; *Cocconeis placentula* Ehr. and its variety *euglypta* (Ehr.) Grun. prevailed on several plants, although, not so frequently, the *Cladophora–Cocconeis* community, well known from literature, was scarce. *Achnanthes minutissima* Kütz. was found in masses on old stems of *Phragmites* and *Typha*; it was usually not abundant on fresh plants except for some cases (single samples from *Phragmites*, *Typha*, *Chara*, *P. pectinatus* and *Myriophyllum*). The other chlorophytes besides filiform genera

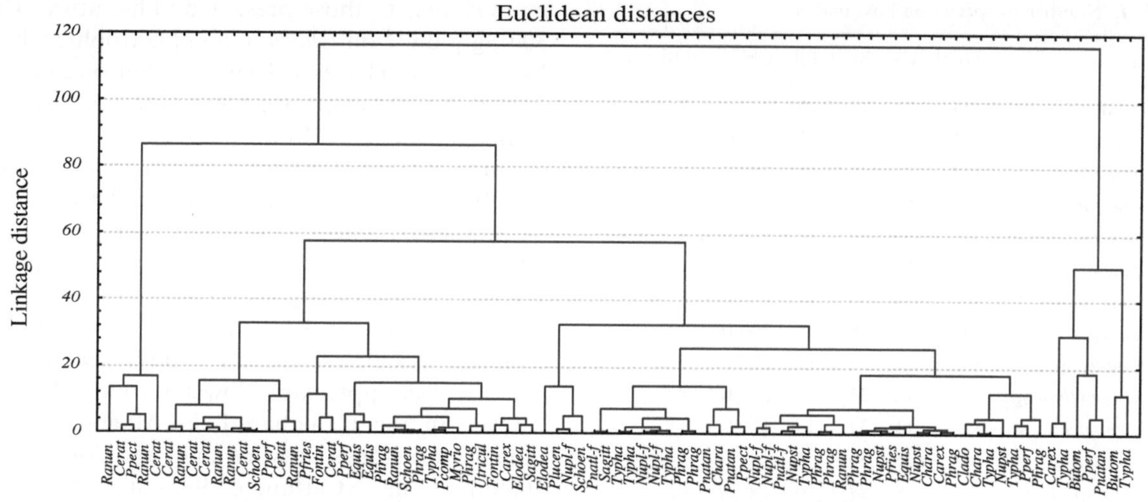

Figure 9. Percentage of cyanobacteria and diatoms in epiphyton biomass.

accounted for less than 5% of biomass. *Phacotus lenticularis* (Ehr.) Stein and *Planctococcus sphaerocystiformis* Korsh. were the most abundant; some host plants as *R. circinatus*, *Chara* and *P. friesii* contained an appreciable amount of desmids in their epiphyton. Table 4 presents the species number of host plants in 1999–2001. *P. lucens*, *P. compressus*, *Myriophyllum* and *Utricularia* were sampled only once; *Phragmites*, *Typha* and *Equisetum*, *Nuphar*, *R. circinatus* and *C. demersum* were represented by 6–12 samples. This circumstance explains the larger species number for most sampled host plants.

Some interesting chlorophyte species were found: *Sorastrum spinulosum* Näg., rare on emergent and submerged plants; *Ducellieria tricuspidata* (Borge) Teil., once on *R. circinatus*; *Gloeotaenium loitlesbergerianum* Hansg. on *R. circinatus* and *Chara*; *Sphaerocystis bilobata* Broady on *M. verticillatum*. The other groups (Eugleno-, Chryso-, Crypto- and Dinophyta) were encountered only occasionally and at low densities.

Discussion

Epiphyton Chl *a* content in L. Verevi exceeds most of the values found in the literature. Majority of available data are presented per unit of area of the host plant. Values closer to ours, <10–40 μg cm^{-2}, were recorded on reed in a eutrophic lake in Germany (Müller, 1995); similar values were also recorded in six meso-eutrophic lakes in Poland (Szczepanska, 1970). In the reservoirs of

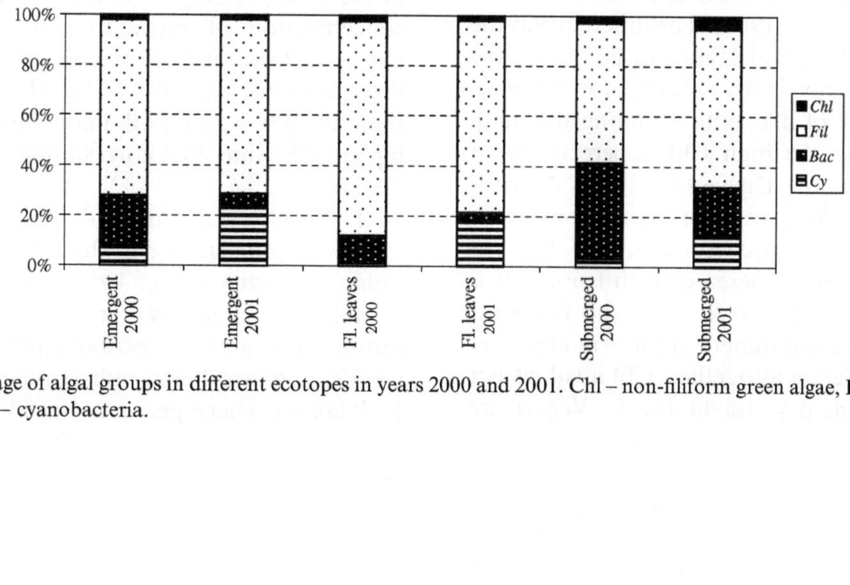

Figure 10. Percentage of algal groups in different ecotopes in years 2000 and 2001. Chl – non-filiform green algae, Fil – green filaments, Bac – diatoms, Cy – cyanobacteria.

Table 4. Number of species on host plants

Taxa	Total	Cy	Bac	Fil	Chl	Others
Phragmites sp.	**126**	**17**	**49**	**20**	**33** (12)	**7**
Typha sp.	**105**	10	**38**	**18**	24 (11)	**8**
P. perfoliatus	**104**	13	34	**18**	**38** (12)	1
Equisetum sp.	97	**17**	30	**21**	26 (10)	3
P. friesii	91	8	32	16	31 (**15**)	4
Ranunculus sp.	91	13	33	17	27 (**17**)	1
Nuphar, stem	90	**14**	30	**18**	26 (12)	2
Chara sp.	87	**14**	36	6	30 (**16**)	1
Fontinalis sp.	84	13	35	13	21 (10)	2
Ceratophyllum sp.	82	12	33	12	21 (10)	4
Schoenoplectus sp.	81	11	37	15	17 (11)	1
P. pectinatus	81	**15**	33	4	28 (10)	1
Nuphar sp., leaf	75	9	27	16	19 (8)	4
Myriophyllum sp.	75	4	29	9	**32** (10)	1
Utricularia sp.	75	9	28	9	28 (7)	1
Carex sp.	74	8	27	11	22 (6)	**6**
Cladophora sp.	65	11	22	12	19 (6)	1
P. natans	65	6	22	12	25 (11)	
Elodea sp.	64	7	25	10	21 (8)	1
Sagittaria sp.	60	3	29	11	16 (10)	1
P. compressus	56	8	28	6	13 (4)	1
P. lucens	50	6	20	6	18 (5)	
Butomus sp.	47	6	21	8	10 (4)	2

Cy – cyanobacteria, Bac – diatoms, Fil – filaments, Chl – chlorophytes, among them species number of desmids in brackets. The largest values are in boldface.

the Volga R., Chl *a* content on four emergent plants was about one magnitude lower and on *Nuphar* 20–40 times lower (Meteleva, 2000) compared with the values for L. Verevi. In the oligotrophic (semidystrophic) L. Pääjärvi, maximum Chl *a* values of epiphyton on *Equisetum* were 5–15 μg cm^{-2} (Kairesalo, 1984). The difference in the trophic state of these two lakes can be one of the causes, since L. Verevi is strongly eutrophic; hence, a very large share of filaments is among the main factors explaining the high Chl *a* percentage of epiphyton (Sigareva & Devyatkin, 1987). According to Cattaneo (1987), mats of filamentous green algae characterize the richest sites, and the importance of large algae increases with trophy. It is also possible that the large share of filaments hinders grazing, or consumers of filiform algae are absent from L. Verevi altogether. Chl *a* values per unit of host plant dry weight for L. Verevi are more similar to those presented in literature: 100–300 μg g dw^{-1} on *Chara* in an oligotrophic lake in New Zealand (Hawes & Schwarz, 1996); on *Littorella* in Danish lakes: epiphyton Chl *a* in the hypertrophic L. Bryrup fluctuated from 540 to 350 00 μg g dw^{-1}; in three eutrophic lakes the average values were 110–1090 μg g dw^{-1} (Sand-Jensen, 1990).

Biomass values similar to our data, 20–300 mg cm^{-2}, were recorded in reservoirs of the Ukraine on solid inorganic substrate (Shevchenko, 1994). In a channel in the Ukraine, epiphyton biomass was 0.7–44 mg dw^{-1} on *P. perfoliatus*, while it was appreciably lower on reed (Kuz'ko, 2000a). In the Canadian Lake Magog, average biomass on stones in May, July and August was 1088 mm^3 cm^{-2} (Cattaneo, 1987); in L. Okeechobee (Florida), the values of biomass on *Eleocharis, Scirpus* and *Vallisneria* were between 0.05 and 175 mm^3 g dw^{-1} (Carrick & Steinman, 2001). In L. Verevi, epiphyton biomass was 10.58 mg cm^{-2} on *Equisetum* and 8.32 on reed (July 1999). In 10 Estonian lakes in summer 1998 and 1999, the highest registered biomass was 170.5 on reed, 121.8 on *Equisetum* and 32.2 mg cm^{-2} on *Carex* (Reunanen, 2000).

The difference in Chl *a* percentage per biomass unit between the years 2000 and 2001 was not statistically significant; still, the larger mean Chl *a* share in 2001 can be associated with almost two-fold lower radiation intensity in May and June 2000 (data from Estonian Meteorological Institute, Tõravere Weather Station). Epiphyton samples were taken from the beginning of July. The influence of light is of importance for the Chl *a* content in phytoplankton volume, and low nutrient concentration can give low chlorophyll content in the cells (Tolstoy, 1977). In water of L. Verevi, concentration of nutrients (N_{tot} and P_{tot}) was markedly lower in 2000 (mean value on the surface layer from the end of April to the end of August) than in 2001, and N:P mass ratio substantially higher (25.5 and 16.2 correspondingly in 2000 and 2001).

The seasonal dynamics of epiphyton on reed in 2000 was quite similar to that observed in lakes in Poland (Szczepanska, 1970), with a maximum in autumn. There were no obvious fluctuations during the vegetation period. Epiphyton dynamics on *Equisetum* differed substantially from that in L. Pääjärvi where peaks occurred in spring or in summer (Kairesalo, 1984). Epiphyton dynamics on *Typha* must be more closely related to weather conditions: the peak in August followed a very rainy July; also, the biomass of *Gloeotrichia pisum*, one of the dominants (lacking on *Phragmites* and *Equisetum*), was the highest in August. Some authors describe an epiphyton minimum in summer (Cattaneo, 1987; Müller, 1994). Comparison of seasonal dynamics is complicated due to differences in the location of lakes and the longer vegetation period in Middle Europe or America.

Szczepanska (1970) described the higher Chl *a* content on old reed stems compared with young stems; this difference persisted during the whole vegetation period. In L. Verevi, epiphyton on old stems showed higher values only at the beginning of vegetation period until the end of May; the very rich diatom flora on old stems did not increase: nor were any filaments added to epiphyton. The possible reason of the richness of diatoms, particularly *Achnanthes minutissima*, on old stems can be the spongy surface of stems and, possibly, better availability of Si. Pioneer algae on young stems are adnate *Cocconeis* cells followed by other diatoms. Real epiphytic filaments of *Oedogonium* occur in June, while metaphytic Zygnematales occur somewhat later.

Epiphyton on different plant species offered particular interest in our study. However, we described the differences but not the factors controlling the differences. Comparison of the nutrient content in host plants (some data of submerged plants are analysed by Mäemets and Freiberg, this issue) revealed obvious seasonal differences within one species as well as between different years and even between close growing sites on one and the same sampling day. The richest in nutrients was water moss (*Fontinalis*), although, its epiphyton was not very rich. Thus, we can associate epiphyton mainly with the structure of plants. Plants with fine branches have a large surface area for epiphytes to attach, sufficient space for metaphyton among the branchlets, and the richest epiphytic flora. Soft and smooth stems are more suitable for cyanobacteria and less suitable for diatoms.

Excretion of toxic substances that inhibit the growth of algae have been suggested to be responsible for the low epiphyton densities observed frequently on *Chara* spp. (Wium-Andersen et al., 1982; cit. Blindow, 1987). Blindow notes that her results are not in accordance with the possible effect of toxic substances, excreted by characeans, on epiphyton density. The low or moderate amount of epiphyton on *Chara* in L. Verevi may be caused by incrusting rather by allelopathy. Still, it is not excluded that the low abundance of cyanobacteria on most plant species is related to allelopathy.

Regarding ecotopes, epiphyton biomass is usually similar for emergent and floating-leaved plants; for submerged plants epiphyton biomass is on average 20 times higher (Kuz'ko, 2000b). In lakes, the amount of epiphytic algae increases significantly, when floating leaved plants are replaced by submerged plants (Cattaneo et al., 1998). In L. Verevi, epiphyton biomass on *Nuphar* leaves is several times lower than that of the stems, which is evidently related to shading effect of large leaves. Submerged plants are far richer in epiphyton and, taking into account their prevalence in the macrophyte flora, it is clear, that their epiphyton plays an important role in the production of the lake.

Acknowledgements

This work was supported by target financed project of Ministry of Education "The influence of the stratification to the biological matter circulation of the lakes" No. 0370208s98 and by grants of Estonian Science Foundation Nos. 3579 & 4835. The authors would like to thank our colleagues Helle Mäemets and Lilian Freiberg for the aid in collecting material, and Mrs. Ester Jaigma for revision of the English text of this paper.

References

Blindow, I., 1987. The composition and density of epiphyton on several species of submerged macrophytes – the neutral substrate hypothesis tested. Aquatic Botany 29: 157–168.

Burkholder, J. M. & R. G. Wetzel, 1989. Epiphytic microalgae on natural substrata in a hardwater lake: seasonal dynamics of community structure, biomass and ATP content. Archiv für Hydrobiologie 83: 1–56.

Burkholder, J. M. & R. G. Wetzel, 1990. Epiphytic alkaline phosphatase on natural and artificial plants in an oligotrophic lake: re-evaluation of the role of macrophytes as a phosphorus source for epiphytes. Limnology and Oceanography 35: 736–746.

Carrick, H. J. & A. D. Steinman, 2001. Variation in periphyton biomass and species composition in Lake Okeechobee,

Florida (USA): distribution of algal guilds along environmental gradients. Archiv für Hydrobiologie 152: 411–438.
Cattaneo, A., 1987. Periphyton in lakes of different trophy. Canadian Journal of Fisheries and Aquatic Sciences 44: 296–303.
Cattaneo, A., G. Galanti, S. Gentinetta & S. Romo, 1998. Epiphytic algae and microinvertebrates on submerged and floating-leaved macrophytes in an Italian lake. Freshwater Biology 39: 725–740.
Gross, E. M., D. Erhard & E. Leu 2002. Allelopathic activity of submersed macrophytes. Intern. Conference on Limnology of Shallow Lakes. Balatonfüred, Hungary. Abstracts, p. 70.
Guidelines for the Baltic Monitoring Programme for the Third Stage 1984. The Baltic Marine Biologists. Publ. 1, 2nd edn.
Hart, E. A. & J. R. Lovvorn, 2000. Vegetation dynamics and primary production in saline, lacustrine wetlands of a Rocky Mountain basin. Aquatic Botany 66: 21–39.
Havens, K. E., T. L. East, R. H. Meeker, W. P. Davis & A. D. Steinman, 1996. Phytoplankton and periphyton responses *in situ* experimental nutrient enrichment in a shallow subtropical lake. Journal of Plankton Research 18: 551–566.
Hawes, I. & A. M. Schwarz, 1996. Epiphytes from a deep-water characean meadow in an oligotrophic New Zealand lake: species composition, biomass and photosynthesis. Freshwater Biology 36: 297–314.
Hepinstall, J. A. & R. L. Fuller, 1994. Periphyton reactions to different light and nutrient levels and the response of bacteria to these manipulations. Archiv für Hydrobiologie 131: 161–173.
Hickman, M. & D. M. Clarer, 1973. Methods for measuring the primary productivity and standing crops of an epiphyte algal community attached to *Scirpus validus* Vahl. Internationale Revue der gesamten Hydrobiologie 58: 893–901.
Iványi, E., D. Erhard, E. M. Gross, 2002. Allelopathic activity of *Ceratophyllum demersum* and *Najas marina*. Intern. Conference on Limnology of Shallow Lakes. Balatonfüred, Hungary. Abstracts, 88 p.
Jeffrey, S. W. & G. F. Humphrey, 1975. New spectrophotometric equations for determining chlorophylls a, b, c_1 and c_2 in higher plants, algae and natural phytoplankton. Biochem. Physiol. Pflanzen (BPP) 167: 191–194.
Jernakoff, P. & J. Nielsen, 1997. The relative importance of amphipod and gastropod grazers in *Posidonia sinuosa* meadows. Aquatic Botany 56: 183–202.
Kairesalo, T., 1984. The seasonal succession of epiphytic communities within an *Equisetum fluviatile* L. stand in Lake Pääjärvi, Southern Finland. Internationale Revue der gesamten Hydrobiologie 4: 475–505.
Kairesalo, T., A. Uusi-Rauva, 1983. Phosphorus release by an emergent macrophyte: significance to epiphyton. Proc. Int. Symp. Aquat. Macrophytes, Nijmegen, 18–23 September, 1983, pp. 101–110.
Kassim, T. I & H. A. Al-Saadi, 1995. Seasonal variation of epiphytic algae in a marsh area (southern Iraq). Acta Hydrobiologica 37: 153–161.
Kuz'ko, O. A., 2000a. Pervichnaya produktsiya epifitnykh gruppirovok vodoroslej v kanale Dnepr-Donbass. Vth All-Russian Conference on Aquatic Plants "Hydrobotany 2000". Abstracts. Borok: 46–47 [Primary production of algal epiphyton groups in Dnepr-Donbass Channel (in Russian)].
Kuz'ko, O. A., 2000b. Epifitnye gruppirovki vodoroslej pojmennykh ozer ust'evoj oblasti Dnepra. Vth All-Russian Conference on Aquatic Plants "Hydrobotany 2000". Abstracts. Borok, 47–48 [Algal epiphyton groups of floodplain lakes of Dnepr estuary area (in Russian)].
Marvan, P., J. Komárek, H. Ettl & J. Komárková, 1978. Dynamics of algal communities. Ecological Studies 28: 315–336.
Meteleva, N. Y., 2000. Soderzhanie khlorofilla v perifitone vodokhranilishch Verkhnej Volgi. Vth All-Russian Conference on Aquatic Plants "Hydrobotany 2000". Abstracts. Borok, 54–55 [Chlorophyll content in periphyton of Upper Volga Reservoirs (in Russian)].
Mäemets, H. & L. Freiberg, 2005. Long- and short-term changes of the macrophyte vegetation in strongly stratified hypertrophic Lake Verevi. Hydrobiologia 547: 175–184.
Müller, U., 1994. Seasonal development of epiphytic algae on *Phragmites australis* (Cav.) Trin. ex Sten. in a eutrophic lake. Archiv für Hydrobiologie 129: 273–292.
Müller, U., 1995. Vertical zonation and production rates of epiphytic algae on *Phragmites australis*. Freshwater Biology 34: 69–80.
Ott, I., T. Kõiv, P. Nõges, A. Kisand, A. Järvalt & E. Kirt, 2005. General description of partly meromictic hypertrophic Lake Verevi, its ecological status, changes during the past eight decades and restoration problems. Hydrobiologia 547: 1–20.
Paterson, M. J., D. L. Findlay, A. G. Salki, L. L. Hendzel & R. H. Hesslein, 2002. The effects of Daphnia on nutrient stoichiometry and filamentous cyanobacteria: a mesocosm experiment in a eutrophic lake. Freshwater Biology 47: 1217–1233.
Putz, R., 1997. Periphyton communities in Amazonian black- and whitewater habitats: community structure, biomass and productivity. Aquatic Sciences 59: 74–93.
Reunanen, M., 2000. Perifüüton mõnedes Eesti teravalt kihistunud järvedes. Bakalaureusetöö. Manuscript in Tartu University. Tartu, 52 pp. [Periphyton in some steeply stratified Estonian lakes (in Estonian, English summary)].
Sand-Jensen, K., 1990. Epiphyte shading: its role in resulting depth distribution of submerged aquatic macrophytes. Folia Geobotanica Phytotaxonomica 25: 315–320.
Shevchenko, T. F., 1994. Fitoperiphyton Kievskogo i Kakhovskogo vodokhranilishch. Gidrobiol. Zh. 30(4): 13–21 [Phytoperiphyton of the Kiev and the Kakhovka Water Reservoirs (in Russian)].
Sigareva, L. E. & V. G. Devyatkin, 1987. Soderzhanie fotosinteticheskikh pigmentov v perifitone Rybinskogo vodokhranilishcha. In Monakov, A.V. (ed.), Fauna i biologiya presnovodnykh organizmov. Trudy Instituta Vnutrennykh vodoemov 54(57): 3–18 [Content of photosynthetic pigments in periphyton of the Rybinsk reservoir (in Russian)].
Szczepanska, W., 1970. Periphyton of several lakes of the Mazurian lakeland. Polish Archive of Hydrobiology 17: 397–418.
Tolstoy, A. 1977. Chlorophyll *a* as a measure of phytoplankton biomass. Acta Universitatis Uppsaliensis (Uppsala) 416: 30.
Vymazal, J., C. B. Craft & C. J. Richardson, 1994. Periphyton response to nitrogen and phosphorus additions in Florida Everglades. Archiv für Hydrobiologie 103: 75–97.

Hydrobiologia (2005) 547:151–162
I. Ott & T. Kõiv (eds), Lake Verevi, Estonia – A Highly Stratified Hypertrophic Lake
DOI 10.1007/s10750-005-4156-8

Vertical distribution of zooplankton in a strongly stratified hypertrophic lake

Kaidi Kübar[1,2,*], Helen Agasild[1], Taavi Virro[2] & Ingmar Ott[1]
[1]*Estonian Agricultural University, Institute of Zoology and Botany, Võrtsjärv Limnological Station, Rannu 61101, Estonia, Tartu County*
[2]*University of Tartu, Institute of Zoology and Hydrobiology, Vanemuise 46, Estonia, Tartu*
*(*Author for correspondence: E-mail: kotkas@kotkas.ee)*

Key words: zooplankton, vertical distribution, seasonal dynamics, stratification

Abstract

The vertical and temporal distribution of metazooplankton in the small hypertrophic, strongly stratified, temperate Lake Verevi (Estonia) was studied during 1998–2001. The zooplankton of Lake Verevi is characteristic of hypertrophic lakes, with a small number of dominant species, rotifers being the main ones, and juveniles prevailing among copepods. In 1999–2001, the average abundance of metazooplankton in the lake was 1570×10^3 ind m^{-3}; in the epilimnion 2320×10^3 ind m^{-3}, in the metalimnion 2178×10^3 ind m^{-3}, and in the hypolimnion 237×10^3 ind m^{-3}. The average biomass of metazooplankton was 1.75 g m^{-3}; in the epi-, meta- and hypolimnion, accordingly, 2.16, 2.85 and 0.26 g m^{-3}. The highest abundances – $19{,}136 \times 10^3$ ind m^{-3} and $12{,}008 \times 10^3$ ind m^{-3} – were registered in the lower half of the metalimnion in 24 May and 5 June 2001, respectively. Rotifer *Keratella cochlearis* f. *typica* (Gosse, 1851) was the dominating species in abundance. In biomass, *Asplanchna priodonta* Gosse, 1850, among the rotifers, and *Eudiaptomus graciloides* (Lilljeborg, 1888), among the copepods, dominated. According to the data from 2000–2001, the abundance and biomass of both copepods and rotifers were highest in spring. Zooplankton was scarce in the hypolimnion, and no peaks were observed there. During the summers of 1998 and 1999, when thermal stratification was particularly strong, zooplankton was the most abundant in the upper half of the metalimnion, and a distinct peak of biomass occurred in the second fourth of the metalimnion. Probably, the main factors affecting the vertical distribution of zooplankton in L. Verevi are fish, *Chaoborus* larvae, and chemocline, while food, like phytoplankton, composition and abundance may affect more the seasonal development of zooplankton.

Introduction

Processes and interactions, including the role of zooplankton in the whole ecosystem, in the shallow, well-mixed lakes, have been intensively investigated in the last decades. A growing attention has been paid also to those in stratified lakes. Matter circulation is particularly complicated in the stratified lakes, and there occur specific biotic communities. Many deep but formerly non-stratified lakes have become stratified due to eutrophication. For predicting of the future changes in their ecosystem, it is extremely important to understand the processes taking place in the metalimnion of the lakes. In the Estonian lakes, increase of oxygen-deficient zone and even more distinct stratification were observed in the last decades (Ott, 1996; Ott et al., 1997); therefore, the processes in the metalimnion acquire steadily increasing importance. The metalimnion can be the most nutrient- or food-rich layer in some stratified lakes (Wetzel, 1983). In Lake Verevi, the summer stratification is formed quickly due to the weak spring water circulation, even lacking completely in some years; therefore, the epilimnion remains poor in nutrients. As a result of the temperature gradient, the water density is different in the different layers; this impedes sedimentation and concentrates organic matter in the metalimnion. Nutrients accrue to the metalimnion also from the hypolimnion by diffusion.

Distribution of zooplankton in the water column of a lake is determined mostly by light, temperature, the concentration and availability of food (Miracle, 1977; Andronikova, 1989; Carpenter & Kitchell, 1993; Kizito & Nauwerck, 1995), also by diurnal vertical migration for avoiding predation, and seasonal vertical distribution induced by life cycles. Fish are the determining predators for zooplankton (Lammens, 1990; Van Donk et al., 1990; Herzig, 1994; Jeppesen et al., 1996; Stransfield et al., 1997), while the meroplanktic *Chaoborus* larvae have a remarkable role, too (Leibold, 1990). The herbivorous zooplankton is located mostly in the epilimnion at the maximum of phytoplankton. In stratified lakes, the vertical distribution of herbivorous zooplankton could be different, depending mostly on the more structured distribution of phytoplankton. In deeper layers, the food of zooplankters consists mainly of detritus, bacteria, and heterotrophic protozoans (Kizito & Nauwerck, 1995).

The very first data on the zooplankton of Lake Verevi were presented in the manuscripts of Prof. H. Riikoja based on the samples collected from July 1928 to October 1929. The next exhaustive investigations on L. Verevi were carried out in 1985 (May–October), 1988–1989 (June–April), 1991 (January–October), and 1993 (March–December). The results on zooplankton have been presented by Timm & Mäemets (1991) and Kübar (1994). More than 50 taxa of zooplankton have been identified in the pelagial of L. Verevi, including over 26 species of rotifers, 15 cladocerans, and 9 copepods (Timm & Mäemets, 1991; Kübar, 1994; Olt, 2001). *Chaoborus* larvae also occur in the plankton. *Keratella cochlearis* (Gosse, 1851), *Polyarthra* sp. and nauplii dominate in abundance, while *Asplanchna* sp., *Eudiaptomus graciloides* (Lilljeborg, 1888) and the cyclopoid copepods dominate in biomass.

The vertical distribution of zooplankton is still not much investigated in the Estonian lakes. As the zooplankton, especially crustaceans, occupy a functionally central position in the matter circulation of a lake (Viitasalo, 1994; Lampert, 1997), it is important to know its distribution in the water column.

The aim of the present work is to describe the vertical and temporal distribution of zooplankton estimating the role of metazooplankton in the hypertrophic stratified Lake Verevi.

Materials and methods

Lake Verevi is a small (12.6 ha, maximum depth 11 m, mean depth 3.6 m; for more details see Ott et al., 2005a), hard water, hypertrophic lake located in the town of Elva, South-Eastern Estonia. It is characterized by weak water exchange (0.63 times per year), strong stratification, and anoxic hypolimnion.

Samples were collected during four years from 1998 to 2001. In 1998 and 1999, L. Verevi was sampled in summer from 8 layers (2 in epi-, 4 in meta- and 2 in hypolimnion). In 2000, the lake was sampled during the whole vegetation period, more frequently in spring and autumn, from 3 layers (epi-, meta- and hypolimnion), except 23 October, when only epi- and hypolimnion was studied. In 2001, the lake was investigated during spring from 3 layers (epi- meta- and hypolimnion) until 26 April, and since 30 April from 4 layers (epi-, 2 meta- and hypolimnion layers). On 29 March, 2 layers (epi- and hypolimnion) were sampled. The number of samples collected from the water column was flexible following the formation of lake stratification and degradation.

Altogether 116 samples of metazooplankton were collected from the lake in the centre of the larger southern part of the lake, at its deepest point. The depth of the layers was chosen taking into account the vertical profiles of temperature and oxygen – the change of temperature by 1.5 °C m^{-1} was considered as the upper boundary of the metalimnion (Nõges & Nõges, 1998). Van Dorn sampler (volume 2 l) was used. 10 l of water from every layer were filtered through 48 μm plankton net; as an exception, 85 μm net was used in 1998. It is generally accepted that the use of plankton nets with mesh sizes larger than 50 μm leads to under-estimation of rotifer numbers

(Bottrell et al., 1976; Ruttner-Kolisko, 1977a; Virro, 1989). This induces undetermined error in the estimates of the abundance and biomass of rotifers, as well as total zooplankton in 1998 data. The real values were probably higher. Samples were preserved with Lugol's solution.

The metazoan zooplankters were identified and counted using a stereomicroscope (MBS-9) with 32–56 × magnification and Bogorov's counting chamber. The individual wet weights of rotifers were estimated from average lengths according to Ruttner-Kolisko (1977b). The lengths of crustaceans were converted to wet weight according to Studenikina & Cherepakhina (1969) for nauplii, and Balushkina & Winberg (1979) for other groups.

Results

Abundance, biomass and dominating zooplankters

In 1999–2001, the average abundance of metazooplankton for the whole water column was 1570×10^3 ind m^{-3}; the abundance in the epilimnion was 2320×10^3 ind m^{-3}, in the metalimnion 2178×10^3 ind m^{-3}, and in the hypolimnion 237×10^3 ind m^{-3}. The average biomass for the whole water column was 1.75 g m^{-3}; in the epi-, meta-, and hypolimnion, 2.16, 2.85 and 0.26 g m^{-3}, accordingly. Rotifers dominated in the epi- and hypolimnion, both in abundance and biomass, while in the metalimnion, rotifers dominated in abundance, and copepods in biomass (Fig. 1). Rotifer *Keratella cochlearis* f. *typica* (Gosse, 1851) was the dominating species in abundance, giving on an average 59, 55 and 26% of the total metazooplankton abundance in the epi-, meta- and hypolimnion, respectively. In the biomass, rotifer *Asplanchna priodonta* Gosse, 1850 (35, 25 and 14% of the metazooplankton biomass in the epi-, meta- and hypolimnion, respectively), and copepod *Eudiaptomus graciloides* (15, 15 and 23%) dominated.

In the summers of 1998 and 1999, the abundance of metazooplankton was the highest in the epilimnion and in the upper layers of the metalimnion, without any sharp peak. In the biomass, a distinct peak occurred in the second fourth of the metalimnion (Fig. 2).

The seasonal dynamics of metazooplankton was investigated in 2000–2001. Abundance and biomass were the highest in spring (Fig. 3) in the lower half of the metalimnion; the numbers reached $19{,}136 \times 10^3$ ind m^{-3} (24 May 2001) and $12{,}008 \times 10^3$ ind m^{-3} (5 June 2001). *Keratella cochlearis* f. *typica* was dominating in both dates, making 72 and 88% of the total abundance of zooplankton, respectively. The highest biomass (8.47 g m^{-3}) occurred in 3 May 2000 in the lower

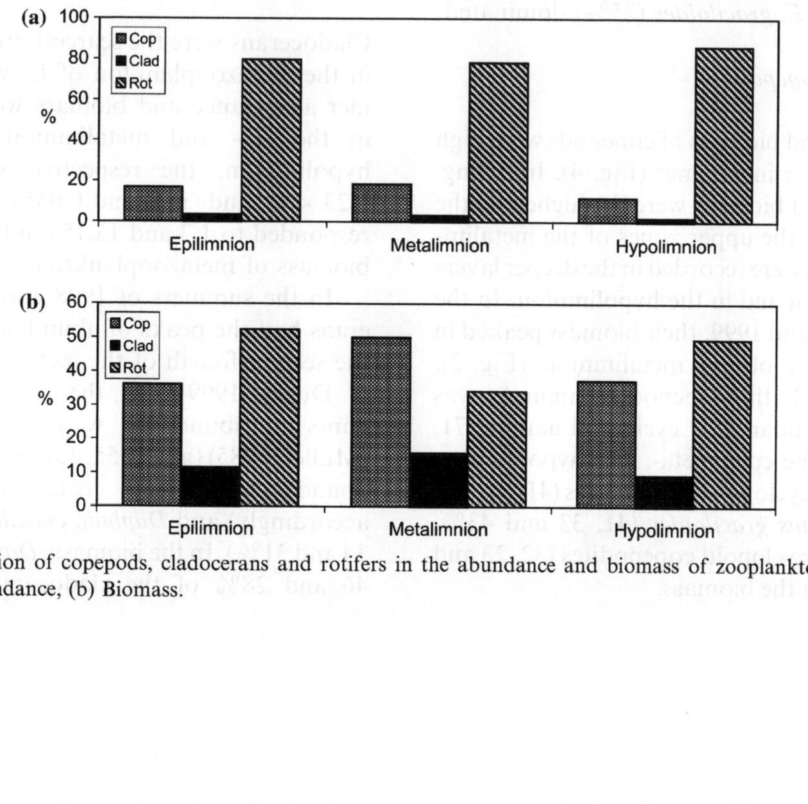

Figure 1. The proportion of copepods, cladocerans and rotifers in the abundance and biomass of zooplankton in epi-, meta- and hypolimnion. (a) Abundance, (b) Biomass.

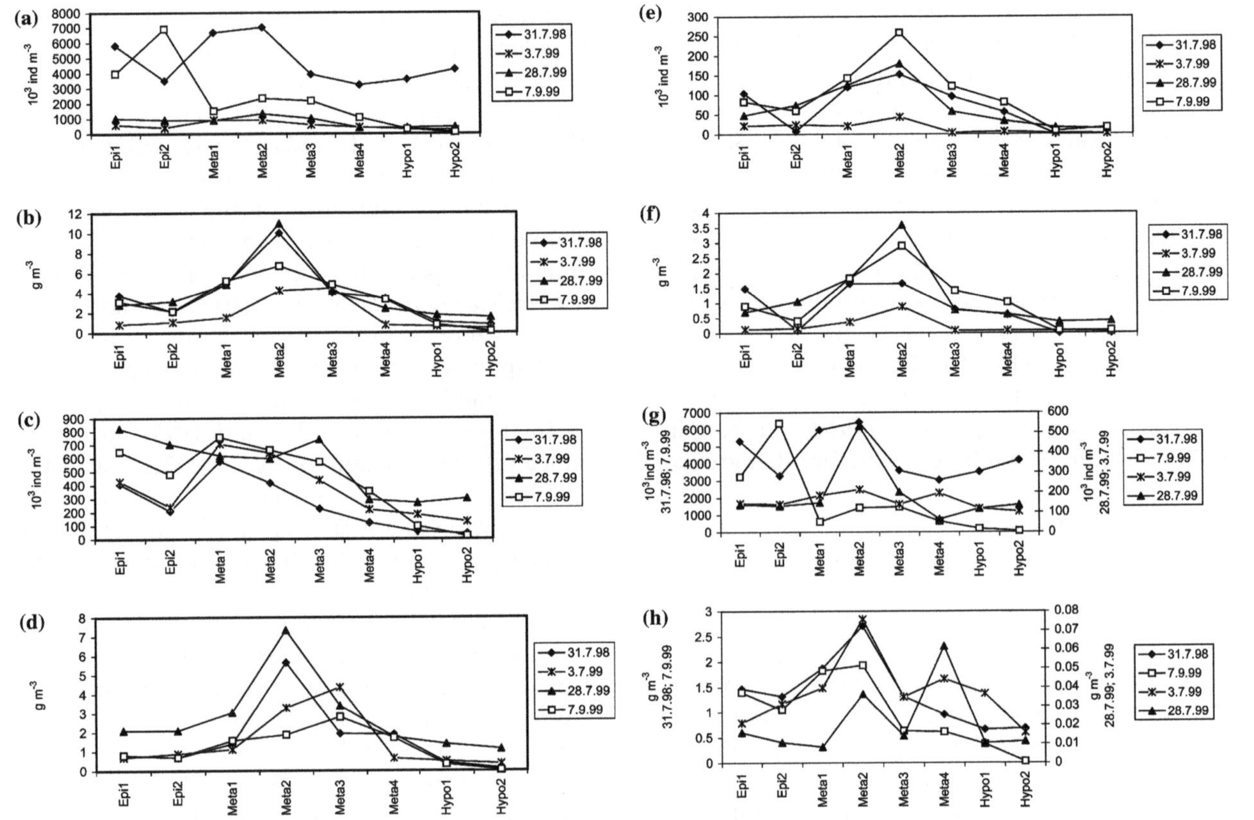

Figure 2. Abundance and biomass of zooplankton in the different layers of the epi-, meta- and hypolimnion during the maximum summer stratification in the years 1998 and 1999. (a) Abundance, (b) Biomass, (c) Abundance of copepods, (d) Biomass of copepods, (e) Abundance of cladocerans, (f) Biomass of cladocerans, (g) Abundance of rotifers, (h) Biomass of rotifers.

half of the metalimnion, where *Asplanchna priodonta* (32%) and *E. graciloides* (25%) dominated.

Distribution of Copepoda

The abundance and biomass of copepods were high in spring and lower in summer (Fig. 4). In spring, the abundance and biomass were the highest in the epilimnion and in the upper zones of the metalimnion; lower values were recorded in the deeper layers of the metalimnion and in the hypolimnion. In the summers of 1998 and 1999, their biomass peaked in the second fourth of the metalimnion (Fig. 2). During 1999–2001, the copepod community was dominated in abundance by cyclopoid nauplii (71, 51, and 69% in the epi-, meta-, and hypolimnion, accordingly) and cyclopoid copepodites (41, 33, and 43%). *Eudiaptomus graciloides* (41, 33 and 43%, accordingly) and cyclopoid copepodites (32, 23 and 30%) prevailed in the biomass.

Distribution of Cladocera

Cladocerans were the scarcest group by abundance in the metazooplankton of L. Verevi. Their summer abundance and biomass were relatively high in the epi- and metalimnion (Fig. 5). In the hypolimnion, the respective values were only 3.23×10^3 ind m^{-3} and 0.0351 g m^{-3}, which corresponded to 1.3 and 13.1% in the abundance and biomass of metazooplankton.

In the summers of 1998 and 1999, the cladocerans had the peaks of abundance and biomass in the second fourth of the metalimnion (Fig. 2).

During 1999–2001, the main cladoceran dominants in abundance were *Bosmina longirostris* (Müller, 1785) (giving 55, 43 and 51% of cladoceran abundance in epi-, meta- and hypolimnion, accordingly) and *Daphnia cucullata* Sars, 1862 (27, 34 and 31%). In the biomass, *Daphnia cucullata* (33, 46 and 28% of the cladoceran biomass in the

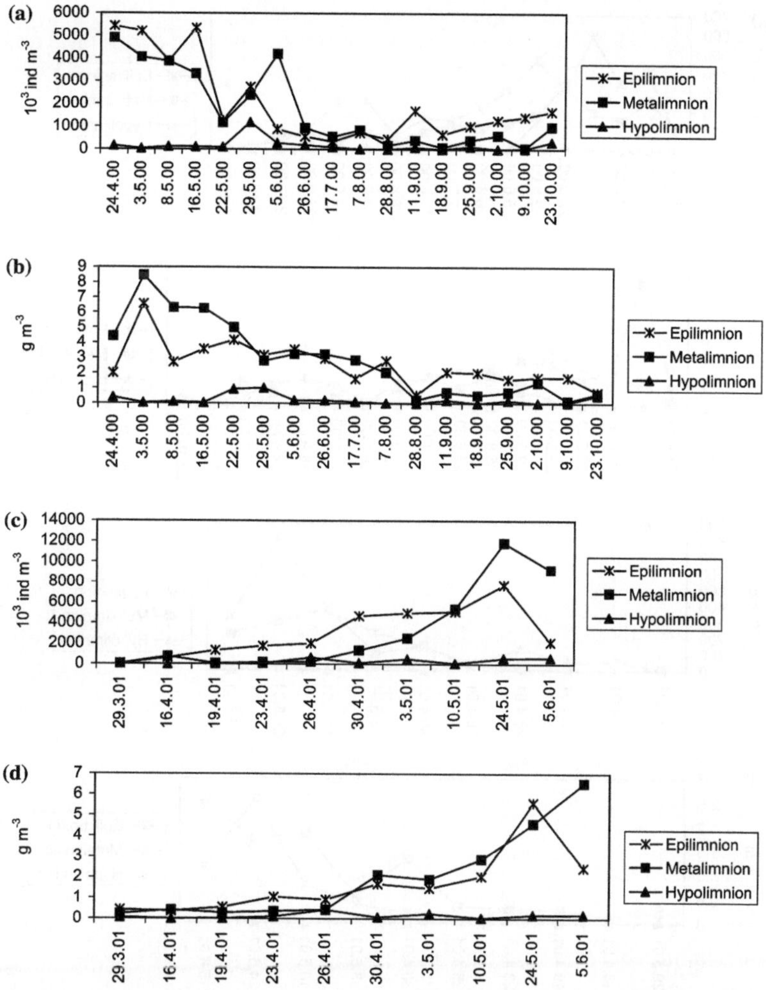

Figure 3. Abundance and biomass of zooplankton in epi- (samples from two layers), meta- (four layers) and hypolimnion (two layers) in 2000 and 2001. (a) Abundance in 2000, (b) Biomass in 2000, (c) Abundance in 2001, (d) Biomass in 2001.

epi-, meta- and hypolimnion, accordingly), *Diaphanosoma brachyurum* (Liévin, 1848) (29, 28 and 39%) and *Bosmina longirostris* (12, 21 and 27%) dominated.

Distribution of Rotifera

The average abundance and biomass of rotifers were relatively high in the epi- and metalimnion in spring, decreased in summer, and increased slightly in autumn (Fig. 6). There was no clear pattern in the spatial distribution of the rotifer abundance in the summers of 1998 and 1999; however, the maximum was always either in the second or in the third layer of the metalimnion. In two cases out of four, their abundance unexpectedly increased in direction from the fourth layer of the metalimnion to the bottom layer of the hypolimnion. A peak of the rotifer biomass was observed in the second fourth of the metalimnion (Fig. 2). *Keratella cochlearis* f. *typica* was the most abundant species in 1999–2001 (67, 62 and 31% of the rotifer abundance in the epi-, meta- and hypolimnion, respectively), while *Asplanchna priodonta* dominated in the biomass (70, 64 and 43%, accordingly).

Distribution of Chaoborus larvae

In early spring, *Chaoborus* larvae were not encountered in the plankton samples. They appeared in the hypolimnion since 16 May 2000 and 26 April 2001. In the metalimnion, *Chaoborus* larvae were found every sampling time since 7

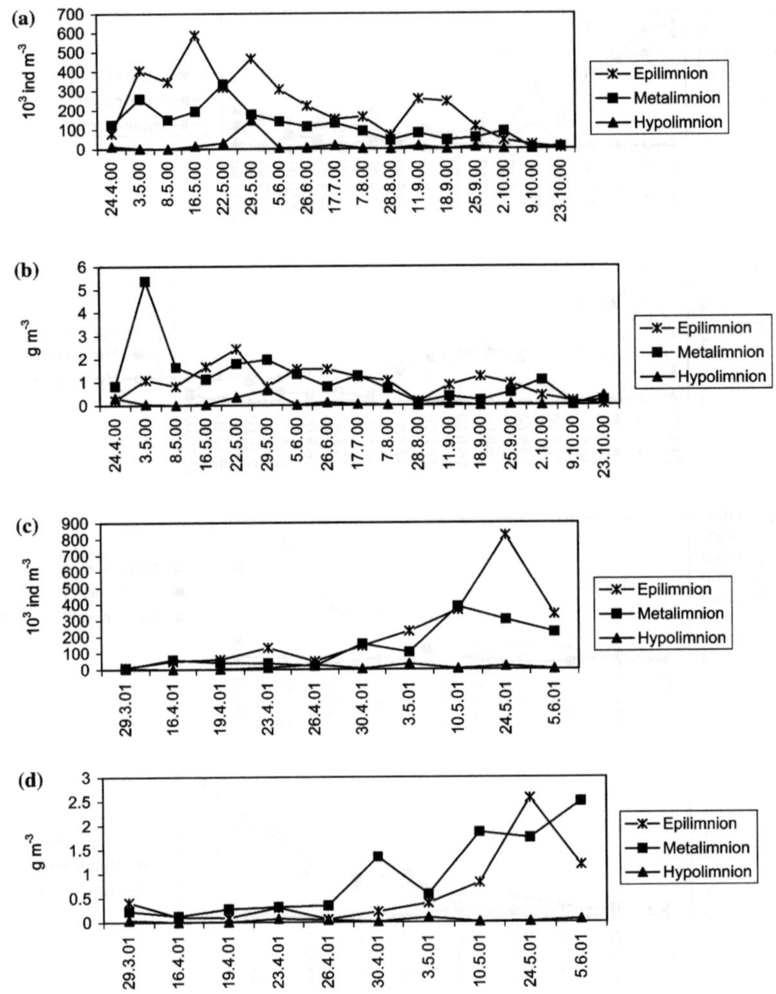

Figure 4. Abundance and biomass of copepods in epi-, meta- and hypolimnion in 2000 and 2001. (a) Abundance in 2000, (b) Biomass in 2000, (c) Abundance in 2001, (d) Biomass in 2001.

August 2000. Their highest abundance in the metalimnion was on 9 October 2000 (2600 ind m^{-3}) and in the hypolimnion on 25 September 2000 (1700 ind m^{-3}).

Relationships with phytoplankton and bacteria

Spearman correlation analysis was used to estimate the relationships between metazooplankton and its food sources: phytoplankton and bacteria. Positive correlation was found between phytoplankton biomass and zooplankton abundance and biomass ($r=0.24$, $p<0.05$ and $r = 0.47$, $p < 0.0001$, respectively). The additional analysis by separate layers (i.e. epi-, meta- and hypolimnion) showed significant correlation between phytoplankton and zooplankton biomass only in metalimnion ($r = 0.46$, $p < 0.01$). In the whole water column, negative correlation between total number of bacteria and zooplankton abundance and biomass ($r = -0.56$, $p<0.0001$ and $r=-0.63$, $p<0.0001$, respectively) was found.

Discussion

The zooplankton features of Lake Verevi are typical of hypertrophic lakes. At high levels of trophy, the ecosystem generally becomes poor (Kira, 1993). Small number of dominants, high share of rotifers in zooplankton abundance, and the prevailing of rotifers even in biomass in the

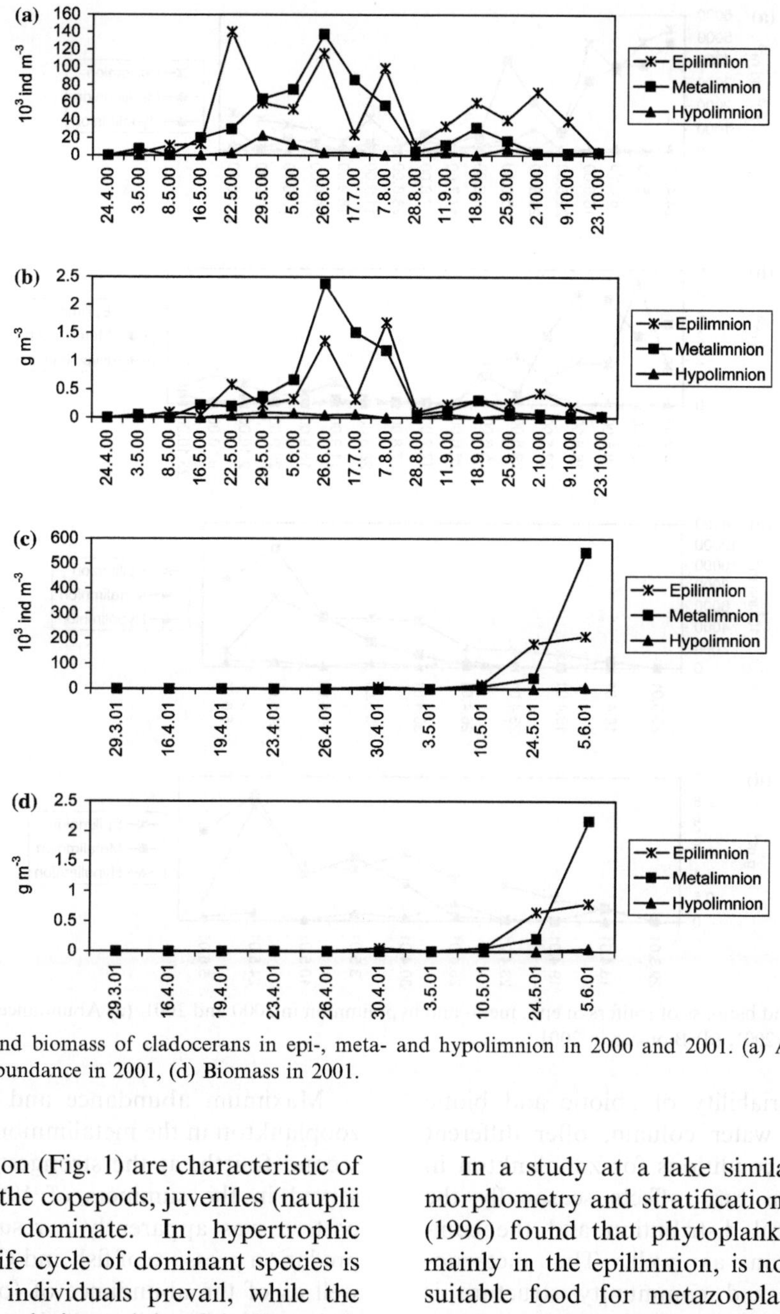

Figure 5. Abundance and biomass of cladocerans in epi-, meta- and hypolimnion in 2000 and 2001. (a) Abundance in 2000, (b) Biomass in 2000, (c) Abundance in 2001, (d) Biomass in 2001.

epi- and hypolimnion (Fig. 1) are characteristic of L. Verevi. Among the copepods, juveniles (nauplii and copepodites) dominate. In hypertrophic water bodies, the life cycle of dominant species is short, and young individuals prevail, while the adults are strongly decimated by fish (Ponyi & Zankai, 1982; Haberman, 1998; Nicholls & Tudorancea, 2001).

Metazooplankters inhabit mainly the epilimnion and metalimnion, offering them suitable living conditions. The anoxic hypolimnion does not qualify as a habitat for many zooplankters. This unsuitable zone expands in the process of eutrophication.

In a study at a lake similar to L. Verevi in morphometry and stratification, Williamson et al. (1996) found that phytoplankton, concentrating mainly in the epilimnion, is not always the most suitable food for metazooplankton. Depending on the concentrations and ratios of nutrients, different groups of phytoplankton can dominate (Tilman, 1977, 1982; Sommer, 1989; Bulgakov & Levich, 1999). A large proportion of algae, unsuitable as food for zooplankton, is characteristic of hypertrophic water bodies (Sommer, 1989).

Phytoplankton cluster analysis (Kangro et al., 2005) showed variability in dominating groups both in space and time. Microhabitats,

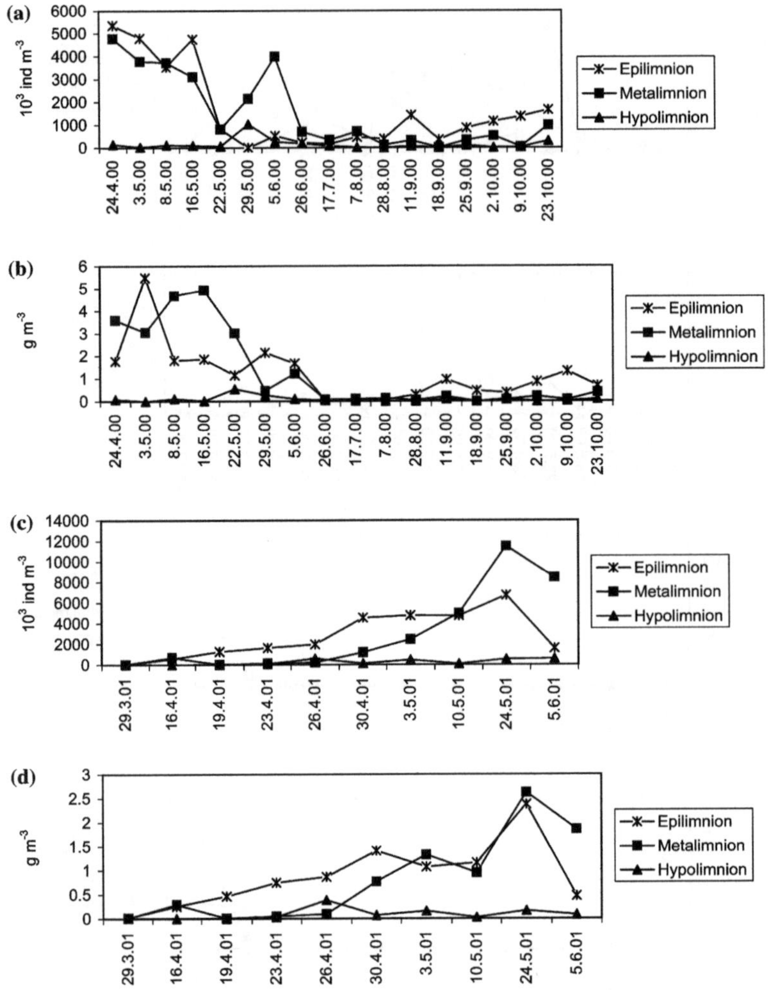

Figure 6. Abundance and biomass of rotifers in epi-, meta- and hypolimnion in 2000 and 2001. (a) Abundance in 2000, (b) Biomass in 2000, (c) Abundance in 2001, (d) Biomass in 2001.

created by the variability of abiotic and biotic factors along the water column, offer different living and feeding conditions for zooplankton in L. Verevi. This in turn affects seasonally the zooplankton vertical distribution and the community composition as well. The metazooplankton survival and community structure is largely influenced by food supply from algae and bacteria (Rothhaupt, 1990; Hofmann & Höfle, 1993; Cordova et al., 2001). Stronger relationship between phytoplankton and zooplankton occurs in metalimnion in L. Verevi. The negative correlation with bacteria also indicates the active use of that food source by zooplankton in L. Verevi (see also Tammert et al., 2005).

Maximum abundance and biomass of metazooplankton in the metalimnion, particularly in its second fourth, at the strongest gradient, was observed in the summers of 1998 and 1999. This pattern was apparently a result of avoiding the predation pressure of fish and *Chaoborus* larvae, as well as of the abundance of food. Converging of metazooplankters in a distinct zone of the metalimnion has been described earlier for rotifers (Miracle & Armengol-Diaz, 1995), as well as for cladocerans (Williamson et al., 1996). Anoxic conditions were established in the depth of 2 m in 1998 (the water level was lowered by 0.7 m during the reconstruction of shores for swimming pool) and 3–3.5 m in 1999 (Fig. 7). Exactly, on that layer the zooplankton maxima were located. In

2000, when the water level was restored, anoxia occurred in the depths of 5–5.5 m, and zooplankton biomass was not concentrated in very narrow layer (Fig. 4). This indicates the synergetic influence of light penetration and water density. Probably, in 1998 and 1999, zooplankters tried to avoid the zone of intensive illumination (i.e., increased exposure to predators), but dense water and hypoxic conditions in the depth of 2–3.5 m blocked the migration.

Bürgi et al. (1999) have also shown that daphnids avoid the layers with high densities of non-grazeable interfering algae. Possible location of the maximum abundance and biomass is evidently species-specific; e.g., some rotifer species prefer thermocline, while others follow the shift of oxycline (Armengol-Diaz et al., 1993). It has been stated that stratification enables the co-existing species to find the most suitable habitat (=water layer) (Arvola et al., 1987; Armengol et al., 1998). It is possible that stratification with moderate gradients in spring creates more niches for biota than the summer stratification, where the temperature and oxygen gradients can be too harsh for zooplankton, seeking food and escaping predators.

High abundance of the *Chaoborus* larvae in pelagic zone of a lake is an important factor regulating the vertical distribution of zooplankton (Leibold, 1990). These larvae occur first of all in stratified water bodies. Their distribution is determined rather by stratification than by the

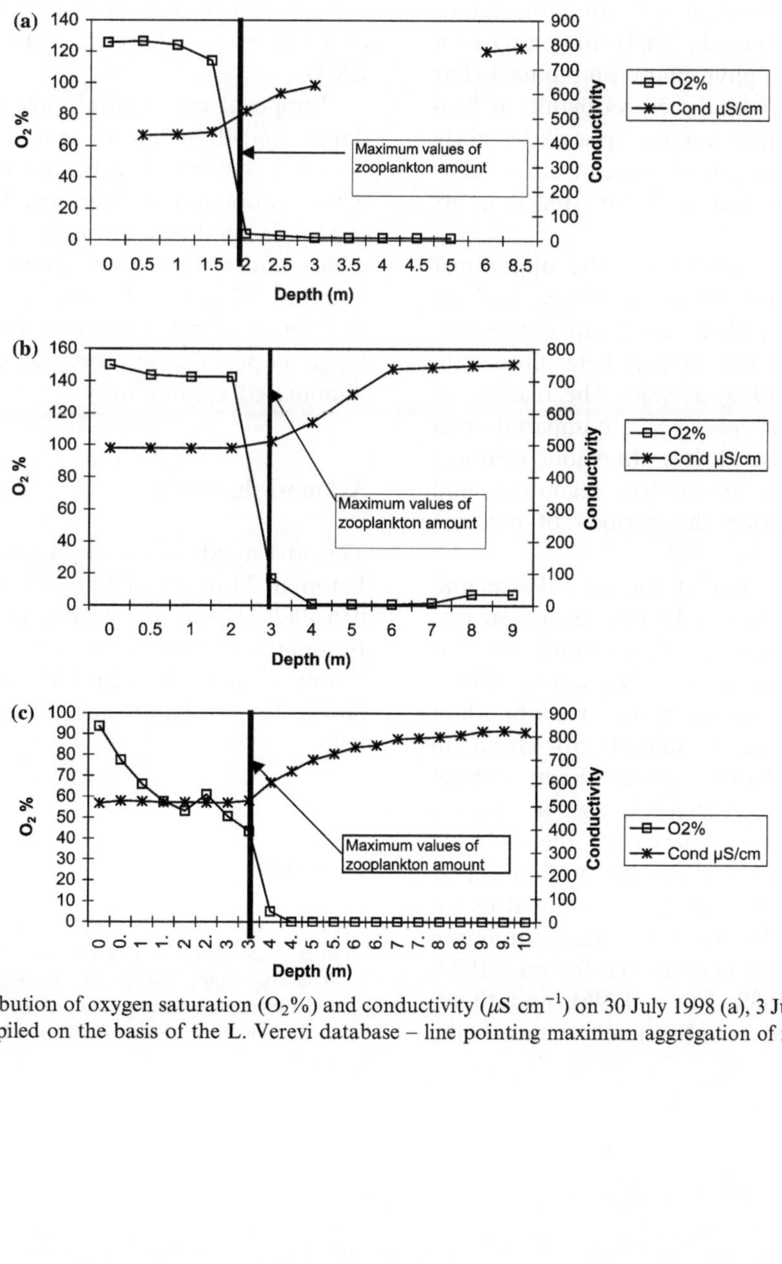

Figure 7. Vertical distribution of oxygen saturation (O_2%) and conductivity (μS cm^{-1}) on 30 July 1998 (a), 3 July 1999 (b) and 28 July 1999 (c). Figure is compiled on the basis of the L. Verevi database – line pointing maximum aggregation of zooplankton.

level of trophy (Liljendahl-Nurminen et al., 2002). Scarcity of zooplankton at daytime in the hypolimnion, and also in the metalimnion in autumn, is explainable, besides other factors, by the presence of *Chaoborus* in L. Verevi.

The high abundance and biomass of metazooplankton in the epi- and metalimnion in spring arose from suitable conditions. There is no fish pressure in spring, as it was established in 2001. The only fish species found in the pelagic zone was the 10 cm roach (Järvalt et al., 2005). Moreover, there is enough small-sized phytoplankton as food (Kangro et al., 2005). Therefore, total phytoplankton biomass is likely to be more affected by grazing in spring period than in summer, when larger non-grazeable forms start to dominate (Vanni & Temte, 1990). Clear water was observed in the upper part of the water column in the end of June in 2000 (Kangro et al., 2005). In comparison with the maximum of phytoplankton biomass (Ott et al., 2005b), the highest metazooplankton biomass in the epi- and metalimnion appeared slightly later. Such time lag is typical: the consumer needs some time for reproduction before reaching its maximum biomass.

Rotifer *Keratella cochlearis* – the uppermost dominant in zooplankton of L. Verevi had its highest abundance in May–June; similar seasonal dynamics has been reported elsewhere (Godenau, 1978; Bosselmann, 1979) as well. The reason for the dominance of *K. cochlearis*, compared with other rotifers, can be its low threshold resource level, the greatest ability to store resources, and ration their use during the periods of extreme resource scarcity (Kirk, 2002).

The spring maximum of the abundance and biomass of rotifers is well-known. Some species have a similar maximum in autumn and very few species in summer (Kizito & Nauwerck, 1995). Affinity of rotifers to cooler water, and abundant presence of small algae suitable for food in spring are the probable reasons. Even a brief circulation of water gives advantage to rotifers capable to adapt most quickly to the changed environmental conditions (Kizito & Nauwerck, 1995). Both *Keratella cochlearis* dominating in abundance and *Asplanchna priodonta* dominating in biomass are eurythermic (Hofmann, 1977; Berzins & Pejler, 1989; Virro, 1996), and therefore, they prevailed in all three zones of the water column. Increase of the abundance of cladocerans also plays certain role in the summer decrease of rotifers in L. Verevi, especially in the epi- and metalimnion, as they compete with rotifers for food (Fradkin, 1995). Among cladocerans, the small-sized ubiquists *Bosmina longirostris*, *Daphnia cucullata* and *Diaphanosoma brachyurum* prevailed in L. Verevi.

The abundance and biomass of metazooplankton decreased in summer and autumn in L. Verevi. This can be explained by the life cycles of the zooplankters, the changes in the living conditions of zooplankton, especially of the prevailing rotifers, and by pressure of fish. It is known that many rotifer species prefer lower water temperatures (Kizito & Nauwerck, 1995). In L. Verevi, the fish predation was substantial in summer, when young roaches and perches were abundant in the pelagial of the lake (Järvalt et al., 2005; Nõges & Kangro, 2005).

Temporal and spatial variations of the abundance and biomass of copepods in L. Verevi generally coincided with the distribution of the whole zooplankton. The smaller in size juvenile copepods gathered mostly in the epilimnion, while adults were more numerous in the metalimnion. This can be explained mainly by the avoidance of fish predation (Lampert, 1992), as bigger copepods are more vulnerable in the better illuminated epilimnion.

Acknowledgements

This investigation was supported by core grants of Estonian Ministry of Education Nos. 0370208s98, 0362482s03 and by grants of Estonian Science Foundation Nos 3579 and 4835. We thank Toomas Kõiv, Katrin Olt, Katrin Ott, Diana Sarik, Tarmo Timm and Lea Tuvikene for their help.

References

Andronikova, I. N., 1989. Strukturno-funktsionalnaya organizatsiya zooplanktona ozernykh ekosistem raznykh troficheskikh tipov. Theses doc. Leningrad, 39 pp. [Structural and Functional Arrangement of Zooplankton in Lake Ecosystems of Various Trophic Types. In Russian].

Armengol, X., A. Esparcia & M. R. Miracle, 1998. Rotifer vertical distribution in a strongly stratified lake: a multivariate analysis. Hydrobiologia 387/388: 161–170.

Armengol-Diaz, J., A. Esparcia, E. Vicente & M. R. Miracle, 1993. Vertical distribution of planktonic rotifers in a karstic meromictic lake. Hydrobiologia 255/256: 381–388.

Arvola, L., K. Salonen, R. I. Jones, A. Heinänen & I. Bergström, 1987. A three day study of the diel behaviour of plankton in a highly humic and steeply stratified lake. Archiv für Hydrobiologie 109: 89–106.

Balushkina, E. V. & G. G. Winberg, 1979. Zavisimost mezhdu massoi I dlinoi tela u planktonnykh zivotnykh Obshchiye osnovy izucheniya vodnykh ekosistem. In Winberg, G. G. (ed.), Obshchie osnovy izucheniya vodnykh ekosistem. Nauka. Leningrad: 169–172 [Relation Between Body Mass and Length in Planktonic Animals. In Russian].

Berzins, B. & B. Pejler, 1989. Rotifer occurrence in relation to temperature. Hydrobiologia 175: 223–231.

Bosselmann, S., 1979. Population dynamics of *Keratella cochlearis* in Lake Esrom. Archiv für Hydrobiologie 87: 152–165.

Bottrell, H. H., A. Duncan, Z. M. Gliwicz, E. Grygierek A. Herzig, A. Hillbricht-Ilkowska, H. Kurasawa, P. Larsson & T. Weglenska, 1976. A review of some problems in zooplankton production studies. Norwegian Journal of Zoology 24(4): 419–456.

Bulgakov, N. G. & A. P. Levich, 1999. The nitrogen:phosphorus ratio as a factor regulating phytoplankton community structure. Archiv für Hydrobiologie 146(1): 3–22.

Bürgi, H. R., C. Heller, S. Gaebel, N. Mookerji & J. V. Ward, 1999. Strength coupling between phyto and zooplankton in Lake Lucerne (Switzerland) during phosphorus abatement subsequent to a weak eutrophication. Journal of Plankton Research 21: 485–507.

Carpenter, S. R. & J. F. Kitchell (eds.) 1993. The Trophic Cascade in Lakes. University Press, Cambridge.

Cordova, S. E., J. Giffin & K. L. Kirk, 2001. Food limitation of planktonic rotifers: field experiments in two mountain ponds. Freshwater Biology 46: 1519–1527.

Fradkin, S. C., 1995. Effects on interference and exploitative competition from large-bodied cladocerans on rotifer community structure. Hydrobiologia 313/314: 387–393.

Godeanu, S. P., 1978. Specificity of the zooplankton in several lakes from northern Germany with different degree of eutrophication. Verhandlungen. Internationale Vereinigung für Theoretische und Angewandte Limnologie 20: 963–968.

Haberman, J., 1998. Zooplankton of Lake Võrtsjärv. Limnologica 28: 49–65.

Herzig, A., 1994. Predator–prey relationships within the pelagic community of Neusiedler See. Hydrobiologia 275/276: 81–96.

Hofmann, W., 1977. The influence of abiotic environmental factors of population dynamics in planktonic rotifers. Archiv für Hydrobiologie. Beiheft Ergebnisse der Limnologie 8: 77–83.

Järvalt, A., T. Krause & A. Palm, 2005. Diel migration and spatial distribution of fish in a small stratified lake. Hydrobiologia 547: 197–203.

Jeppesen, E., E. A. Madsen, J. P. Jensen & N. J. Anderson, 1996. Reconstructing the past density of planktivorous fish and trophic structure from sedimentary zooplankton fossils: a surface sediment calibration data set from shallow lakes. Freshwater Biology 36: 115–127.

Kangro, K., R. Laugaste, P. Nõges & I. Ott, 2005. Long-term changes and seasonal development of phytoplankton in a strongly stratified, hypertrophic lake. Hydrobiologia 547: 91–103.

Kira, T., 1993. Major environmental problems in world lakes. In De Bernardi R., R. Pagnotta & A. Pugnetti (eds), Strategies for Lake Ecosystems Beyond 2000. Memorie dell'Istituto Italiano di Idrobiologia 52: 1–7.

Kirk, K. L., 2002. Competition in variable environments: experiments with planktonic rotifers. Freshwater Biology 47: 1089–1096.

Kizito, Y. S. & A. Nauwerck, 1995. Temporal and vertical distribution of planktonic rotifers in a meromictic crater lake, Lake Nyahirya (Western Uganda). Hydrobiologia 313/314: 303–312.

Kübar, K., 1994. Arbi ja Verevi järve zooplankton. Diploma paper. Manuscript at the Institute of Zoology and Hydrobiology of the University of Tartu, 81 pp. [Zooplankton of Lake Arbi and Lake Verevi. In Estonian, English summary].

Lammens, E. H. R. R., 1990. The relation of biotic and abiotic interactions to eutrophication in Tjeukemeer, The Netherlands. Hydrobiologia 191: 29–37.

Lampert, W., 1992. Zooplankton vertical migrations: implications for phytoplankton–zooplankton interactions. Archiv für Hydrobiologie. Beiheft Ergebnisse der Limnologie 35: 69–78.

Lampert, W., 1997. Zooplankton research: the contribution of limnology to general ecological paradigms. Aquatic Ecology 31: 19–27.

Leibold, M. A., 1990. Resources and predators can affect the vertical distributions of zooplankton. Limnology and Oceanography 35(4): 938–944.

Liljendahl-Nurminen, A., J. Horppila, P. Eloranta, T. Malinen & L. Uusitalo, 2002. The seasonal dynamics and distribution of *Chaoborus flavicans* larvae in adjacent lake basins of different morphometry and degree of eutrophication. Freshwater Biology 47: 1283–1295.

Miracle, M. R., 1977. Migration, patchiness, and distribution in time and space of planktonic rotifers. Archiv für Hydrobiologie Beiheft Ergebnisse der Limnologie 8: 19–37.

Miracle, M. R. & X. Armengol-Diaz, 1995. Population dynamics of oxiclinal species in lake Arcas-2 (Spain). Hydrobiologia 313/314: 291–301.

Nicholls, K. H. & C. Tudorancea, 2001. Species-level and community-level data analyses reveal spatial differences and temporal change in the crustacean zooplankton of a large Canadian lake (Lake Simcoe, Ontario). Journal of Limnology 60(2): 155–170.

Nõges, T. & K. Kangro, 2005. Primary production of phytoplankton in a strongly stratified temperate lake. Hydrobiologia 547: 105–122.

Nõges, P. & T. Nõges, 1998. Stratification of Estonian lakes studied during hydrooptical expeditions in 1995–97. Proceedings of the Estonian Academy of Sciences. Biology. Ecology 47: 268–281.

Olt, K., 2001. Zooplanktoni vertikaalne jaotus Verevi järves ja Nohipalu Valgjärves suvise stratifikatsiooni tingimustes. Diploma paper. Manuscript at the Institute of Zoology and Hydrobiology of the University of Tartu, 59 pp. [Vertical Distribution of Zooplankton in Lake Verevi and Lake Nohipalu Valgjärv in Conditions of Summer Stratification. In Estonian, English summary].

Ott, I., 1996. Relationship between organic matter and summer phytoplankton species composition. Eutrophication in planktonic food web dynamics and elemental cycling. International PELAG symposium, p. 54.

Ott, I., T. Kõiv, P. Nõges, A. Kisand, A. Järvalt & E. Kirt, 2005a. General description of partly meromictic hypertrophic Lake Verevi, its ecological status, changes during the past eight decades and restoration problems. Hydrobiologia 547: 1–20.

Ott, I., R. Laugaste, S. Lokk & A. Mäemets, 1997. Plankton changes in Estonian small lakes in 1951–1993. Proceedings of the Estonian Academy of Sciences. Biology. Ecology 46(1/2): 58–79.

Ott, I., A. Rakko, D. Sarik, P. Nõges & K. Ott, 2005b. Sedimentation rate of seston during the formation of temperature stratification after ice break-up in the partly meromictic Lake Verevi. Hydrobiologia 547: 51–61.

Ponyi, J. E. & N. P. Zankai, 1982. Population dynamics, biomass and biomass production of *Eudiaptomus gracilis* (G. O. SARS) in two water areas of different trophic state of L. Balaton (Hungary). Acta Hydrochimica et Hydrobiologica 10: 597–610.

Rothhaupt, K. O., 1990. Population growth rates of two closely related rotifer species: effects of food quality, particle size, and nutritional quality. Freshwater Biology 23: 561–570.

Ruttner-Kolisko, A., 1977a. Comparison of various sampling techniques, and results of repeated sampling of planktonic rotifers. Archiv für Hydrobiologie. Beiheft Ergebnisse der Limnologie 8: 13–18.

Ruttner-Kolisko, A., 1977b. Suggestions for biomass calculation of planktonic rotifers. Archiv für Hydrobiologie. Beiheft Ergebnisse der Limnologie 8: 71–76.

Sommer, U., 1989. Toward a Darwinian ecology of plankton. In Sommer, U. (ed.) Plankton Ecology Succession in Plankton Communities. Springer Verlag, New York, Berlin, Heidelberg: 1–8.

Stransfield, J. H., M. R. Perrow, L. D. Tench, A. J. D. Jowitt & A. A. L. Taylor, 1997. Submerged macrophytes as refuges for grazing Cladocera against fish predation: observations on seasonal changes in relation to macrophyte cover and predation pressure. Hydrobiologia 342/343: 229–240.

Studenikina, E. I. & M. M. Cherepakhina, 1969. Srednii ves osnovnykh form zooplanktona Azovskogo morya. Gidrobiologicheskii Zhurnal 5: 89–91 [Mean Weight of Basic Zooplankton Forms of the Azov Sea. In Russian].

Tilman, D., 1977. Resource competition between planktonic algae: an experimental and theoretical approach. Ecology 58: 338–348.

Tilman, D., 1982. Resource Competition and Community Structure. Princeton Univ. Press, Princeton, N.J.

Timm, M. & A. Mäemets, 1991. Zooplankton. In Timm, H. (ed.), Verevi järve seisund. [State of Lake Verevi]. Estonian Academy of Sciences, Institute of Zooloogy and Botany, Tartu: 91–94 [Zooplankton. In Estonian].

Van Donk, E., M. P. Grimm, R. D. Gulati, P. G. M. Heuts, W. A. de Kloet & L. Van Liere, 1990. First attempt to apply whole-lake food-web manipulation on a large scale in The Netherlands. Hydrobiologia 200/201: 291–301.

Vanni, M. J. & J. Temte, 1990. Seasonal patterns of grazing and nutrient limitation of phytoplankton in a eutrophic lake. Limnology and Oceanography 35: 697–709.

Viitasalo, M., 1994. Seasonal succession and long-term changes of mesozooplankton in the Northern Baltic Sea. Finnish Marine Research 263: 3–39.

Virro, T., 1989. Sravnenie metodov sbora planktonnykh kolovratok (Rotatoria) na primere Chudskogo ozera. Proceedings of the Academy of Sciences of the Estonian SSR. Biology 38: 119–122 [The Comparison of Sampling Methods of Planktonic Rotifers (Rotatoria) on the Example of Lake Peipsi. In Russian].

Virro, T., 1996. Taxonomic composition of rotifers in Lake Peipsi. Hydrobiologia 338: 125–132.

Wetzel, R. G., 1983. Limnology (2nd ed.). Saunders College Publishing, Philadelphia.

Williamson, C. E., R. W. Sanders, R. E. Moeller & P. L. Stutzman, 1996. Utilization of subsurface food resources for zooplankton reproduction: Implications for diel vertical migration theory. Limnology and Oceanography 41(2): 224–233.

Hydrobiologia (2005) 547:163–174
I. Ott & T. Kõiv (eds), Lake Verevi, Estonia – A Highly Stratified Hypertrophic Lake
DOI 10.1007/s10750-005-4157-7

Vertical and seasonal dynamics of planktonic ciliates in a strongly stratified hypertrophic lake

Priit Zingel[1,2]
[1]*Centre for Limnology, Institute of Agricultural and Environmental Sciences, Estonian Agricultural University, Rannu, 61101 Tartumaa, Estonia*
[2]*Institute of Veterinary Medicine and Animal Sciences, Estonian Agricultural University, 1 Kreutzwaldi St., 51014 Tartu, Estonia*
E-mail: pzingel@zbi.ee

Key words: ciliates, protozooplankton, vertical distribution, seasonal dynamics

Abstract

Seasonal population dynamics and the vertical distribution of planktonic ciliates in a hypertrophic and strongly stratified temperate lake were studied from April to October in 2000 and from April to June in 2001. In the epi- and metalimnion the ciliate abundance peaked in spring and late summer, reaching maximum values in the metalimnion (86 cells ml^{-1}) on 7th August 2000. In the epilimnion, the highest biomass content (414 μg C l^{-1}) was observed on 8th May 2000. In the hypolimnion only a late summer peak occurred and the ciliate numbers were always lower than in the epi- and metalimnion. Five groups dominated the community of ciliates: *Oligotrichida, Gymnostomatea, Prostomatida, Hymenostomata* and *Peritrichia*, and the community composition varied greatly with depth. In the epilimnion the ciliate numbers were dominated by oligotrichs but small algivorous prostomatids, peritrichs and gymnostomes were also numerous. In the metalimnion these groups were gradually replaced by scuticociliates and mixotrophic *Coleps* spp. In the hypolimnion scuticociliates and species known as benthic migrants dominated. In the epilimnion and upper metalimnion in spring large herbivores and in summer small bacterivores were more numerous.

Introduction

Ciliates (*Ciliophora*) are one of the largest groups of protozoans – over 7000 species of ciliates have been described and they can be found in almost every aquatic environment. Studies in the two last decades have highlighted the importance of planktonic ciliates in freshwater ecosystems. Ciliates have an important role in the 'microbial loop' (Azam et al., 1983) as they prey upon bacteria and microflagellates. They consume also pico- and nanoalgae (Sherr & Sherr, 1984; Fenchel, 1987; Gonzales et al., 1990; Kisand & Zingel, 2000) that are not efficiently grazed by larger metazooplankters. There is plenty of evidence that planktonic ciliates are an important food resource for large metazoan zooplankton (Stoecker & Capuzzo, 1990; Dolan & Coats, 1991; Gifford, 1991). Thus, planktonic ciliates may be a critical link between microbial and macroscopic components of pelagic food webs. In lakes, planktonic ciliates can at times constitute over 50% of the biomass of zooplankton (Zingel, 1999).

In the past years the number of studies that deal with freshwater ciliates have increased rapidly. Still there is lack of detailed studies about the temporal and vertical distribution of ciliates in strongly stratified lakes. During the formation of the metalimnion, an adaptation of organisms to changeable conditions will take place. In summer stagnation the vertical distribution of organisms differs essentially from that in the moment of circulation. In the relatively narrow metalimnetic layer, a temporary microbial loop will form, in which aerobic, microaerobic and anaerobic conditions occur. In addition, vertical gradients of temperature, oxygen and radiation will develop inside the metalimnion, and cause a succession of microniches and -environments in time and space. A great variety of microbes will allow for the functioning of a microbial loop. This metalimnetic circulation of matter consists of microalgae, bacteria, heterotrophic nanoflagellates, ciliates and other microzooplankton; the organic matter produced by this loop can in principle return into classical matter circulation (Steenbergen et al., 1993). Little is known about the role of ciliates in the described pattern.

The aim of our study was to describe the vertical and temporal distribution and community structure of planktonic ciliates in a hypertrophic temperate lake.

Materials and methods

Lake Verevi is a small (12.6 ha; mean depth 3.6 m; maximum depth 11 m) hypertrophic lake (tot P > 100 μg l^{-1}; tot N > 1500 μg l^{-1}; chl *a* > 40 μg l^{-1}) with very small water exchange, situated in Southern Estonia. It is characterized as a strongly stratified water body with an anoxic hypolimnion. The ice cover lasts usually from November to April. During the vegetation period, the Secchi depth usually does not exceed 1 m. For more thorough description of lake and study methods see Ott et al., 2005.

The lake was sampled from April to October in 2000 and from April to June in 2001, altogether in 12 and 9 occasions, respectively. In the year 2000, at every occasion eight subsamples were collected from different depths for ciliate counts: two from the epi- and hypolimnion and four from the metalimnion. In October, the deepest hypolimnion layer was not sampled. In 2001, three or four subsamples were collected: one from the epi- and hypolimnion and one (in April) or two (May and June) from the metalimnion.

The samples from the surface were taken directly into a bottle, the others using a special vacuum probe (similar to the one used by Guerrero et al., 1985). A masterflex pump (model N 7533–60) with an 'easy-load' pump-head (model 7518–12) was used for pumping water to the surface through a Ø 8 mm hose. The lower end of the vertical hose was connected to a 7-cm long horizontal tube in order to get water from horizontal layers as precisely as possible. The flow of the device was ~2 l min^{-1}. Collected samples were preserved and fixed with acid Lugols solution. The ciliate biomass and community composition were determined using the Utermöhl (1958) technique. Samples were stored at 4 °C in the dark. Volumes of 50 ml were settled for at least 24 h in plankton chambers. Ciliates were enumerated and identified with an inverted microscope (mainly Olympus IX50) at 400–1000 × magnification. The entire content of each Utermöhl chamber was surveyed. Ciliates were usually identified to genus on the basis of several sources (Kahl, 1930–1932, 1935; Patterson & Hedley, 1992; Foissner & Berger, 1996). The first 20 measurable specimens encountered for each taxon were measured. Biovolumes of each taxa were estimated by assuming geometric shapes and converted to carbon weight using a factor of 190 fg C μm^{-3} (Putt & Stoecker, 1989). For statistical analysis nonparametric methods were used. For bacterial and metazooplankton data used in analyses see Tammert et al., 2005 and Kübar et al., 2005.

Results

In 2000, the mean abundance of ciliate protozoa in L. Verevi was 29.7 cells ml^{-1} and in 2001 27.6 cells ml^{-1}. The mean biomass was 110 μgC l^{-1} and 116 μgC l^{-1} in 2000 and 2001, respectively. Throughout the investigation period the ciliates species composition consisted mainly of five groups: oligotrichs, gymnostomes, scuticociliates, peritrichs and prostomatids (Fig. 1). The highest abundance (85.9 cells ml^{-1}) was registered on 7 August 2000 in the metalimnion (Fig. 2) and the highest biomass (414 μg C l^{-1}) on 8 May 2000 in the epilimnion (Fig. 3). In spring 2001, the highest abundance (68.6 cells ml^{-1}) and highest biomass (406 μgC l^{-1}) were recorded on 7 May (Fig. 4).

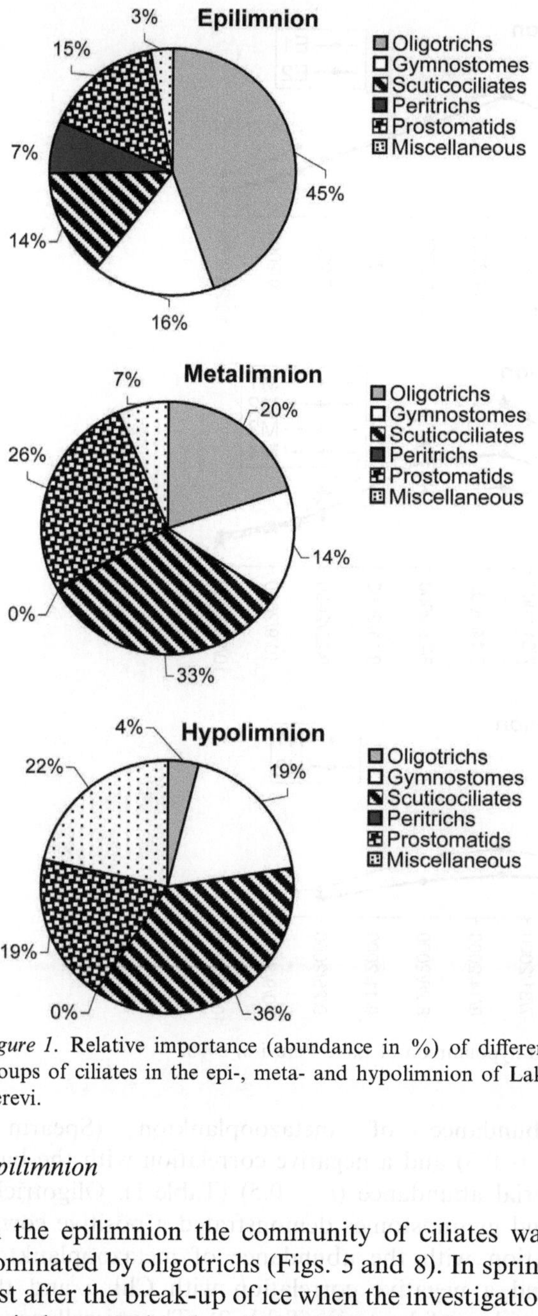

Figure 1. Relative importance (abundance in %) of different groups of ciliates in the epi-, meta- and hypolimnion of Lake Verevi.

Epilimnion

In the epilimnion the community of ciliates was dominated by oligotrichs (Figs. 5 and 8). In spring just after the break-up of ice when the investigation period started, the dominating species were large-sized (Ø > 50 μm) herbivores like *Pelagostrombidium* spp., *Limnostrombidium* spp., *Rimostrombidium* spp., *Codonella cratera* and *Tintinnidium fluviatile*. They were most numerous in the surface layer, reaching up to values 39.4 cells ml^{-1} on 8 May 2000 (at that time the total ciliate abundance was 75.4 cells ml^{-1}) and 37.9 ml^{-1} on 7 May 2001 (68.6 cells ml^{-1}). In spring the other numerous species were *Mesodinium* sp., *Askenasia* sp., *Coleps hirtus*, *Urotricha* spp. and *Uronema* sp. In the end of May the abundance in epilimnion dropped, remaining around values of about 20 cells ml^{-1}, also in June. In August, the ciliate number rose again. At that time in addition to herbivores numerous small-sized (Ø ~20 μm) bacterivores (*Halteria* sp., *Rimostrombidium* sp.) were present. In some occasions small prostomatids (mostly *Urotricha* sp.) and peritrichs (*Vorticella* sp. and *Epistylis* sp.) were also quite abundant. In late summer and autumn the ciliate number slowly decreased.

Metalimnion

In the upper metalimnion oligotrichs, scuticociliates and prostomatids (Fig. 6) dominated. The share of oligotrichs decreased with increasing depth. In the deepest layers they were present only in spring and autumn. Peritrichs were present only in the upmost part of metalimnion. In the deep metalimnion scuticociliates (*Cyclidium* sp., *Uronema* sp., *Cinetochilum margaritaceum*) and prostomatids (*Coleps spetai*, Coleps sp.) clearly dominated. In spring, the greatest numbers were registered in the upper metalimnion, and the abundance decreased with increasing depth. As in the epilimnion, the abundance dropped after the spring peak in the end of May and stayed low (<20 cells ml^{-1}) through June. In August a second peak of ciliates occurred. In that period the picture was reversed – the upper metalimnion contained least ciliates and the highest abundances were recorded in the deep metalimnion. That peak consisted mainly of small scuticociliates and prostomatids (*Coleps* spp.), in upper layers also by small oligotrichs (*Halteria* sp.). After the second peak the ciliate number decreased slowly also in the metalimnion.

Hypolimnion

In the hypolimnion the dynamics of ciliates demonstrated a high stability, the abundance ranging between 3.2 and 35.3 cells ml^{-1}. It was visible that the spring peak (made mostly up by large-sized herbivores) did not occur in the hypolimnion. But the second summer peak occurred despite the fact that the number of ciliates stayed lower than in the

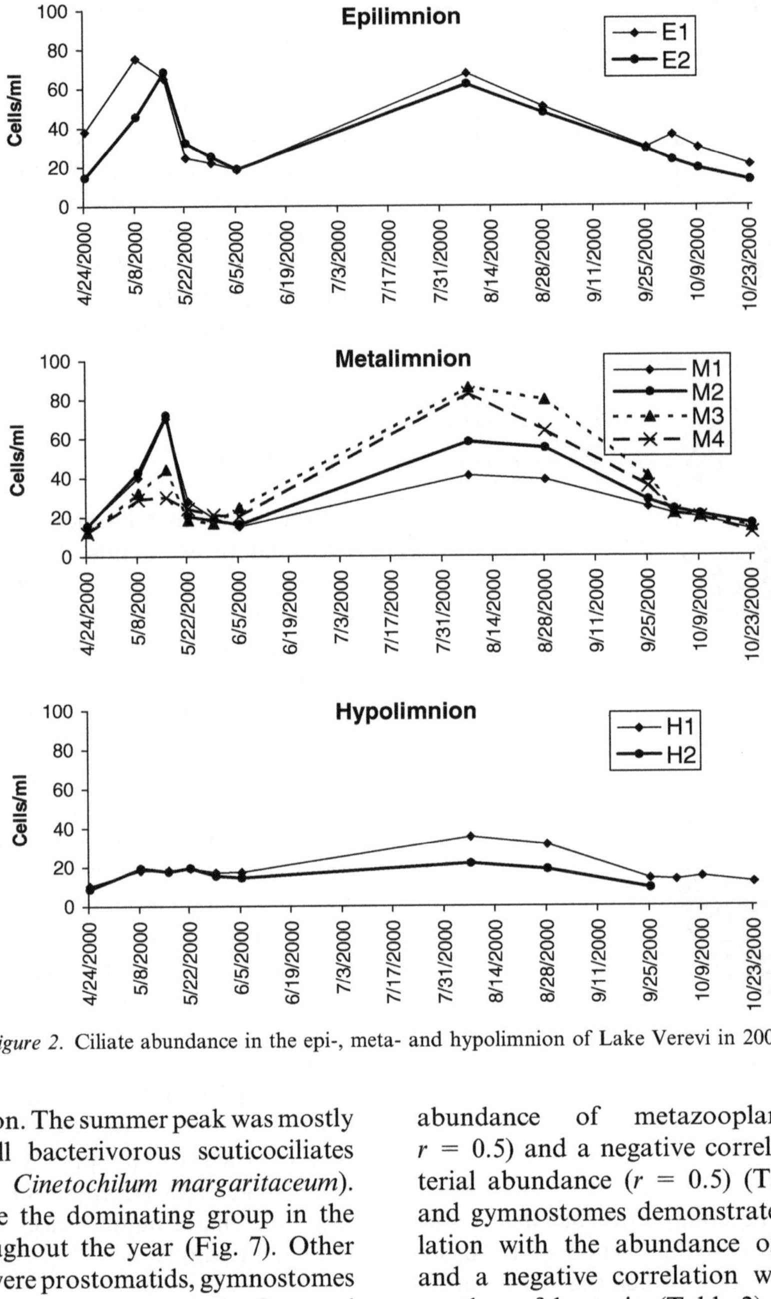

Figure 2. Ciliate abundance in the epi-, meta- and hypolimnion of Lake Verevi in 2000.

meta- and epilimnion. The summer peak was mostly made up by small bacterivorous scuticociliates (*Uronema* sp. and *Cinetochilum margaritaceum*). Scuticociliates were the dominating group in the hypolimnion throughout the year (Fig. 7). Other important groups were prostomatids, gymnostomes and a miscellaneous group composed of several different species of the orders *Heterotrichida, Odontostomatida, Hypotrichia and Hymenostomatea.* Oligotrichs were found only in some spring and autumn samples. While the ciliate numbers were the lowest in the hypolimnion, the species diversity was the highest in the hypolimnion.

The ciliate abundance in the epiliminion had a significant ($p < 0.05$) positive correlation with the abundance of metazooplankton (Spearman $r = 0.5$) and a negative correlation with the bacterial abundance ($r = 0.5$) (Table 1). Oligotrichs and gymnostomes demonstrated a positive correlation with the abundance of metazooplankton and a negative correlation with Chl *a* and the number of bacteria (Table 2). The miscellaneous ciliate group was in a positive correlation with the number of bacteria.

Discussion

Our study demonstrated that ciliate numbers and the community structure vary greatly both

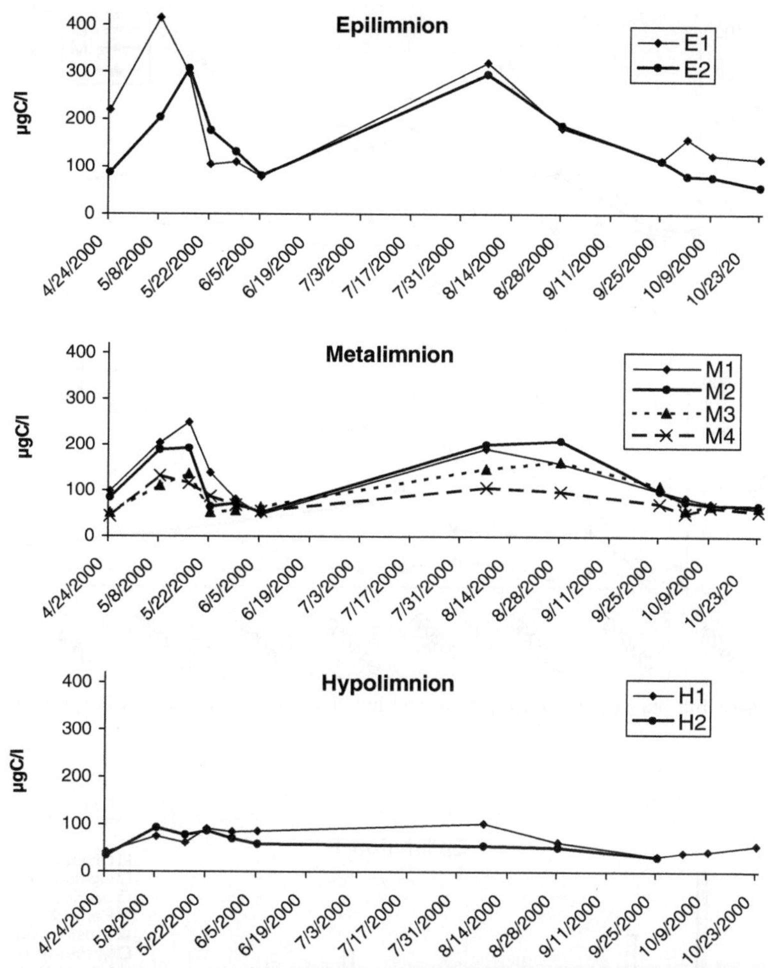

Figure 3. Ciliate biomass in the epi-, meta- and hypolimnion of Lake Verevi in 2000.

vertically and temporally. The decline in ciliate numbers in the beginning of summer, which in our study was observed in the epi- and metalimnion, coincided with a decline in zooplankton abundance. Metazooplankton is known to prey intensively on ciliates (Sorokin & Paveljeva, 1972; Maly, 1975; Berk et al., 1977; Porter et al., 1979; Heinbokel & Beers, 1979) and can easily affect their numbers. But it has also been suggested (e.g. Beaver & Crisman, 1982) that limited food resources may control the ciliate community composition and abundance rather than metazoan grazing. This may be the case also in Lake Verevi, where some ciliate groups seem to be clearly bottom-up controlled. The metazooplankters and larger herbivorous ciliates (oligotrichs and gymnostomes) seemed to depend on similar food resources as they were positively, not negatively correlated. It is unlikely that there exists a resource overlap between metazooplankters and small ciliates as the latter feed mainly on bacterio- and autotrophic picoplanton. As herbivorous species were generally more common in surface layers, it is understandable that the correlation between ciliates and metazooplankters was significant only in the epilimnion.

In the epilimnion the most numerous were oligotrichs. Oligotrichs are generally known to be a common component of lacustrine protozooplankton, especially in the epilimnion (Laybourn-Parry et al., 1990; Müller et al., 1991). In spring, their numbers were dominated by large-sized taxa (>50 μm Ø) including both heterotrophic and mixotrophic species that feed mainly on

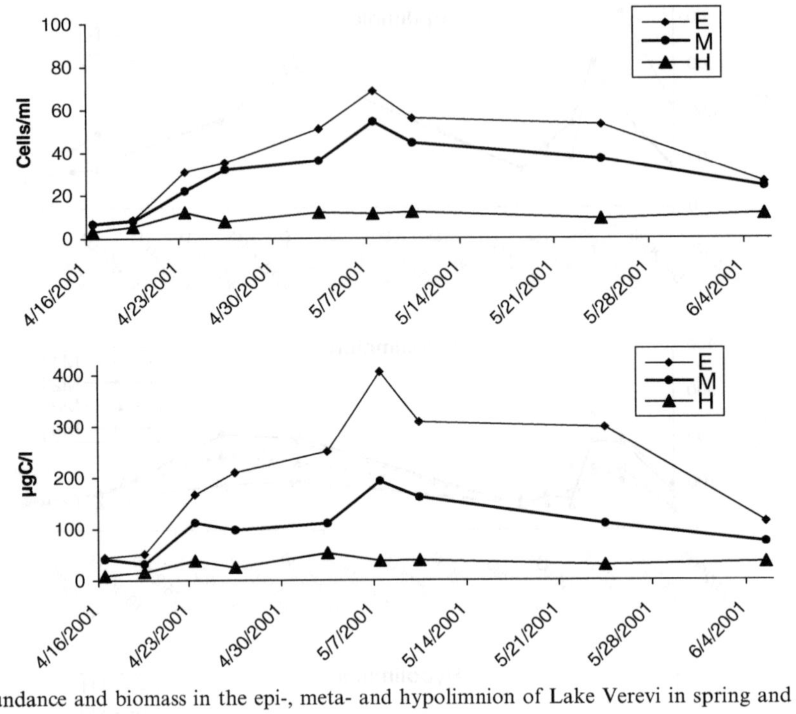

Figure 4. Ciliate abundance and biomass in the epi-, meta- and hypolimnion of Lake Verevi in spring and early summer 2001.

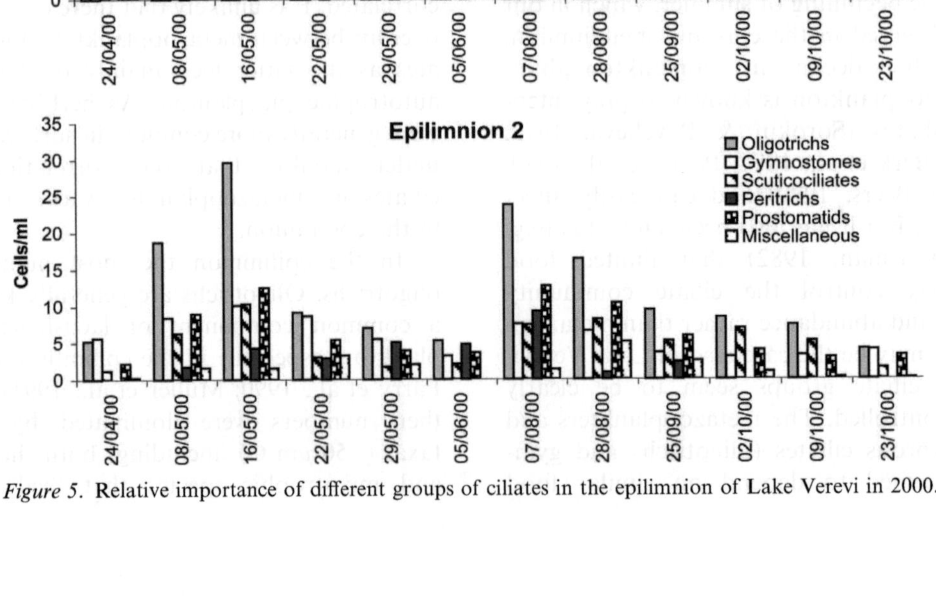

Figure 5. Relative importance of different groups of ciliates in the epilimnion of Lake Verevi in 2000.

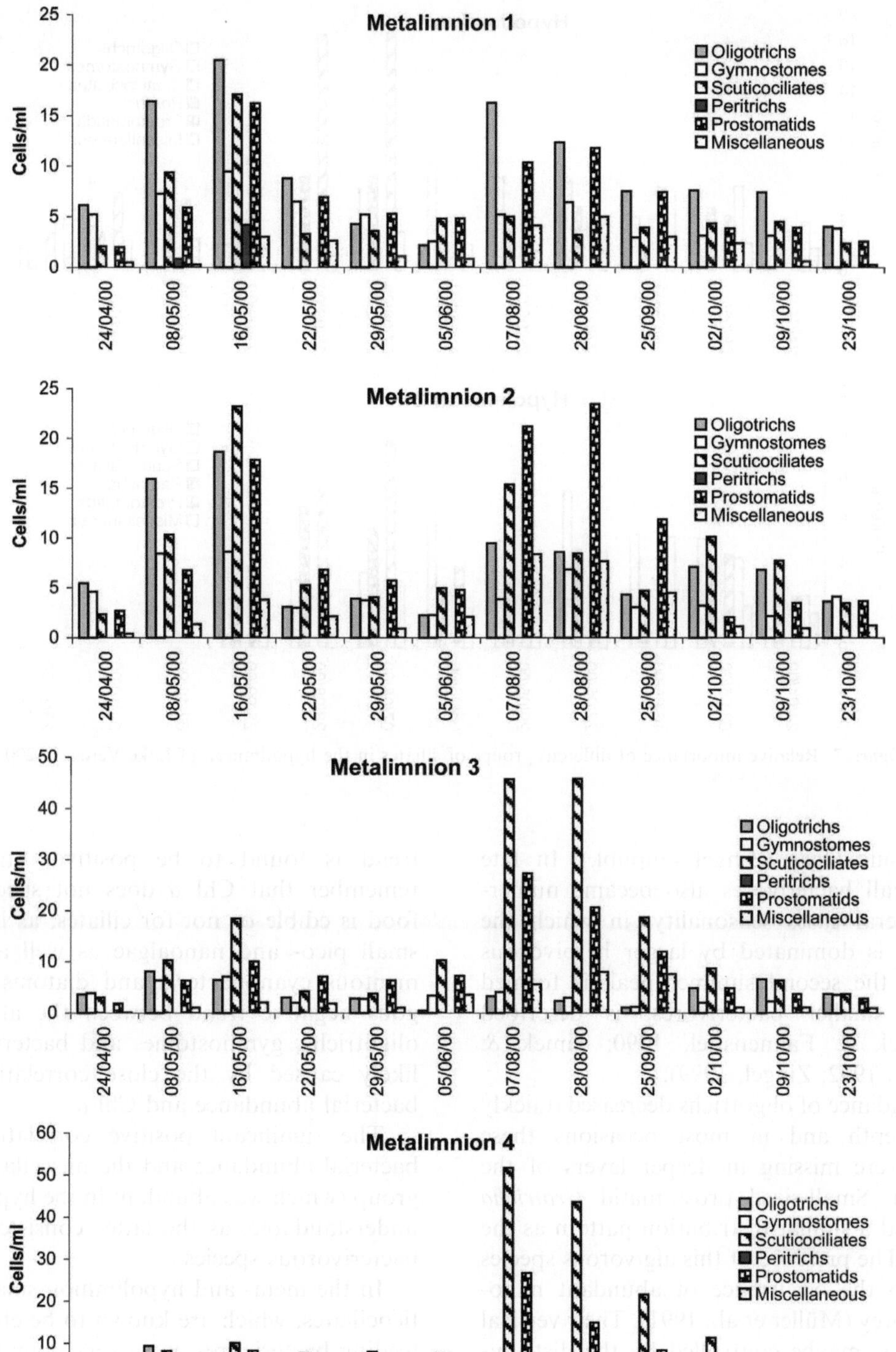

Figure 6. Relative importance of different groups of ciliates in the metalimnion of Lake Verevi in 2000.

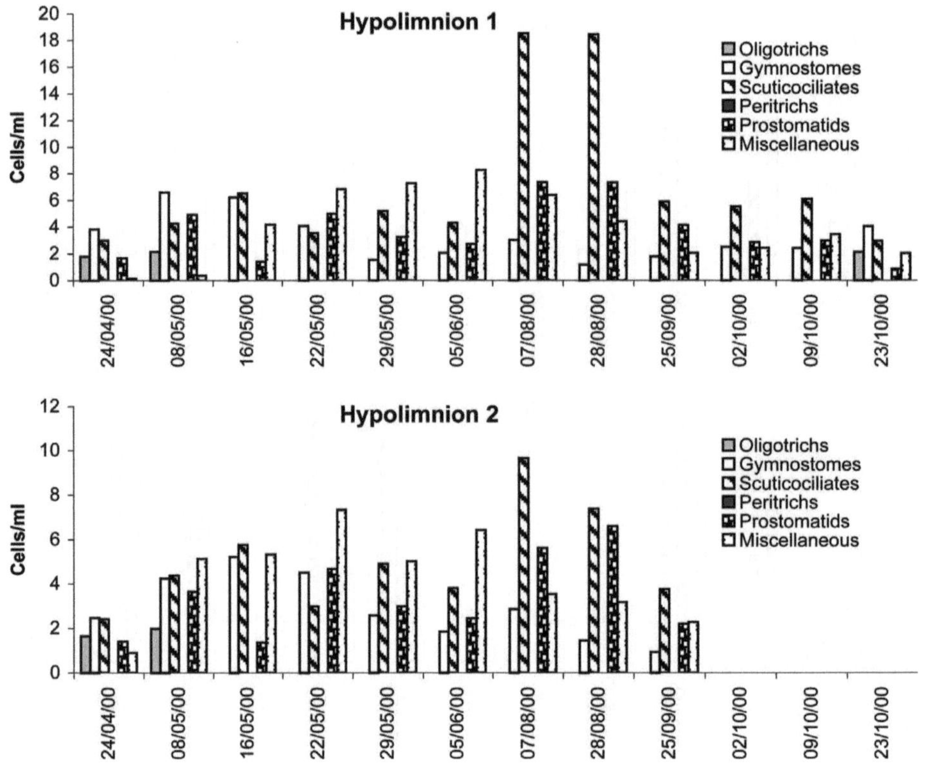

Figure 7. Relative importance of different groups of ciliates in the hypolimnion of Lake Verevi in 2000.

nanoplanktonic algae (Zingel, unpubl.). In late summer, small bacterivores also became numerous. In several cases seasonality, in which the spring peak is dominated by larger herbivorous ciliates and the second summer peak is formed mostly by smaller bacterivores, is described (e.g., Carrick & Fahnenstiel, 1990; Šimek & Staškrabová, 1992; Zingel, 1999).

The abundance of oligotrichs decreased quickly with the depth and in most occasions these organisms were missing in deeper layers of the metalimnion. Small-sized prostomatid *Urotricha* demonstrated a similar distribution pattern as the oligotrichs. The presence of this algivorous species is related to the occurrence of abundant nanoplanktonic prey (Müller et al., 1991). Their vertical distribution is maybe controlled by the distribution of convenient food organisms.

In our study, we found a significant negative correlation between oligotrichs and gymnostomes and Chl *a* (such a trend applied also to metazooplankton). This is surprising, as in most cases that trend is found to be positive. Still we must remember that Chl *a* does not specify whether food is edible or not for ciliates, as it consists of small pico- and nanoalgae as well as large filamentous cyanobacteria and diatoms. An analogous negative trend between the abundance of oligotrichs, gymnostomes and bacteria was most likely caused by the close correlation between bacterial abundance and Chl *a*.

The significant positive correlation between bacterial abundance and the miscellaneous ciliate group (which was abundant in the hypolimnion) is understandable, as the latter consisted mainly of bacterivorous species.

In the meta- and hypolimnion small-sized scuticociliates, which are known to be effective filter-feeding bacterivores, were very abundant. Several other studies also report high numbers of scuticociliates near the metalimnion (Pace, 1982; Zingel & Ott, 2000). In the meta- and hypolimnion, the abundance of metazooplankton that potentially prey on these small-sized ciliates, usually decreases.

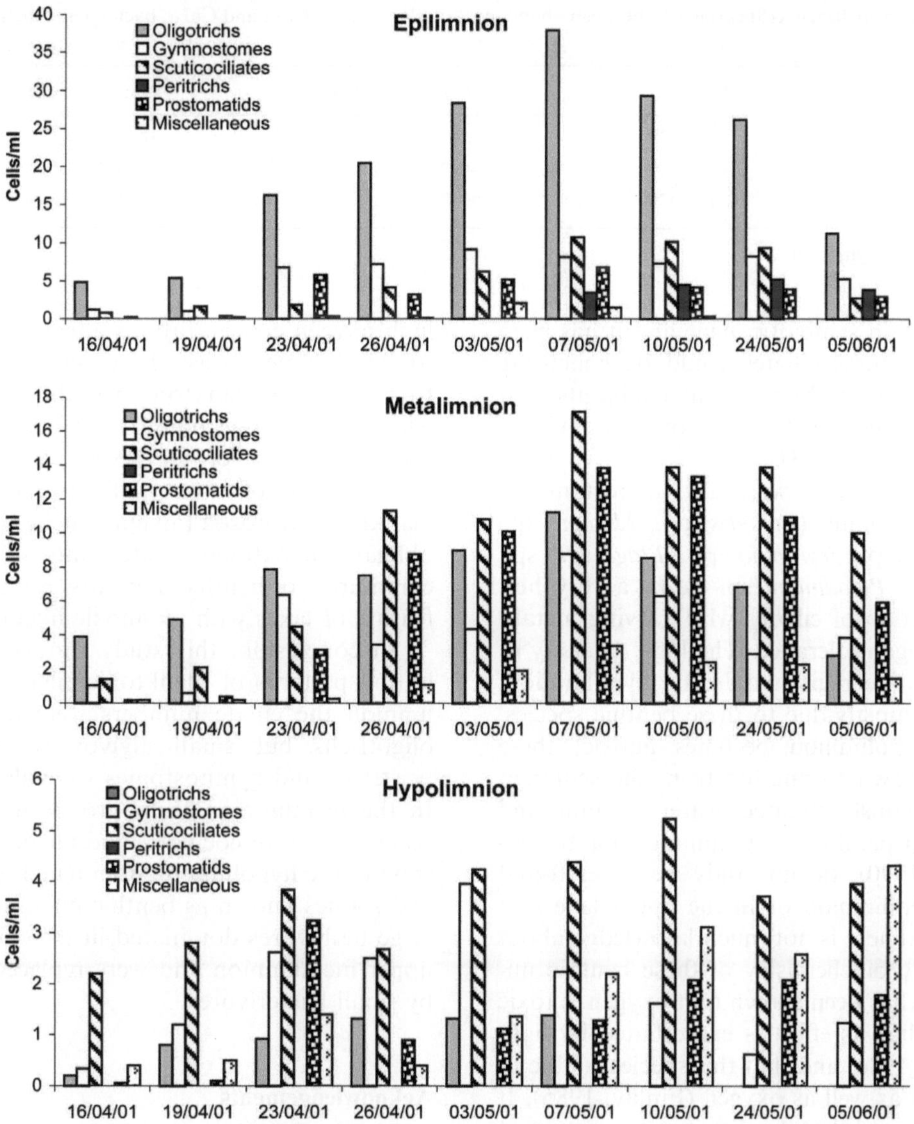

Figure 8. Relative importance of different groups of ciliates in the epi-, meta- and hypolimnion of Lake Verevi in 2001.

This may explain the higher abundances of scuticociliates in the deeper layers. It has also been suggested (Müller et al., 1991) that scuticociliates tend to concentrate in the oxycline, where bacterial productivity is high. In the epilimnion, the dominating bacterivores were in most cases small-sized oligotrichs (*Rimostrombidium* spp., *Halteria* sp.), which may act as competitors against scuticociliates. In deeper layers small-sized oligotrichs were gradually replaced by scuticociliates throughout the year. In addition the epilimnion contributed a community of vorticellas, which are known to have high cell-specific ingestion rates (Šimek et al., 1995) and feed mainly on planktonic bacteria. As peritrichs, small oligotrichs were almost absent in deeper layers. Their association with the surface water layer has been usually explained by the species structure of colonial microcyanobacteria that act as their support (Müller et al., 1991; Carrias et al., 1998). But it is also reasonable to consider the possibility that the above-described pattern may also be influenced by competition for food resources.

In the metalimnion high densities of prostomatids *Coleps* spp. were observed. These species are

Table 1. Correlation coefficients (Spearman *R*) between abundances of planktonic ciliates and Chl *a*, bacteria and metazooplankton in Lake Verevi

Abundance of ciliates	Chl *a*	Abundance of bacteria	Abundance of metazooplankton
Epilimnion	-0.273^{NS}	-0.445	0.513
Metalimnion	-0.024^{NS}	0.078^{NS}	0.093^{NS}
Hypolimnion	0.171^{NS}	0.076^{NS}	-0.143^{NS}

$p < 0.05$; ^{NS} = not significant.

known to contain symbiotic algae and it has been suggested that these ciliates could be benefiting from oxygen supplied by the algal symbionts, and may thus easily cope with a low-oxygen environment (Esteve et al., 1988).

In the hypolimnion were also important so-called benthic migrants (*Loxodes* spp., *Metopus* sp., *Caenomorpha* spp., *Frontonia* spp., *Plagiopyla* sp., *Epalxella* spp., *Pelodinium* sp. etc.) – a very heterogeneous group of ciliates with varying metabolism and oxygen tolerance. The high diversity of species, which we registered in the hypolimnion, was achieved mostly due to these benthic species. When the hypolimnion becomes anoxic, these species are known to migrate from the sediment into the previously vacated water column and form a distinct planktonic community (Laybourn-Parry et al., 1990). In our study we never found them in the epilimnion or in the upper layers of metalimnion. There is not much knowledge about the respiratory biochemistry of these benthic migrants. Still it has been shown that oxygen is toxic to *Loxodes*, that its effect is exacerbated by light (Finlay et al., 1986) and that this species is able to respire nitrate as well as oxygen (Finlay, 1985). It has been suggested that the distribution pattern of this species as well as other benthic migrants could be explained as a compromise between avoiding high oxygen concentrations and meeting the need for aerobic metabolism in a zone where predation from metazooplankton will be minimal and where at the same time food is readily available. Laybourn-Parry et al. (1990) suggest that the development of two distinct communities of planktonic protozoa (an epilimnetic community of obligate planktonic ciliates and a hypolimnetic community of benthic migrants) is a characteristic feature of lakes with an anoxic hypolimnion.

In conclusion, this study showed clear distribution patterns of planktonic ciliates. In the epilimnion the ciliate numbers were dominated by oligotrichs but small algivorous prostomatids, peritrichs and gymnostomes were also numerous. In the metalimnion these groups were gradually replaced by scuticociliates and mixotrophic *Coleps* spp.; in the hypolimnion dominated scuticociliates and species known as benthic migrants. In spring, large herbivores dominated in the epilimnion and upper metalimnion and were replaced in summer by small bacterivores.

Acknowledgements

We would like to express our gratitude to the whole staff of the Võrtsjärv Limnological Station, who

Table 2. Correlation coefficients (Spearman *R*) between abundances of different ciliate groups and Chl *a*, bacteria and metazooplankton in Lake Verevi

Abundance of ciliates	Chl *a*	Abundance of bacteria	Abundance of metazooplankton
Oligotrichs	-0.491	-0.472	0.468
Gymnostomes	-0.471	-0.501	0.442
Scuticociliates	0.163^{NS}	0.169^{NS}	0.238^{NS}
Peritrichs	-0.372^{NS}	-0.266^{NS}	0.183^{NS}
Prostomatids	-0.038^{NS}	-0.098^{NS}	0.275^{NS}
Miscellaneous	0.357^{NS}	0.477	-0.205^{NS}

$p < 0.05$; ^{NS} = not significant.

made this study possible. This research was partly supported by the ESF Grant Nos. 3579, 4835, and the target financed research project No. 0370208 and the Nessling foundation Grant No. 99084.

References

Azam, F., T. Fenchel, J. G. Field, J. S. Gray, L. A. Meyer-Reill & F. Thingstad, 1983. The ecological role of water-column microbes in the sea. Marine Ecology Progress Series 10: 257–263.

Beaver, J. R. & T. L. Crisman, 1982. The trophic response of ciliated protozoans in freshwater lakes. Limnology and Oceanography 27: 246–253.

Berk, S. G., D. C. Brownlee, D. R. Heinle, H. J. Kling & R. R. Colwell, 1977. Ciliates as a food source for marine planktonic copepods. Microbial Ecology 4: 27–40.

Carrias, J. F., C. Amblard & G. Bourdier, 1998. Seasonal dynamics and vertical distribution of planktonic ciliates and their relationships to microbial food resources in the oligo-mesotrophic Lake Pavin. Archiv Für Hydrobiologie 143: 227–255.

Carrick, H. J. & G. L. Fahnenstiel, 1990. Planktonic protozoa in Lakes Huron and Michigan: seasonal abundance and composition of ciliates and dinoflagellates. Journal of Great Lakes Research 16: 319–329.

Dolan, J. R. & D. W. Coats, 1991. Preliminary prey digestion in a predacious estuarine ciliate and the use of digestion data to estimate digestion. Limnology and Oceanography 36: 558–565.

Esteve, I., J. M. N. Gaju, H. McKhann & L. Margulis, 1988. Green endosymbiont of Coleps from Lake Cisó identified as *Chlorella vulgaris*. Symbiosis 3: 197–210.

Fenchel, T., 1987. Ecology of Protozoa. The Biology of Free-living Fagotrophic Protists. Springer-Verlag, Berlin, 197 pp.

Finlay, B. J., 1985. Nitrate respiration by protozoa (*Loxodes* spp.) in the hypolimnetic nitrate maximum of productive freshwater pond. Freshwater Biology 15: 333–346.

Finlay, B. J., T. Fenchel & S. Garfdiner, 1986. Oxygen perception and oxygen toxicity in the freshwater ciliated protozoan Loxodes. Journal of Protozoology 33: 157–165.

Foissner, W. & H. Berger, 1996. A user-friendly guide to the ciliates (Protozoa, Ciliophora) commonly used by hydrobiologists as bioindicators in rivers, lakes, and waste waters, with notes on their ecology. Freshwater Biology 35: 375–482.

Gifford, D. J., 1991. The protozoan–metazoan trophic link in pelagic ecosystems. Journal of Protozoology 38: 81–86.

Gonzales, J. M., E. B. Sherr & B. F. Sherr, 1990. Size-selective grazing on bacteria by natural assemblages of estuarine flagellates and ciliates. Applied and Environmental Microbiology 56: 583–589.

Guerrero, R., C. Pedros-Alio, T. M. Schmidt, & J. Mas, 1985. Phototrophic sulphur bacteria in two Spanish lakes: vertical distribution and limiting factors. Limnology and Oceanography 30: 807–819.

Heinbokel, J. F. & J. R. Beers, 1979. Studies on the functional role of tintinnids in the Southern California Bight. 3. Graizing impact of natural assemblages. Marine Biology 52: 23–32.

Kahl, A., 1930. Urtiere oder Protozoa I: Wimpertiere oder Ciliata (Infusoria) 1. Allgemeiner Teil und Prostomata. Tierwelt Deutschlands 18: 1–180.

Kahl, A., 1931. Urtiere oder Protozoa I. Wimpertiere oder Ciliata (Infusoria) 2. Holotricha auber den im 1. Teil behandelten Prostomata. Tierwelt Deutschlands. 21: 181–398.

Kahl, A., 1932. Urtiere oder Protozoa I: Wimpertiere oder Ciliata (Infusoria) 3. Spirotricha. Tierwelt Deutschlands 25: 399–650.

Kahl, A., 1935. Urtiere oder Protozoa I: Wimpertiere oder Ciliata (Infusoria) 4. Peritricha und Chonotricha. Tierwelt Deutschlands 30: 651–886.

Kisand, V. & P. Zingel, 2000. Dominance of ciliate grazing on bacteria during spring in a shallow eutrophic lake. Aquatic Microbial Ecology 22(2): 135–142.

Kübar, K., H. Agasild, T. Virro & I. Ott, 2005. Vertical distribution of zooplankton in a strongly stratified hypertrophic lake. Hydrobiologia 547: 151–162.

Laybourn-Parry, J., J. Olver, A. Rogerson & P. L. Duvergé, 1990. The temporal and spatial patterns of protozooplankton abundance in a eutrophic temperate lake. Hydrobiologia 203: 99–110.

Maly, E. J., 1975. Interactions among the predatory rotifer *Asplancha* and two prey *Paramecium* and *Euglena*. Ecology 56: 346–358.

Müller, H., A. Schöne, R. M. Pinto-Coelho, A. Schweizer & T. Weisse, 1991. Seasonal succession of ciliates in Lake Constance. Microbial Ecology 21: 119–138.

Ott, I., T. Kõiv, P. Nõges, A. Kisand, A. Järvalt & E. Kirt, 2005. General description of partly meromictic hypertrophic Lake Verevi, its ecological status, changes during the past eight decades and restoration problems. Hydrobiologia 547: 1–20.

Pace, M. L., 1982. Planktonic ciliates: their distribution, abundance and relationship to microbial resources in a monomictic lake. Canadian Journal of Fisheries & Aquatic Sciences 39: 106–116.

Patterson, D. J. & S. Hedley, 1992. Free-living Freshwater Protozoa. A Color Guide. Wolfe Publishing Ltd., England, 223 pp.

Porter, K. G., M. L. Pace & J. F. Battey, 1979. Ciliate protozoans as links in freshwater planctonic food chains. Nature 277: 563–565.

Putt, M. & D. K. Stoecker, 1989. An experimentally determined carbon:volume ratio for marine 'oligotrichous' ciliates from estuarine and coastal water Js. Limnology and Oceanography 34: 1097–1103.

Sherr, E. B. & B. F. Sherr, 1984. Role of heterotrophic protozoa in carbon and energy flow in aquatic ecosystems. In Klug, M. J. & C. A. Reddy (eds), Current Perspectives in Microbial Ecology. American Society for Microbiology, Washington, 412–423.

Sorokin, Y. I. & E. B. Paveljeva, 1972. On the quantitative characteristics of the pelagic ecosytem of Dalnee Lake (Kamchatka). Hydrobiologia 40: 519–552.

Steenbergen, C. L. M., J. P. Sweerts & T. E. Cappenberg, 1993. Microbial biogeochemical activities in lakes: stratification and eutrophication. 4. Aquatic Microbiology. An Ecological Approach. Blackwell Science Publication, 69–101.

Stoecker, D. K. & J. M. Capuzzo, 1990. Predation on protozoa: its importans to zooplankton. Journal of Plankton Research 12: 891–908.

Šimek, K. & V. Straškrabová, 1992. Bacterioplankton production and protozoan bacterivory in a mesotrophic reservoir. Journal of Plankton Research 14: 773–787.

Šimek, K., J. Bobkova, M. Macek, J. Nemoda & R. Psenner, 1995. Ciliates grazing on picoplankton in a eutrophic reservoir during summer phytoplankton maximum: a study at the species and community level. Limnology and Oceanography 40: 1077–1090.

Tammert, H., V. Kisand & T. Nõges, 2005. Bacterioplankton abundance and activity in a small hypertrophic stratified lake. Hydrobiologia 547: 83–90.

Utermöhl, H., 1958. Zur Vervollkommung der quantitativen Phytoplankton Methodik. Mitteilungen Internationale Vereiningung für Theoretische und Angewandte. Limnologie 9: 1–38.

Zingel, P., 1999. Pelagic ciliated protozoa in a shallow eutrophic lake: community structure and seasonal dynamics. Archiv für Hydrobiologie 146: 495–511.

Zingel, P. & I. Ott, 2000. Vertical distribution of planktonic ciliates in strongly stratified temperate lakes. Hydrobiologia 435: 19–26.

Hydrobiologia (2005) 547:175–184
I. Ott & T. Kõiv (eds), Lake Verevi, Estonia – A Highly Stratified Hypertrophic Lake
DOI 10.1007/s10750-005-4158-6

Long- and short-term changes of the macrophyte vegetation in strongly stratified hypertrophic Lake Verevi

Helle Mäemets* & Lilian Freiberg
Institute of Zoology and Botany, Estonian Agricultural University, Võrtsjärv Limnological Station, 61101 Rannu, Tartu County, Estonia
*(*Author for correspondence: E-mail: helle@zbi.ee)*

Key words: biomass, *Ceratophyllum demersum*, replace of dominants, re-establishing, tissue N and P

Abstract

The aim of study was to bring out changes in the macrophyte vegetation, caused by eutrophication, short-term lowering of the water level and the following restoration of equilibrium in L. Verevi. Also biomass and N and P content of shoots of main submergent species were studied in 1999–2001, to follow the temporal and specific differences. Due to strong eutrophication, the type of the lake changed from a *Myriophyllum-Potamogeton*-Charophyta lake to a *Ceratophyllum-Lemna trisulca* lake in 1984–1988, obviously owing to the formation of loose organic-rich sediment. Water lowering by 0.7 m during summer months of 1998 facilitated mineralization of sediments, as a consequence of which a mass development of *Ranunculus circinatus* and a temporary increase in the abundance and biomass of other nutrient-demanding species took place during following years. Our data suggest differences in nutrient supply and release of submerged species and the need for more species-related approach to this group. The problem of nutrient supply of unrooted plants at the time of stratification arises. Regarding the increase of biomass of *Ceratophyllum demersum* in second half of summer, we suppose that one part of nutrients for this growth may derive from freshly decayed filamentous algae or vascular plants.

Introduction

The small strongly stratified L. Verevi (12.6 ha, hard-water, more particularly characterized in Ott et al., 2005) has been rich in macrophytes at least during the last century. Emergent and submerged species dominate, while scattered stands of floating-leaved and floating plants are frequent and large filamentous algae occur abundantly. Macrovegetation has occupied 35–50% of the lake's area in different years. Over 50 macrophyte species have been recorded, among them 30 species of hydrophytes. The diverse bottom of the littoral zone (sand, gyttja, calcareous gyttja and dy) supports a diverse vegetation. Hard water favours high productivity of macroalgae, *Potamogeton* spp., *Elodea canadensis*, *Chara* spp. and other macrophytes which have high or intermediate extraction capacity of HCO_3^- (Madsen & Sand-Jensen, 1991) as the source of carbon. Supply from sediments is the advantage for rooted macrophytes over the phytoplankton in a situation where nutrients are scarce in water during the stratification period. However, also unrooted *Ceratophyllum* and large filamentous algae are abundant in such lakes and their nutrient supply is poorly studied. Changes in the opulent macrovegetation as the habitat, feeding and hiding area, are important for the biota of the littoral as well as for nutrient dynamics of whole lake (Diehl & Kornijów, 1998; Scheffer & Jeppesen, 1998; Feldman, 2001). Covering densely a large area, the submerged macrovegetation is a particularly essential integral component of the lacustrine ecosystem (Structuring Role, 1998).

The species composition of the macrovegetation probably determines the specific features of matter circulation, at least in the littoral zone. The aim of our study was to bring out long- and short-term changes in the macrovegetation and their possible reasons. An attempt was made to associate these changes with nutritional differences between the species.

Materials and methods

Three sets of literature and unpublished data, at three time levels, were available for our purposes:

(1) changes in the species composition in the 20th century;
(2) phenomena followed to the 0.7 m water lowering in summer months in 1998;
(3) seasonality in the biomass and nutrient content of macrophytes, studied in the most recent years.

To find out long-term changes, species abundance estimates according to Braun-Blanquet scale (1964) for the years 1929, 1957, 1988 and 2003 (Riikoja, 1940; Eesti järved, 1968; Mäemets Aime, 1991; our data), and vegetation distribution schemes (for 1957 compiled by H. Tuvikene, for 1988 compiled by A. Mäemets) were used. From the newest data 2003 was chosen, regarding the state as least influenced by the water lowering in 1998. Short-term changes resulting from water lowering were followed in 1998–2003 on the basis of: (1) a yearly description (1998–2003) and mapping (1998–2001) the vegetation in July when the species composition, abundances and depth limits were registered; (2) sampling of the shoots of main submerged plants with SCUBA (ducking equipment) from 33 quadrats (0.5 × 0.5 m) in midsummer 1999, 2000 and 2001 (Fig. 1). Five samples of filamentous algae taken in June 2000 and July 2001 included only their floating (not attached) part. For observation of seasonal changes in biomass and nutrient content, altogether 11 samples were taken from the same stands at the beginning of June and in mid-July in 2000. In 2001, four samples after every 2 weeks (total 20) between 24.05 and 21.07 were taken from stands in transition area between the wide southern part and the narrow northern part of the lake (Fig. 1) (*Ceratophyllum demersum* and *Potamogeton friesii* at the western shore, *Fontinalis antipyretica* and *Ranunculus circinatus* at the eastern shore). All above-ground part of plants (both for the dominating and other species) were collected, and their air-dry weight (=air-dry biomass = ADW) was determined without the crumbling (by drying) part of calcareous precipitate. The content of total phosphorus and total nitrogen in the plants was determined from a 2 g mixture (flowers, leaves and shoots) at the Laboratory of Plant Biochemistry of Faculty of Agronomy of EAU). Phosphorus (P) in plant tissues was analysed by Kjeldahl digest, stannous chloride method. Nitrogen (N) in tissues was analysed after H_2SO_4 (+Se) digest with gas-diffusion method. The amounts of nutrients per surface unit of stand area were calculated multiplying their concentration by biomass.

Data on weather conditions were obtained from the Tõravere Meteorological Station located at a distance of 5 km of the lake. Data on hydrochemistry and plankton were drawn from the database of the Võrtsjärv Limnological Station of the Institute of Zoology and Botany of EAU. The analysis of variance (Nonparametrics – Kruskal–Wallis ANOVA) (α = 0.05) was performed with Statistica 5.5 to test the significance of differences in biomass and nutrient content of plant samples.

Results

Long-term changes

Long-term changes are mostly related to abundance; species composition has been more stable (Table 1). However, three recently important species, *Lemna trisulca*, *Potamogeton friesii* and *P. pectinatus*, were not mentioned before 1980s (*Potamogeton* spp. were not identified to the species in 1929). *Myriophyllum spicatum*, being opulent in 1929 (Riikoja, 1940), is absent today, as well as *Stratiotes aloides*. The decline of *Schoenoplectus lacustris* and *Equisetum fluviatile* was marked already in 1957, as well as the increasing abundance of *Potamogeton natans*. The most remarkable is the replacement of the dominating charophytes by *Ceratophyllum demersum* in the 1980s (Fig. 2). Both Charophyta and *Ceratophyllum* occurred opulently during 4–5 years (Mäemets Aime, 1991);

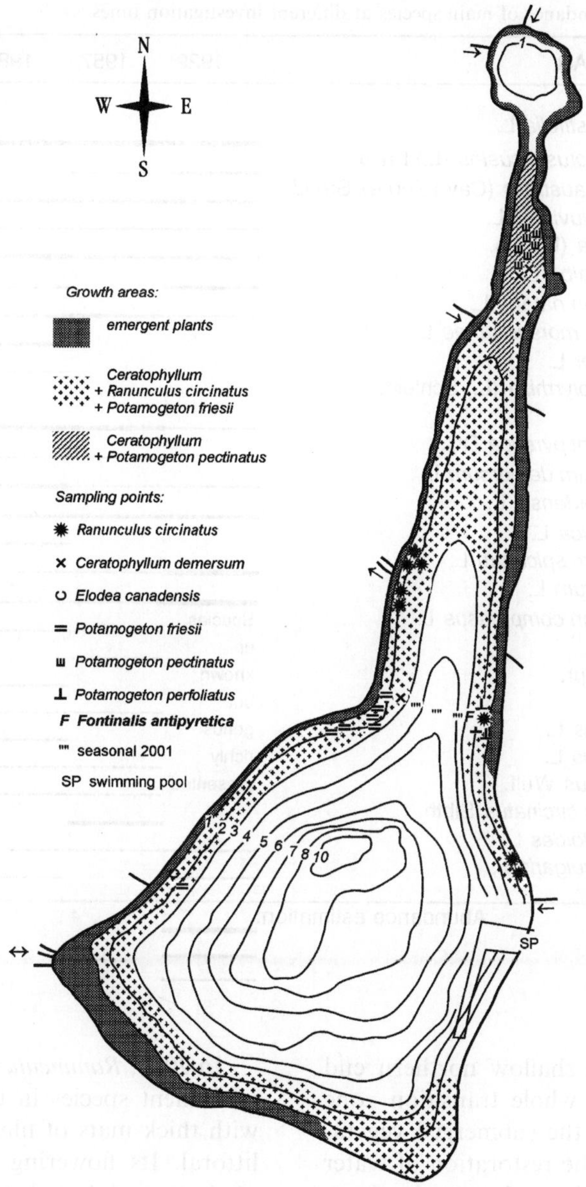

Figure 1. Lake Verevi, vegetation scheme and sampling areas.

later on, *Ceratophyllum* prevailed. Floating plants and large filamentous algae became important in the same period. The occurrence of the other *Potamogeton* species not mentioned above has been quite variable. The growth areas of different plant species have also changed. Only the charophytes (*Nitellopsis obtusa* Desv. in Lois., *Chara aspera* Deth. ex Willd., *Ch. globularis* Thuill., *Ch. tomentosa* L., *Nitella* sp.) have continuously preferred the northern part, rich in springs. The depth limit of the submergent belt was slightly decreasig before 1998, being 3.8 m in 1957 and 3.0 m in 1988.

Changes after the water lowering in 1998

During the 0.7 m water lowering period connected with the restoration of swimming pool in the

Table 1. Abundance of main species at different investigation times

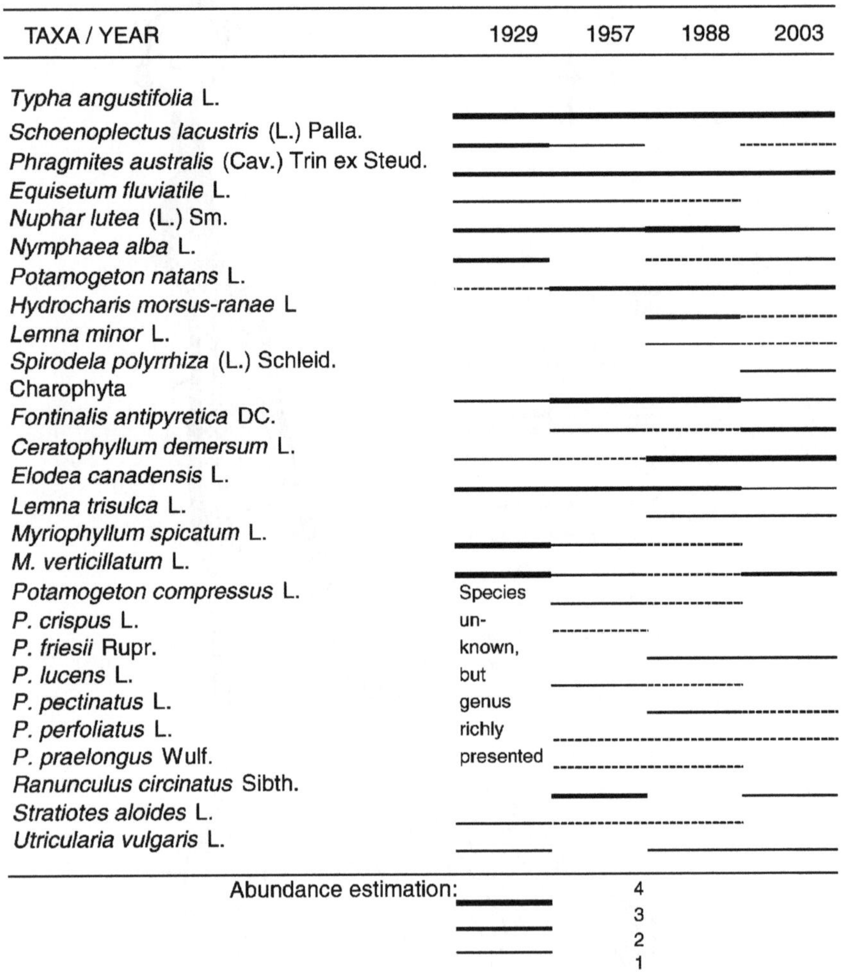

summer months 1998, the shallow northern end of the lake as well as the whole transition zone between the emergent and the submerged vegetation were denuded. After the restoration of water level in summer 1999, the submerged plants reached a depth of 2.0 m; in 2000 2.5 m. In 2001–2003, the maximum depth of submerged plants was 4.0 m, but mostly up to 3.0 m. The emergent plants reached usually a depth of 1 m, in some places 1.5 m. The growth depth of the floating-leaved stands was mostly 1.0–2.0 m.

In 1999, (Table 2) *Potamogeton pectinatus* and *P. friesii* occurred with high abundance. In the northern part, *P. pectinatus – Ceratophyllum* stands were extraordinarily luxuriant: in some samples total ADW of above-ground part exceeded 600 g m^{-2}, the mean was 346 ± 281.7 g m^{-2}.

In 2000, *Ranunculus circinatus* became the most prominent species in the lake, covering, together with thick mats of filamentous algae, most of the littoral. Its flowering started in the first decade of June and lasted over a month. This species dominated in the depth zone of 1–2 m, while in deeper water it was accompanied by *Ceratophyllum*, charophytes, *Potamogeton friesii*, *Elodea canadensis* and *Fontinalis antipyretica* occurred in many places, while some other *Potamogeton* species, e.g., *P. crispus*, were frequent locally. The maximum total ADW of the *Ranunculus*-dominated samples amounted to 973 g m^{-2}; the mean of submerged plants was 408 ± 214.8 g m^{-2}. The ADW of filamentous algae, mainly *Cladophora glomerata* (L.) Kütz., was about (based on two by hand samples from water surface) about 300 g m^{-2} at the begin-

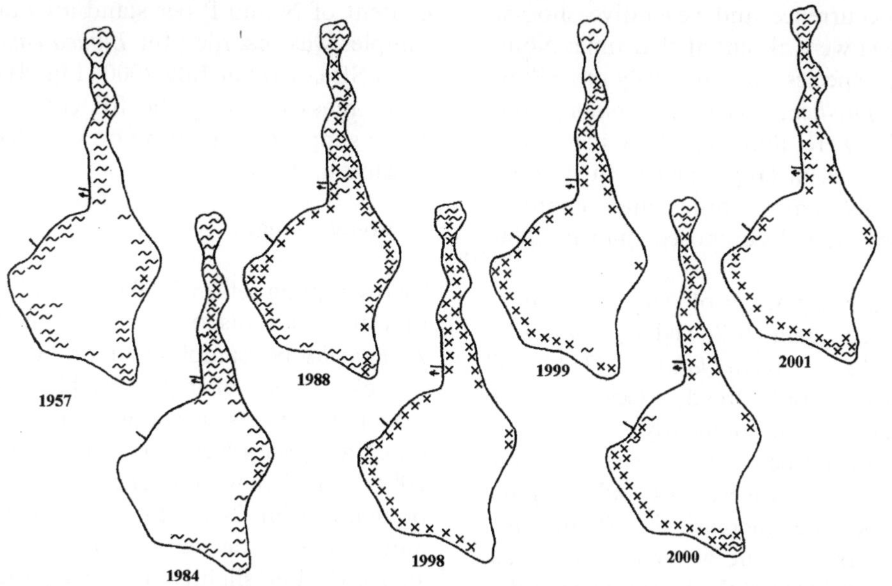

Figure 2. Distribution schemes of *Ceratophyllum* and charophytes in different years: ~ – Charophyta; x – *Ceratophyllum demersum*.

ning of mat-formation. *Potamogeton friesii* was decaying in the second half of July.

In summer 2001, some *Potamogeton* species were less represented (Table 2), while charophytes, *Elodea* and *Fontinalis* occupied the same position as before. The mass flowering of *Ranunculus* began in the second half of June (the spring was cooler than that of 2000), but *Potamogeton friesii* decayed already in midsummer and had less inflorescences than in the previous year. Although *Ranunculus* was opulent on the surface of plant mass again, it was *Ceratophyllum* that actually prevailed in ADW of samples. Maximum total ADW in such samples was about 600 g m^{-2}; while ADW of floating filamentous algae only 20–30 g m^{-2}.

The investigation of species composition in July of 2002 revealed a strong decline of *Ranunculus*

Table 2. Abundances of main submerged taxa after water lowering*

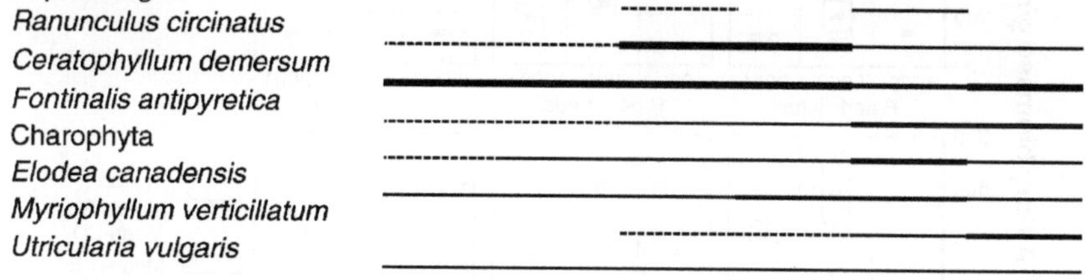

* Legend given at Table 1

with scattered occurrence and vegetative shoots. Filamentous algae were absent at that time. None of the submerged species was obviously prevailing in 2002 – *Ceratophyllum*, *Fontinalis*, *Potamogeton friesii* and *Elodea* were abundant. In some places charophytes occurred in large masses. In spite of the extraordinarily warm spring and summer, *P. friesii* appeared less decomposed than in 2000 and 2001.

In 2003, (at the beginning of August) *Ceratophyllum* dominated again, followed by *Fontinalis* and *Myriophyllum verticillatum* (Table 2). *Potamogeton friesii* was already decaying and filamentous algae occurred opulently.

The changes in the belts of the emergent and floating-leaved vegetation were observed less particularly. A moderate increase in *Phragmites* stands in first years after the water lowering was remarkable. But in 2002–2003, *Typha angustifolia* had replaced many of them. *Potamogeton natans* had recovered its frequence in 2002, and *Nymphaea alba* occurred more abundantly than in the preceding decades (Tables 1 and 2).

In the sampling years of 1999, 2000 and 2001, a statistically significant change was found in the midsummer air-dry biomass of *P. pectinatus* ($p = 0.027$) (Fig. 3). The plants of this species were conspicuously weaker and less productive in 2000 and 2001. The mean biomass of *Ceratophyllum* decreased in 1999–2001 (552.8 → 346.3 → 332.5 g m^{-2}), and that of *Ranunculus* decreased in 2000–2001 (514.9 → 261.9 g m^{-2}), but the changes were statistically not significant ($p > 0.05$). In all three sampling years, only the midsummer N% and P% in *Potamogeton perfoliatus* (Fig. 4) were statistically different ($p < 0.05$). The highest content of N and P per stand area among all our samples was recorded for *Elodea canadensis* taken from SE corner in July 2000 (Fig. 4). The opulent plant mass covering the SE corner (cleansed in 1998) supported the formation of black mud there already in 2000.

Seasonal changes

Two sampling sets in 2000 (9.06 and 17.07) showed that the amounts of N and P in *P. friesii* and *Ranunculus* per sampling quadrat decreased during 5 weeks despite of increasing biomass. Therefore, seasonal changes in the nutrient content of four different species were more thoroughly studied in 2001. Together with a general trend of increase in biomass from 24.05 to 21.07 (Fig. 5) the N% and P% were decreasing for *Ceratophyllum* and *P. friesii*, but slightly increasing for *Ranunculus*. Especially remarkable increase in ADW of *Ceratophyllum* and *Ranunculus* occurred in July. N% and P% for *Fontinalis* remained at a high level (Fig. 5). The amounts of N and P calculated per stand area unit, were in accordance with increasing biomass for *Ceratophyllum* and *Ranunculus*, but without any trend for *P. friesii*. The total amount of N and P, calculated for these four species per stand surface unit, increased during the observed period 2.5 and 2.6 times, respectively, mainly at the expense of increment in *Ranunculus* and *Ceratophyllum*. (Table 3).

Discussion

Owing to strong eutrophication in the 1970s and 1980s, described in the present issue, L. Verevi

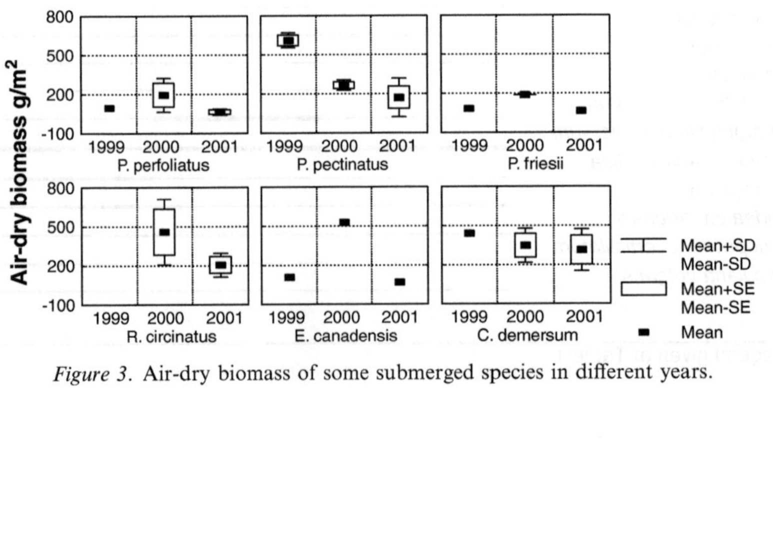

Figure 3. Air-dry biomass of some submerged species in different years.

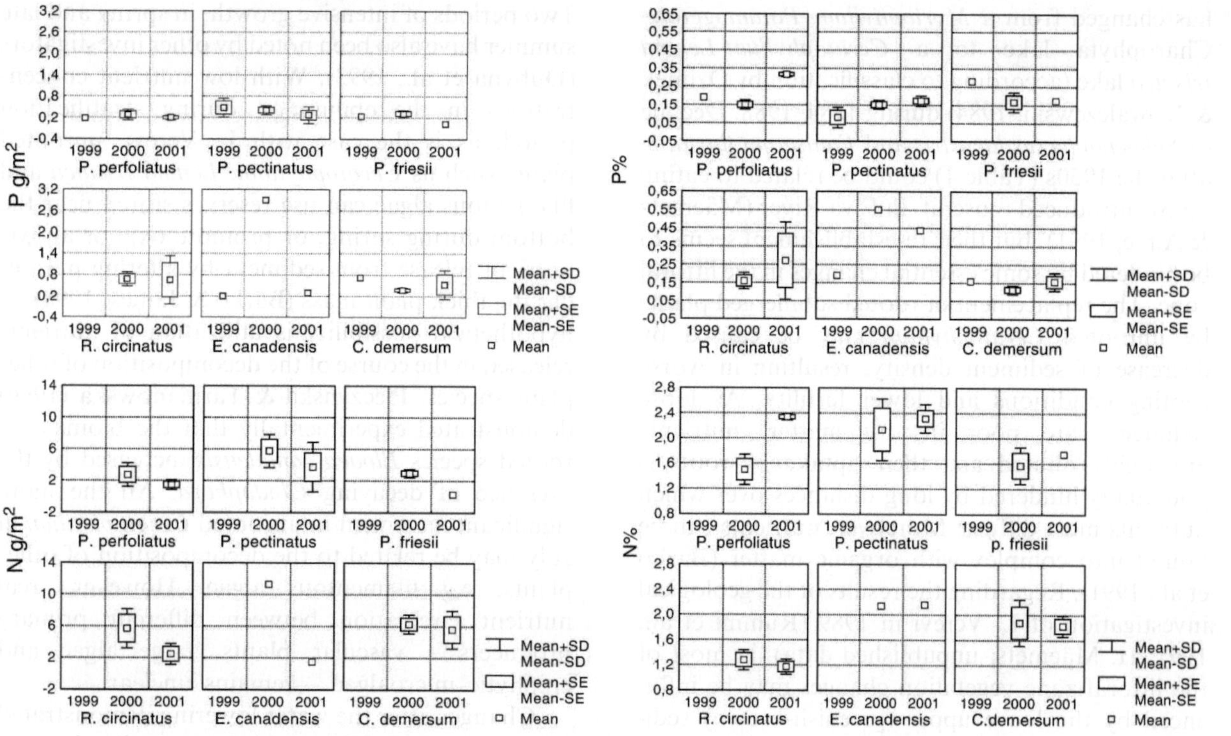

Figure 4. Tissue phosphorus and nitrogen concentrations (in air-dry matter) and amount of phosphorus and nitrogen per stand area unit of some submerged species in different years.

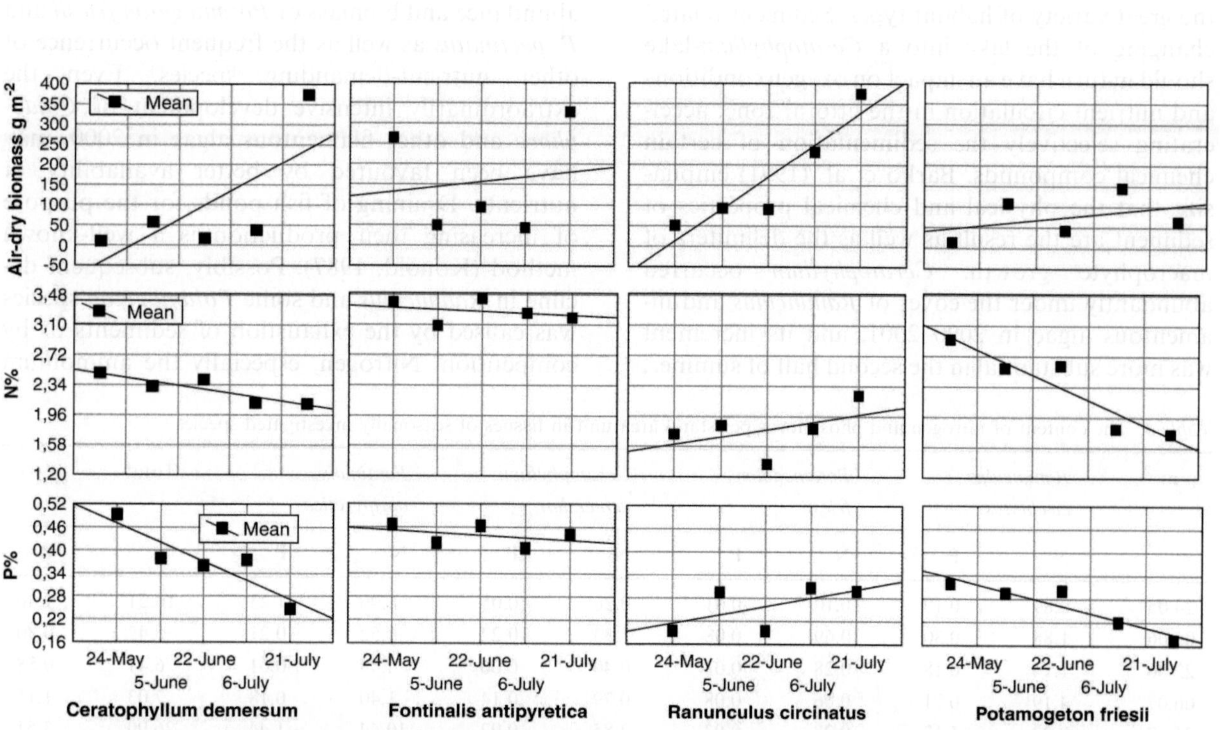

Figure 5. Air-dry biomass and nutrient concentrations of some submerged species during seasonal investigation.

has changed from a *Myriophyllum–Potamogeton–*Charophyta lake to a *Ceratophyllum-Lemna trisulca* lake (according to classification by Ozimek & Kowalczewski, 1984) during 1984–1988. Decline of *Schoenoplectus lacustris* and *Equisetum fluviatile* after the 1950s (Table 1) could be related to eating up by introduced muscrat in Elva River (Mäemets & Aime, 1991), but their re-establishment seems to be hindered by some essential changes in the littoral zone. The replacement of rooted submerged plants by unrooted *Ceratophyllum* may be caused by decrease of sediment density, resulting in worse rooting conditions and lower fertility. As loose sediments are poor in dry matter, nutrients are highly diluted and their uptake by roots is potentially hindered by long distances over which nutrients must diffuse. Moreover, nutrients can be bound into complex with organic matter (Barko et al., 1991). Regarding the results of the geological investigation of L. Verevi in 1989 (Rummi et al., 1991; H. Mäemets, unpublished data), in most of the littoral zone vegetation changes may be influenced by the loose upper (greenish-brown) sediment layer. The extinction of *Myriophyllum spicatum* may be associated with this circumstance. The general species richness was preserved due to the great variety of habitat types. Sediment related changing of the lake into a *Ceratophyllum*-lake should in turn have an impact on oxygen conditions and nutrient circulation in the littoral zone, accelerating selectively the sedimentation of certain chemical compounds. Barko et al. (1991) emphasize that the physical and chemical properties of sediment are the result as well as the delimiters of macrophyte growth. *Ceratophyllum* occurred abundantly under the cover of *Ranunculus* and filamentous algae in 2000–2001, and its increment was more substantial in the second half of summer. Two periods of intensive growth, in spring and late summer have also been noted by other investigators (Dubyna et al., 1993). With low nutrient concentrations in the epilimnion during stratification period, as is the case with L. Verevi, unrooted plants such as *Ceratophyllum*, *Lemna trisulca* and filamentous algae can use reserves stored near the bottom during spring, or promote oxic or anoxic nutrient release from sediment by altering pH under the thick plant mass (Barko & James, 1998). A hypothetical possibility is utilization of nutrients released in the course of the decomposition of other plant species. Pieczyńska & Tarmanowska (1996) demonstrated experimentally that the biomass of rooted species *Elodea canadensis* increased by the presence of decaying *Cladophora*. All the more significant increment in unrooted *Ceratophyllum* in July may be related to the decomposition of other plants, e.g filamentous algae. However, real nutrient circulation between different primary producers – vascular plants, large algae and epiphytic microalgae – remains unclear.

Changes after the water lowering demonstrated a strong impact of the aeration, mineralization and thickening of bottom sediments, resulting in the mass development of *Ranunculus circinatus*, high abundance and biomass of *Potamogeton friesii* and *P. pectinatus* as well as the frequent occurrence of other nutrient-demanding species. Even the extraordinarily intensive development of *Cladophora* and other filamentous algae in 2000, may have been favoured by better availability of nutrients. Draining of fish ponds for the purpose of increasing their production is a well-known method (Konold, 1987). Possibly, subsequent decline in *Ranunculus* and some *Potamogeton* species was caused by the exhaustion of sediments or by competition. Nitrogen, especially the ammonium

Table 3. The content of nitrogen and phosphorus per stand area unit in tissues of seasonally investigated species

$g\ m^{-2}$	*Ranunculus circinatus*		*Potamogeton friesii*		*Ceratophyllum demersum*		*Fontinalis antipyretica*		Total	
	N	P	N	P	N	P	N	P	N	P
24.05	0.89	0.10	0.10	0.01	0.26	0.05	8.96	1.23	10.21	1.40
05.06	1.88	0.30	0.69	0.08	1.33	0.22	1.52	0.21	5.42	0.80
22.06	1.24	0.18	0.28	0.03	0.40	0.06	4.49	0.61	6.41	0.88
06.07	4.19	0.71	0.66	0.08	0.79	0.14	1.40	0.18	7.03	1.11
21.07	8.33	1.12	0.28	0.03	7.85	0.92	10.44	1.44	26.90	3.51

N of the sediment is considered to be more limiting for macrophyte growth compared with phosphorus (Barko et al., 1991). But Figure 4 which presents the data of 3 years display mostly similar percent of the total N in the separate species in 2000 and 2001. Among all investigated species, the highest N% was characteristic for *Fontinalis* (Fig. 5) which has continuously increased in abundance after the water lowering. Supply from water and the sparing metabolism of mosses (Vinogradov et al., 2000) probably promotes its advancement. The decline of well-anchored nutrient-demanding *Potamogeton pectinatus* in recent years has been unambiguous. On the contrary, changes in the state of *Potamogeton friesii* are hard to understand (See in Results). The earlier or later turion formation seems to be depending on some other factors as the temperature. Also the uniform level of the amounts of N and P in *P. friesii* per sampling quadrate during 2 months in 2001 appears intriguing (Table 3). The preliminary increase of *Phragmites* after the water lowering is in good accordance with opinions that the water level is of great importance for its increase or decline (Andersson, 2001; Schmieder et al., 2002). The replacement of *Phragmites* by *Typha angustifolia*, preferring deeper water, was intensive after the rainy 2001.

The seasonal data should be interpreted with great caution, regarding the possible impacts of the actual state of stands and weather conditions. Due to small number of transect samples (one for every studied species from each sampling) these results are preliminary. The above described differences in the nutrient content of *Ranunculus* and *P. friesii* in 2000 and 2001, were probably related to weather conditions. The correlations with the temperature or percipitation differences of summers are probably "delayed", as the development of plants depends essentially on the conditions in preceding year. These preliminary results of the seasonal investigations suggest the existence of significant differences in matter circulation, which depend on the dominating species. They are in accordance with the conclusions by Wigand et al. (1998) that macrophyte species composition can alter sediment biogeochemistry, resulting in different phosphate, solid-phase phosphorus and metal levels in the pore water. Described changes in the character of macrovegetation, occupying more than a third of the lake's area, must be essential for the whole lake. Unfortunately, investigations following simultaneously other main groups of biota as well as hydrochemical parameters of littoral zone, are almost lacking. Some seasonal samples of ciliates from L. Verevi in 2001 (analyzed by K. Preismann) showed their markedly (to 10 times) larger amount in water sieved from plants (*Ranunculus circinatus*) than in water taken from among plants. The abundance of ciliates increased in midsummer. As many of their species are feeding on bacteria, we may suppose that nutrients released from senescing plants caused the abundance of bacteria and ciliates. The main consequence of our study might be the need for simultaneous complex investigations of littoral and more species-related way of approaching in macrophyte investigations.

Acknowledgements

The study was supported by the core grants of the Ministry of Education Nos. 0370208s98, 0362482s03 and by the Estonian Science Foundation grant N 3579 and 4835. We are very thankful to Dr. Reet Laugaste and Mr. Rene Freiberg for effective help and support, as well as to Ms. Kaili Preismann and other good colleagues from Võrtsjärv Limnological Station, providing us with data about L. Verevi. Dr. Tarmo Timm, Mr. Toomas Kõiv and Ms. Anu Palm helped us reviewing the manuscript, Mrs. Ester Jaigma improved the translation into English.

References

Andersson, B., 2001 Macrophyte development and habitat characteristics in Sweden's large lakes. Ambio 30: 503–513.

Barko, J. W., D. Gunnison & S. R. Carpenter, 1991. Sediment interactions with submersed macrophyte growth and community dynamics. Aquatic Botany 41: 41–65.

Barko, J. W. & W. F. James, 1998. Effects of submerged aquatic macrophytes on nutrient dynamics, sedimentation, and resuspension. In Jeppesen, E. M. Søndergaard, & K. Christoffersen (eds), The Structuring Role of Submerged Macrophytes in Lakes. Springer, New York, 197–214.

Braun-Blanquet, J., 1964. Pflanzensoziologie. Stuttgart. 865 pp.

Carpenter, S. R. & D. M. Lodge, 1986. Effects of submerged macrophytes on ecosystem processes. Aquatic Botany 26: 341–370.

Diehl, S. & R. Kornijów, 1998. Influence of Submerged Macrophytes on Trophic Interactions Among Fish and

Macroinvertebrates. In Jeppesen, E. M. Søndergaard, & K. Christoffersen (eds), The Structuring Role of Submerged Macrophytes in Lakes. Springer, New York, 24–46.

Dubyna, D. V., S. M. Stojko, K. M. Sytnik, L. A. Tasenkevich, Y. R. Shelyag-Sosonko, S. Hejný, Z. Hroudova, S. Husak, G. Otyagelova & O. Érzhabkova, 1993. Makrofity–indikatory izmenenij prirodnoj sredy. Kiev, Naukova dumka, 434 pp. [Macrophytes – the indicators of changes of natural environment. In Russian].

Eesti järved (Estonian Lakes), 1968. Ed. A. Mäemets, Valgus, Tallinn, 547 pp.

Feldman, R.S., 2001 Taxonomic and size structures of phytophilous macroinvertebrate communities in *Vallisneria* and *Trapa* beds of the Hudson River, New York. Hydrobiologia 452: 233–245.

Konold, W., 1987. Oberschwäbischen Weiher und Seen, I-II. Karlsruhe, 634 pp.

Madsen, T. V. & K. Sand-Jensen, 1991. Photosnthetic carbon assimilation in aquatic macrophytes. Aquatic Botany 41: 5–40.

Mäemets, Aime, 1991. Suurtaimestik. In Timm, H. (ed.), State of Lake Verevi (Hydrobiological Researches XVII), Tartu, 95–106. [Macrovegetation. In Estonian].

Ott, I., T. Kõiv, P. Nõges, A. Kisand, A. Järvalt & E. Kirt, 2005. General description of partly meromictic hypertrophic Lake Verevi, its ecological status, changes during the past eight decades and restoration problems. Hydrobiologia 547: 1–20.

Ozimek, T. & A. Kowalczewski, 1984. Long-term changes of the submerged macrophytes in eutrophic Lake Mikoajskie (North Poland). Aquatic Botany 19: 1–11.

Pieczyńska, E. & A. Tarmanowska, 1996. Effect of decomposing filamentous algae on the growth of *Elodea canadensis Michx* (a laboratory experiment). Aquatic Botany 54(4): 313–319.

Riikoja, H., 1940 Zur Kenntnis einiger Seen Ost-Eestis, insbesondere ihrer Wasserchemie. Annales Societatis rebus naturae investigandis in Universitate Tartuensi constitutae 46: 168–329.

Rummi, P., T. Mägi, J. Ütsi, H. Mäemets, A. Lindpere & A. Mäemets, 1991. Põhjasetted. In Timm, H. (ed.), State of Lake Verevi (Hydrobiological Researches XVII). Tartu: 22–33. [Bottom sediments. In Estonian].

Scheffer, M. & E. Jeppesen, 1998. Alternative stable states. In Jeppesen, E. M. Søndergaard, & K. Christoffersen (eds), The Structuring Role of Submerged Macrophytes in Lakes. Springer, New York, 397–406.

Schmieder, K., M. Dienst & W. Ostendorp, 2002. Auswirkung des Extremhochwassers 1999 auf die Flächendynamik und Bestandsstruktur der Uferröhrichte des Bodensees. Limnologica 32: 131–146.

Structuring Role of Submerged Macrophytes in Lakes, 1998. E. Jeppesen, M. Søndergaard, K. Christoffersen (eds.), Springer, New York, 406 pp.

Vinogradov, G. A., E. V. Borisovskaja & E. V. Lapirov, 2000. Osobennosti obmena kaltsiya i magniya u nekotoryh vodnyh rastenii razlichnyh sistematicheskih grup. – *Zhurnal obshchej biologii*, 61, 2: 163–172 [Differences in the metabolism of Ca and Mg of some hydrophytes of different systematic groups. In Russian].

Wigand, C., J. C. Stevenson & J. C. Cornwell, 1998. Effects of different submersed macrophytes on sediment biogeochemistry. Aquatic Botany 56(3–4): 233–244.

Hydrobiologia (2005) 547:185–195
I. Ott & T. Kõiv (eds), Lake Verevi, Estonia – A Highly Stratified Hypertrophic Lake
DOI 10.1007/s10750-005-4159-5

Macrozoobenthos of Lake Verevi

Henn Timm[1,]* & Tõnu Möls[2]
[1]*Estonian Agricultural University, Institute of Zoology and Botany, Võrtsjärv Limnological Station, 61101 Tartu County, Estonia*
[2]*Estonian Agricultural University, Institute of Zoology and Botany, 51014 Tartu, Estonia*
*(*Author for correspondence: E-mail: htimm@zbi.ee)*

Key words: macrozoobenthos, lake, stratification, Estonia

Abstract

An overview on studies of macrozoobenthos in the small, hard-water, stratified and hypertrophic Lake Verevi (South-Eastern Estonia) is given. The list of macroinvertebrates comprises at least 105 taxa. In the open water habitats, the biomass and abundance of macrozoobenthos (except the phantom midge *Chaoborus flavicans*) was rather constant beginning from the epilimnion up to the upper hypolimnion (depth 2–4 m), but very low in the lower hypolimnion (depth 6 m), which was inhabited mainly by *Chaoborus*. Comparison with long-term reference data from other Estonian lakes, belonging to similar limnological types, indicated that the total biomass and abundance (without *Chaoborus*) in the profundal of Verevi were very low.

Introduction

The hard-water, stratified and hypertrophic Lake Verevi (area 12.6 ha, maximum depth 11 m) is one of the best-studied freshwater bodies in Estonia (Mäemets, 1977; Timm, 1991; Ott & Kõiv, 1999). The first samples of macrozoobenthos were collected from open water by the hydrobiological staff of the Institute of Zoology and Botany (IZB) of the former Academy of Sciences in 1957, in order to give a general description of the lake. The next fieldwork of IZB, in which similar sampling techniques were used, followed in 1984 and 1988. The results of earlier studies are presented in the book "State of Lake Verevi" (Timm, 1991). In 1991, IZB conducted an all-year-round bottom-sampling series, accompanied with hydrochemical and other hydrobiological works. In 1996 and 1998, macroinvertebrates of the outer littoral were studied using handnets, the results being published in Timm et al. (1999). During 1998–2001, a new study in open-water areas was conducted, in order to estimate the relationships between macrozoobenthos and water stratification (Table 1).

In an other paper (Timm et al., 2001), we studied whether and how summer thermal stratification determines the taxonomical composition and amount of macrozoobenthos in several small stratified Estonian lakes, among them in Lake Verevi. In this work, a more thorough analysis of the influence of stratification on macrozoobenthos in Lake Verevi is given. A general description of macroinvertebrates of Lake Verevi is presented. On the basis of data of the years 1957–1996, temporal changes in macrozoobenthos are estimated. The biomass and abundance of macrozoobenthos of L. Verevi is compared with that of 90 reference lakes in Estonia.

Materials and methods

Qualitative samples were collected with a triangular sweepnet (edge length 12.5 cm, mesh size 0.3 mm, rod length 1 m) in May and July 1996 as well as in July 1998. Sampling sites were distributed along the shoreline; each time three samples were taken (Fig. 1). When sampling macrophytes, stable patches of *Carex* (living or dead) were

Table 1. Number of samples of macroinvertebrates taken from Lake Verevi in the years 1957–2001

Year	Quantitative	Qualitative	Sampling time
1957	12		Summer
1984	12		Summer
1988	12		Summer
1991	60		All year round
1996		6	Spring and Summer
1998	15	3	Summer
1999	15		Summer
2000	80		Spring to autumn
2001	20		Summer
Total	226	9	

preferred to sites with thick round stems of *Phragmites* and *Typha* or temporary plants. Three to four specimens from each easily identifiable taxon as well as up to 10 specimens of other taxa were fixed in 70% ethanol *in situ*. Sampling was continued until no new taxa were found.

Quantitative samples were taken from soft unvegetated bottoms in July–August, 1998, July–August, 1999; May, June, August and October, 2000; and August, 2001 (Fig. 1).

Sampling was arranged by zones in the following order:

(1) Sampling depth corresponding to: epilimnion, near the metalimnion's upper boundary; (2) the middle part of metalimnion; (3) the upper hypolimnion, near the metalimnion's lower boundary; (4) the lower (deeper) part of hypolimnion (only in 2000 and 2001). Although the depth of the lake extends to 11 m, samples of zoobenthos were not regularly taken from the very limited deepest region. In 1998, in each case, the positions of the metalimnion and oxycline in the water column were established before samples were taken by the team of hydrochemists of the same project. In 1999–2001, the sampling depths

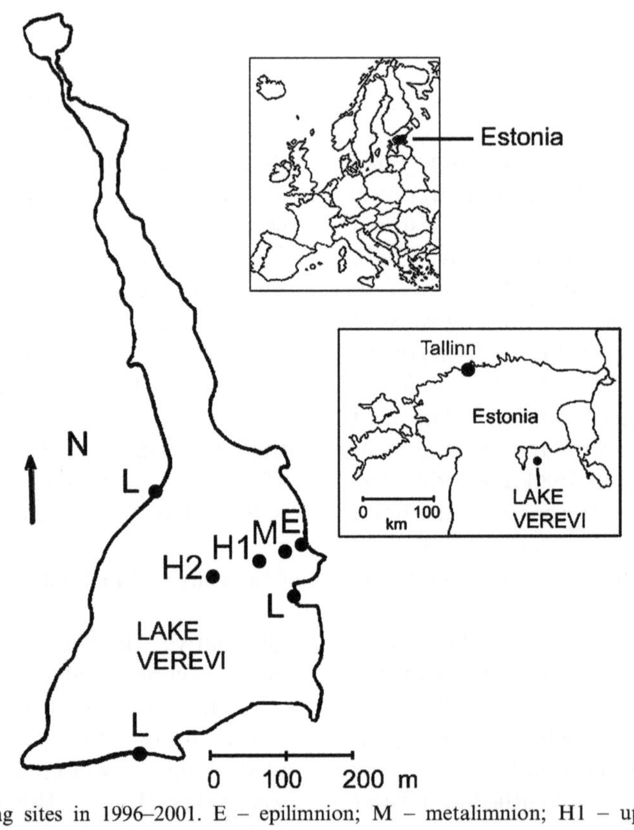

Figure 1. Location of sampling sites in 1996–2001. E – epilimnion; M – metalimnion; H1 – upper hypolimnion; H2 – lower hypolimnion; L – littoral.

were the same as in 1998. We assumed that the location of benthic organisms is more stable than possible small changes in the position of water strata in different years.

The approximate sampling depths were 2 m (epilimnion), 3 m (metalimnion), 4 m (upper hypolimnion) and 6 m (lower hypolimnion). The conditions above the bottom at each sampling site were considered equal to those observed at the same depth in the water column. Epilimnic areas with vegetation were avoided.

The macrozoobenthos was collected with a Boruckij-type sampler (a modification of the Ekman grab, grasp area 225 cm^2, box height 40 cm). Five replicates were taken each time from each zone (130 replicates at all). Macrozoobenthos samples were sieved (mesh size 0.5 mm) *in situ*, sorted alive in laboratory and then fixed in 70% ethanol. Because of the historical comparisons, the wet biomass of macrozoobenthos, measured with a torsion scale (accuracy 1 mg), was used. Where possible, the species of animals were identified (Appendix 2). In most cases, in calculations a higher identification level than genus was used. Genus-level identification did not usually provide a strikingly different description of community patterns than higher-level (e.g., family or order) taxonomic identification (Bowman & Bailey, 1997).

Altogether, the abundance and biomass of 20 taxa of macrozoobenthos were tested to normal distribution. The total abundance, total biomass and number of taxa per sample were also calculated. As chaoborid larvae have clearly a different depth distribution than that of other groups, the total biomass and total abundance without chaoborids were also determined as separate variables.

The reference dataset of the years 1951–2000 included 605 samples from 90 small lakes of the hard-water eutrophic or hypertrophic limnological types (Eesti järved, 1968; unpublished archive of the Limnological Station). The most common bottom substrates in L. Verevi are mud without macrovegetation, muddy sand without vegetation, muddy sand with vegetation, and pure sand with vegetation. All data, including reference data, was united and processed by a general 106 parameter Covariance Analysis model (see Appendix 1). The model consisted of three discrete factors: Status (L. Verevi, reference lake), Macrovegetation (not present, present; nested within Status), and Sediment Type (mud, muddy sand, sand; nested within Status). The Sampling Depth was presented by a second-order polynomial, and the Day Number within the year by a third-order polynomial, both nested within the Status. The Year was presented by a fourth-order polynomial with coefficients nested within Status, Macrovegetation and Sediment. The interaction between Depth and Year was presented by a four-term mixed polynomial nested within Status, Macrovegetation and Sediment. That was included to allow the yearly dependencies differ for different depths.

From all measured biological variables, only 11 were chosen for analysis because only they had a statistically acceptable residual distribution. The following variables were measured: total biomass, total abundance, total biomass without *Chaoborus* biomass, total abundance without *Chaoborus* abundance, biomass of chironomids, abundance of chironomids, biomass of oligochaetes, abundance of oligochaetes, biomass of *Chaoborus*, abundance of *Chaoborus*, number of taxa per sample identified without magnification. Prior to analysis, all abundance and biomass variables were log $(x + 1)$ transformed to make their distribution more normal.

The statistical analysis was carried out with the aid of the SAS system, Release 8.1 (SAS Institute Inc., SAS/STAT Software: Changes and enhancements through release 6.11. Cary, NC: SAS Institute Inc, 1996). The main tools were the estimable contrasts and parametric functions in the framework of the general linear analysis. The estimated model was used to predict values and confidence limits for the variable at different years, but at a fixed status (Verevi or Reference), fixed day (190th day of year), fixed depth (6 m), absence of vegetation and mud sediment. The predicted values were used for picturing year-dependence graphs. Additional parametric functions (SAS estimates) were tailored for testing the statistical significance and the sign of the slope in each year. These were used for marking, with arrows, the time intervals where the increase or decrease of the variable was significant. Specific parametric functions were also constructed to test the differences between Verevi and Reference lakes in each year of 1984–2001.

Results and discussion

Taxa richness

Altogether, 105 benthic macroinvertebrate taxa were found in L. Verevi during 1984–2001. Sixty-seven taxa were identified in quantitative (grab) samples, and 58 taxa in qualitative littoral samples. Chironomids (31 species) and oligochaetes (12 species) were the most taxon-rich groups (see Appendix 2). The samples of 1957 were not identified to species and not preserved, and the identification to species was not complete in some other cases.

Figure 2 indicates a strong relationship of the taxa number to the sampling depth and layers in 2000: the larger the depth (or the lower the layer), the less taxa occurred. In comparison with other seasons, summer seemed to have a negative effect on the taxa number in the metalimnion. On the contrary, in the lower hypolimnion, autumn was the time when the taxa number was the lowest (*Chaoborus flavicans* being the only species).

Compared with other Estonian small lakes, L. Verevi as a nutrient-rich, well-buffered and relatively deep waterbody, excels in a quite high number of macroinvertebrate taxa. In the littoral of 34 Estonian lakes, studied in the years 1994–1996, the number of taxa per 6 qualitative samples varied between 21 and 44. Among them, L. Verevi was on the 2nd to 4th position with 36 taxa (chironomids were not identified to species) (Timm et al., 1999). In comparison with nine other small-stratified South-Estonian lakes, which were studied in 1998–1999, the taxa richness in Lake Verevi was in the third position in the epilimnion, but even the highest in the littoral and metalimnion (Timm et al., 2001). In the hypolimnion, the taxa richness was evenly low in all ten lakes.

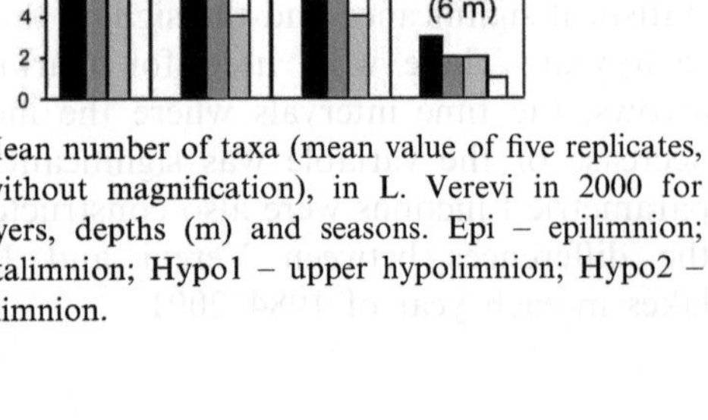

Figure 2. Mean number of taxa (mean value of five replicates, identified without magnification), in L. Verevi in 2000 for different layers, depths (m) and seasons. Epi – epilimnion; Meta – metalimnion; Hypo1 – upper hypolimnion; Hypo2 – lower hypolimnion.

Abundance and biomass: spatial and seasonal variability

In Table 2 the geometric averages of the 11 chosen parameters in four thermal layers during 1998–2001 are presented.

In the case of the total biomass and total abundance, there were no apparent differences between different layers, because the low amount of other taxa in the lower hypolimnion was compensated by *Chaoborus flavicans*. The total biomass and total abundance without *Chaoborus* were significantly lower in the lower hypolimnion than in other layers. Thus the upper hypolimnion was more similar to the meta and the epilimnion than to the lower hypolimnion. The high biomass of chironomids (and the total biomass without *Chaoborus*) in the upper hypolimnion was mainly caused by the presence of large *Chironomus plumosus* larvae. The changes in the mean number of taxa indicated rather an even negative regression, than a sharp decline between the two parts of the hypolimnion.

As integral variables, the total abundance without *Chaoborus* and the total biomass without *Chaoborus* were chosen to characterize the differences between different layers and seasons (Fig. 3). The maximum values were observed in spring, and the minimum ones at the end of summer. Like in Table 2, in 2000 differences between layers were evident only between the lower hypolimnion and all other layers. At the end of summer and in autumn, no animals, except *Chaoborus*, were found in the hypolimnion.

Unlike results of concurrent hydrochemical, plankton, meiozoobenthos and fish studies, presented in this volume (see Järvalt et al., 2005; Kangro et al., 2005; Kübar et al., 2005; Tammert et al., 2005; Zingel, 2005), the total biomass and abundance of animals (except *Chaoborus*) was quite constant beginning from the epilimnion (2 m) up to the upper hypolimnion (4 m), but very low in the lower hypolimnion (6 m).

Table 2. The geometric mean values of the biological parameters with approximately normal distribution during 1998–2001

1998–2001	Epilimnion (2 m)	Metalimnion (3 m)	Upper hypolimnion (4 m)	Lower hypolimnion (6 m)
Total biomass	2.51	1.38	2.16	2.03
Total abundance	575	408	327	525
Total biomass without *Chaoborus*	2.46	1.14	1.73	0.05
Total abundance without *Chaoborus*	567	348	226	33
Biomass of chironomids	0.40	0.43	1.32	0.05
Abundance of chironomids	229	141	150	31
Biomass of oligochaetes	0.40	0.16	0.15	0.01
Abundance of oligochaetes	196	96	84	22
Biomass of *Chaoborus*	0.03	0.19	0.16	1.97
Abundance of *Chaoborus*	25	68	75	504
Mean number of taxa	3.8	2.8	2.3	1.6

Abundance and biomass: long-term variability

The general temporal changes of macrozoobenthos variables are shown in Figures 4 and 5, for reference lakes beginning from 1950 and for L. Verevi from 1984. Figure 4 shows that in the reference lakes there were significant differences in all studied metrics during the whole study period. In particular, biomass and abundance, as well as the number of taxa revealed a significant decrease at the end of the period (the same could be expected in L. Verevi beginning from 1984). At the same time, no significant differences of any variable were found in the macrozoobenthos of L. Verevi in the years 1984–2001 (Fig. 5). This lake, which was hypertrophic already in the mid-1980s (Timm, 1991), had perhaps already reached a relatively stable condition. In the reference lakes, both oligochaetes and *Chaoborus*, but not chironomids or sum values, increased significantly in biomass and abundance during 1960–1975. No reliable explanation of such a change is available. The intensive use of mineral fertilizers in Estonia began in the early1970s (Eesti Loodus, 1995), and therefore could not have caused an enhancement of such low oxygen tolerant groups as oligochaetes (tubificids) and *Chaoborus*. Because of insufficient data about the macrozoobenthos of L. Verevi before 1984, a comparison with reference data was possible only for the period 1984–2001. Likewise, the only bottom type, used in calculations, was soft mud without macrovegetation, standardized to 6 m depth (profundal).

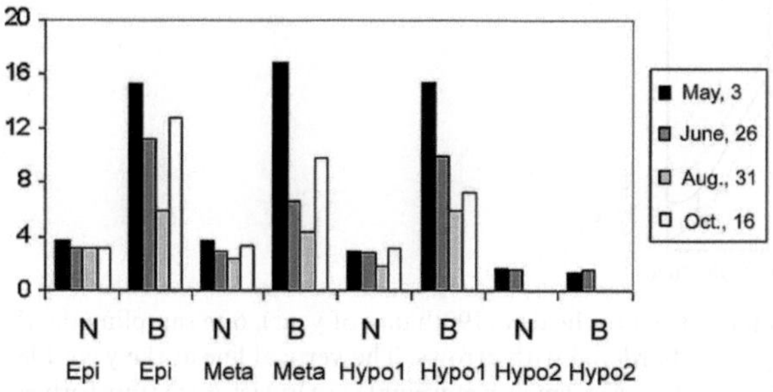

Figure 3. Total abundance without *Chaoborus* (N, ind. m^{-2}) and total biomass without *Chaoborus* (B, wet mg m^{-2}), transformed as log (N+1) and 5 log (B+1) in L. Verevi in 2000. Epi – epilimnion; Meta – metalimnion; Hypo1 – upper hypolimnion; Hypo2 – lower hypolimnion.

The following significant differences between L. Verevi and the reference lakes were observed. The second half of 1980s and the beginning of 1990s were periods when several characteristics of the zoobenthos of L. Verevi were much lower than in reference lakes. Compared with the overall decrease of most of the metrics in the reference lakes in 2000, in Lake Verevi the total biomass and the total abundance, as well as the biomass and abundance of *Chaoborus*, were higher. The relatively large amount of *Chaoborus* probably caused the significantly higher total abundance and biomass in L. Verevi in 2000, compared to other lakes. The number of taxa, identified by naked eye, was the only variable which never differed significantly (Table 3).

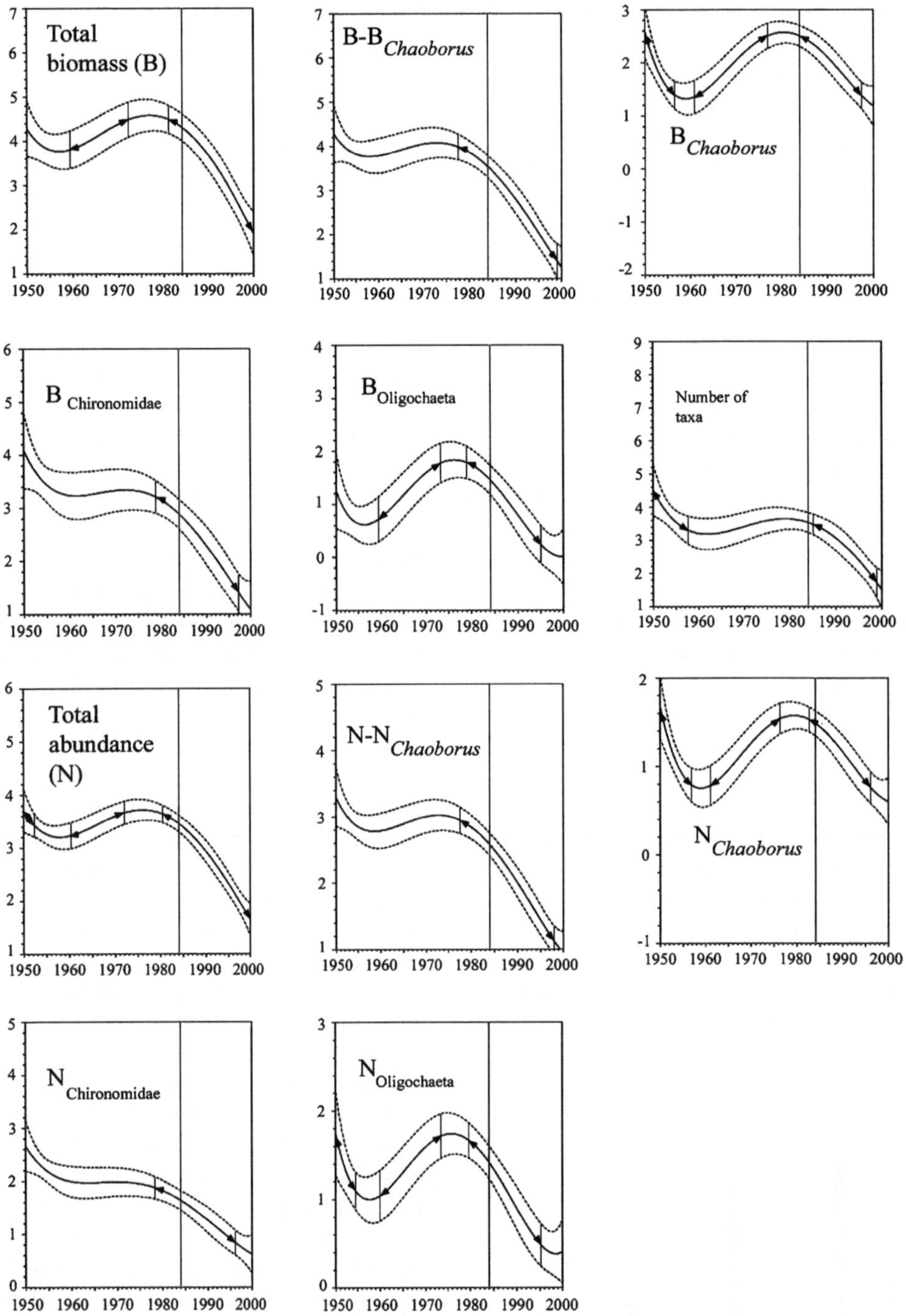

Figure 4. The trends of macrozoobenthos metrics in reference lakes, standardized to the day (190th day of year), 6 m sampling depth, and muddy sediment without vegetation. The areas of significant changes are bordered with arrows. The vertical line at the year 1984 marks the starting-point in Figure 5. Abundance (ind. 675 cm^{-2}) and biomass (mg 675 cm^{-2}, wet weight) are log (x + 1) transformed.

The sampling depth and bottom substrate are the main factors, which influence the distribution of macrozoobenthic taxa (Meriläinen et al., 2000). Hypolimnetic oxygen deficiency, which is the main limiting factor for the benthic community in the profundal, is not necessarily related to the trophic

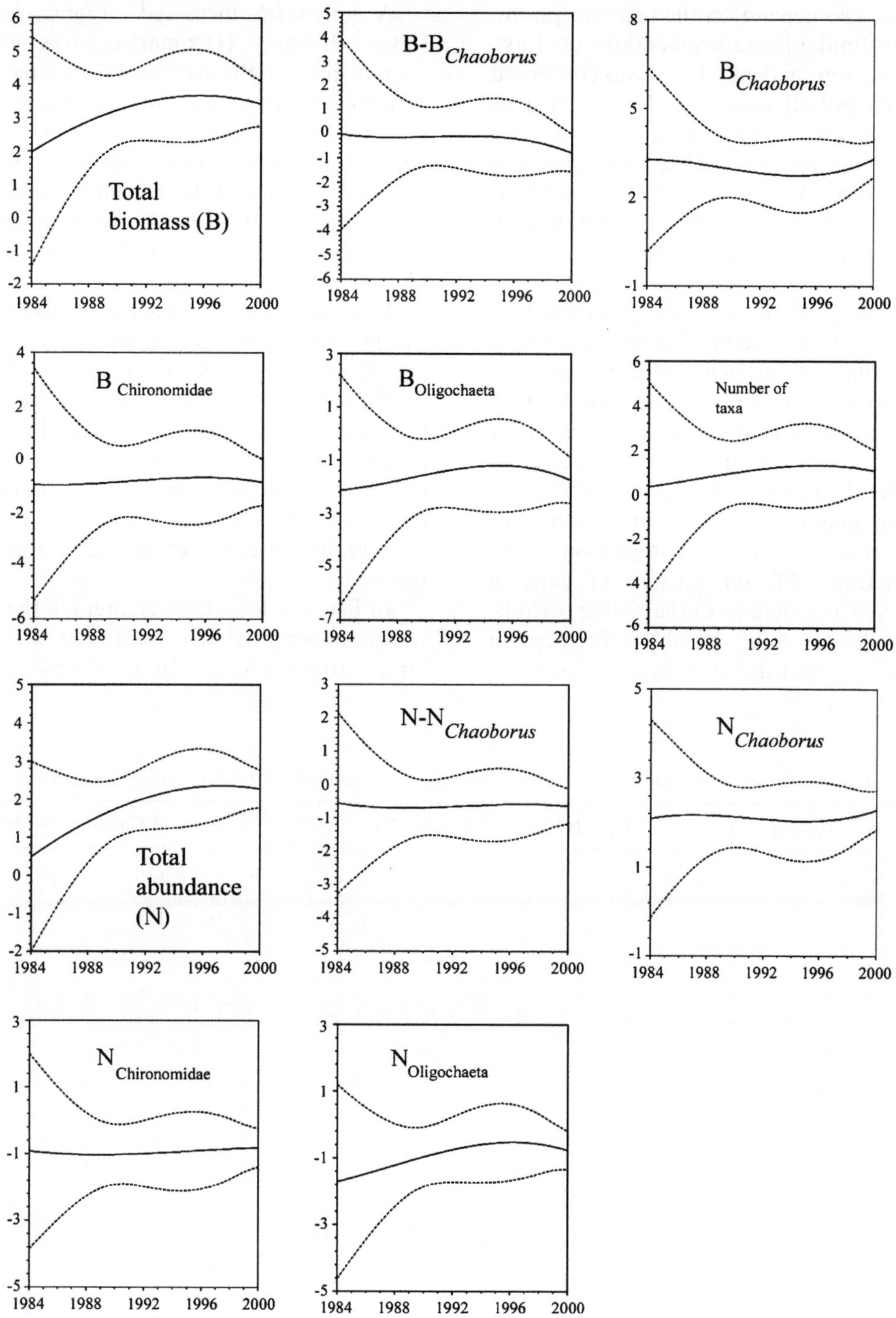

Figure 5. The trends of macrozoobenthos metrics in L. Verevi, standardized to the day (190th day of year), 6 m sampling depth, and muddy sediment without vegetation. Abundance (ind. 675 cm^{-2}) and biomass (mg 675 cm^{-2}, wet weight) are log $(x + 1)$ transformed.

state but may also depend on the lake morphometry. The profundal macrozoobenthos of Lake Nemi (Italy; maximum depth 32 m) was composed of well-known pollution-tolerant forms, such as oligochaetes *Tubifex tubifex* with many relative taxa, *Chironomus plumosus*, *Chaoborus flavicans* and *Procladius* sp. (Bazzanti & Seminara, 1987). It seems that the profundal macrozoobenthos in stratified and polluted lakes is similar, irrespective of their geographical location. On the other hand, the benthic fauna of the very deep Italian lake Albano (160 m) had a depth distribution, which was strongly related to oxygen content in the water column. Seasonal variations generally showed no significant differences in the community parameters between summer/autumn and winter/early spring months (Bazzanti & Seminara, 1995).

Studies on long-term changes of lake macrozoobenthos are quite rare, and comparison of the data is sometimes difficult because of gaps in sampling as well as differences in sampling periods, in sampling devices with variable efficiency, in sample processing and the identification level.

A long-term increased organic pollution in Lake Esrom (Denmark) was considered responsible for the decline in abundance of two *Pisidium* species, and at the same time, the increase of *Potamothrix hammoniensis* (Lindegaard et al., 1997). The observed changes in the profundal fauna of L. Esrom were comparative small during 1932–1995, because the present community in 1995 still included numerous oxyphilous macrozoobenthic taxa.

In Lake Geneva, Switzerland (maximum depth 310 m), 1958–1985 chironomid larvae were studied (Lods-Crozet & Lachavanne, 1994). The profundal zone (>50 m) was colonised only by a few taxa, characteristic to deep-lake fauna. The results still confirmed that the communities had changed in 10 years from an zeta-oligotrophic (1958) to a theta-mesotrophic (1978) chironomid community, thus reflecting the eutrophication of the lake.

In Estonia, long-term changes in the amount of macrozoobenthos have been well studied in two large eutrophic lakes. In L. Võrtsjärv, a general

Table 3. Significance of differences between Lake Verevi and reference lakes in 1984–2001 for 11 biological variables

Year	Total B	Total N	Total B*	Total N*	B Chir	N Chir	B Olig	N Olig	B Chaob	N Chaob	Taxa
1984	.	.	.	.	.	.	.	.	.	.	.
1985	.	.	.	.	.	.	.	.	.	.	.
1986	.	.	.	–	.	.	.	.	.	.	.
1987	.	.	–	–	–	–	.	–	.	.	.
1988	.	–	–	–	–	–	.	–	.	.	.
1989	.	–	–	–	–	–	–	–	.	.	.
1990	.	–	–	–	–	–	–	–	.	.	.
1991	.	.	–	–	–	–	–	–	.	.	.
1992	.	.	–	–	–	–	.	–	.	.	.
1993	.	.	–	–	–	–	.	.	.	+	.
1994	.	.	.	–	.	–	.	.	.	+	.
1995	.	.	.	.	.	.	.	.	.	+	.
1996	.	.	.	.	.	.	.	.	.	+	.
1997	.	.	.	.	.	.	.	.	+	+	.
1998	.	.	.	.	.	.	.	.	+	+	.
1999	.	.	.	–	.	–	.	.	+	+	.
2000	+	+	–	–	–	–	–	–	+	+	.
2001	.	.	–	–	–	–	–	–	+	+	.

Total B – total biomass, Total N – total abundance (* – without Chaoborus), Chir – chironomids, Olig – oligochaetes, Chaob – Chaoborus, Taxa – mean number of taxa, identified by naked eye. Significant differences ($p < 0.01$) are marked with '+' (Verevi > Reference) or '–' (Verevi < Reference); insignificant differences are marked with dots.

increase of *Chironomus plumosus* and oligochaetes was observed during 1973–1996, while the amount of small molluscs (excluding unionids and *Dreissena*) decreased (Kangur et al., 1998). Both trends were explained by increasing eutrophication.

In the open-water area of L. Peipsi-Pihkva, during 1964–1991, the biomass of *Asellus*, Gammaridae (an invader, *Gmelinoides fasciatus*), Valvata, Dreissena and Hirudinea increased significantly, while that of Hydracarina declined (Timm et al., 1996). The decrease of Hydracarina and the increase of Valvata were probably caused by progressing eutrophication. At the same time, the biomass of many groups decreased significantly in the littoral (up to 4 m), because the number of predatory gammarids increased.

Conclusions

According to some macrozoobenthos variables, Lake Verevi proved to be a typical thermally stratified lake, with the lower hypolimnion inhabited mainly by phantom midge (*Chaoborus flavicans*). Unlike results of the concurrent hydrochemical, plankton, meiozoobenthos and fish studies, presented in this publication, the total biomass and abundance of animals (except *Chaoborus*) was quite constant beginning from the epilimnion (2 m) up to the upper hypolimnion (4 m), but very low in the lower hypolimnion (6 m). On the contrary, the number of macrozoobenthic taxa decreased gradually in the direction of epilimnion to hypolimnion. At the same time, the upper littoral of the lake was quite rich in macroinvertebrate taxa, due to relatively high water transparency, alkalinity and nutrient content, as well as rich vegetation with abundant periphyton along the shoreline.

A comparison with long-term reference data of other Estonian lakes belonging to similar limnological types indicated that in the profundal of L. Verevi the total biomass and abundance (without that of *Chaoborus*) were often significantly lower. Hard stratification may be a reason why the abundance and biomass of macrozoobenthos in L. Verevi were rather low, compared with reference lakes, because the lower layers usually support only few species with low or unstable numbers. In 2000, the amount of *Chaoborus* in Lake Verevi had increased so much that the total amount of macrozoobenthos was significantly higher than in the reference lakes.

Acknowledgements

The study was supported by the core Grants of the Ministry of Education Nos. 0370208s98, 0362482s03 and by Grants of the Estonian Science Foundation Nos. 3579, 4483. The sampling, identification and analysis were conducted by a team of the Limnological Station (Institute of Zoology and Botany, Estonian Agricultural University). We are indebted to Drs Külli Kangur and Tarmo Timm for species identification of chironomids and oligochaetes. We also thank Dr Aleksander Maastik for linguistic aid.

References

Bazzanti, M. & M. Seminara, 1987. Profundal macrobenthos in a polluted lake. Depth distribution and its relationship with biological indices for water quality assessment Acta Oecologica. Oecologica Applicata 8: 15–26.

Bazzanti, M. & M. Seminara, 1995. Eutrophication in a deep, meromictic lake (Lake Albano, Central Italy): spatial-temporal patterns of profundal benthic community as a tool for assessing environmental stress in the hypolimnion Limnologica 25: 21–31.

Bowman, M. F. & R. C. Bailey, 1997. Does taxonomic resolution affect the multivariate description of the structure of freshwater benthic macroinvertebrate communities? Canadian Journal of Fisheries and Aquatic Sciences 54: 1802–1807.

Eesti järved, 1968. Tallinn, "Valgus": 532 pp. [Estonian lakes. In Estonian].

Eesti. Loodus, 1995. Compiled by A. Raukas. Publishing Office "Valgus". Tallinn, 606 pp. [Estonia. Nature. In Estonian, Russian and English summary].

Järvalt, A., T. Krause & A. Palm, 2005. Diel migration and spatial distribution of fish in a small stratified lake. Hydrobiologia 547: 197–203.

Kangro, K., R. Laugaste, P. Nõges & I. Ott, 2005. Long-term changes and seasonal development of phytoplankton in a strongly stratified, hypertrophic lake. Hydrobiologia 547: 91–103.

Kangur, K., H. Timm, T. Timm & V. Timm, 1998. Long-term changes in the macrozoobenthos of Lake Võrtsjärv Limnologica 28: 75–83.

Kübar, K., H. Agasild, T. Virro & I. Ott, 2005. Vertical distribution of zooplankton in a strongly stratified hypertrophic lake. Hydrobiologia 547: 157–162.

Lindegaard, C., P. C. Dall & P. M. Jónasson, 1997. Long-term patterns of the profundal fauna in Lake Esrom. In Sand-Jensen, K. & O. Pedersen (eds), Freshwater Biology. Priorities and Development in Danish Research. GEC Gad, Copenhagen, 39–53.

Lods-Crozet, B. & J. B. Lachavanne, 1994. Changes in the chironomid communities in lake Geneva in relation with eutrophication, over a period of 60 years Archiv für Hydrobiologie 130: 453–471.

Mäemets, A., 1977. Eesti NSV järved ja nende kaitse. Valgus, Tallinn: 263 pp [Lakes of the Estonian S.S.R. and their protection. In Estonian].

Meriläinen, J., H. Veijola & J. Hynynen, 2000. Zoobenthic communities in relation to the depth zones in a large boreal lake in Finland Verhandlungen Internationale Vereinigung für theoretische und angewandte Limnologie 27: 985–988.

Ott, I. & T. Kõiv, 1999. Estonian small lakes: special features and changes. EV Keskkonnaministeeriumi Info- ja Tehnokeskus, Eesti Teaduste Akadeemia, Eesti Põllumajandusülikooli Zooloogia ja Botaanika Instituut, Tallinn, 128 pp.

Tammert, H., V. Kisand & T. Nõges, 2005. Bacterioplankton abundance and activity in a small hypertrophic stratified lake. Hydrobiologia 547: 83–90.

Timm, H., (ed.), 1991. Verevi järve seisund. A monograph. Estonian Academy of Sciences, Institute of Zoology and Botany, Tartu 139 pp. State of Lake Verevi. In Estonian, English and Russian summary.

Timm, T., K. Kangur, H. Timm & V Timm, 1996. Macrozoobenthos of Lake Peipsi-Pihkva: long-term biomass changes Hydrobiologia 338: 155–162.

Timm, H., T. Möls, K. Kangur & T. Timm, 1999. Littoral macroinvertebrates in some small lakes of Estonia. In: Biodiversity in Benthic Ecology. Proc. from Nordic Benthological Meeting in Silkeborg, Denmark, 13–14 November 1997. NERI Technical Report, No. 266: 133–139.

Timm, H., T. Möls & T. Timm, 2001. Macro- and meiozoobenthos in some small stratified lakes of Estonia. In: Nordic Benthological Society. Mývatn Research Station. Twin Symposium on Cold Aquatic Environment. Lake Mývatn 13–16 May 2001. Abstracts, 52.

Zingel, P., 2005. Vertical and seasonal dynamics of planktonic ciliates in a strongly stratified hypertrophic lake. Hydrobiologia 547: 163–174.

Appendix 1. Description of the basic model in SAS programming language

```
proc GLM data = Alldata;
    Class Stat Veg Sed;
    model Var = Stat Veg(Stat)|Veg(Stat) Veg(Stat)|Dep(Stat)
        Dep(Stat)|Dep(Stat)|Year(Stat Veg Sed)|Year(Stat Veg Sed)
        Day(Stat)|Day(Stat)|Day(Stat)
        Year(Stat Veg Sed)|Year(Stat Veg Sed)|Year(Stat Veg Sed)|Year(Stat Veg Sed)
                / SS3 solution;
run;
```

Var – biological variable; Stat – status; Veg – macrovegetation; Sed – sediment; Dep – sampling depth; Day – sampling day; Year – sampling year.

Appendix 2. List of macroinvertebrate taxa collected in Lake Verevi during 1984-2001

Nematoda	*Sialis lutaria*	*Hippeutis complanatus*
Mermithidae indet.	Lepidoptera	*Lymnaea stagnalis*
Oligochaeta	Pyralidae indet.	*Physa fontinalis*
Ilyodrilus templetoni	Trichoptera	*Planorbis planorbis*
Limnodrilus hoffmeisteri	*Phryganea grandis*	*Radix auricularia*
Limnodrilus udekemianus	*Leptocerus tineiformis*	*Radix ovata*
Potamothrix bedoti	*Triaenodes bicolor*	*Valvata piscinalis*
Potamothrix hammoniensis	*Cyrnus flavidus*	Diptera
Psammoryctides barbatus	*Grammotaulius nigropunctatus*	*Ablabesmyia monilis*
Tubifex tubifex	*Limnephilus flavicornis*	*Ablabesmyia phatta*
Dero digitata	*Limnephilus politus*	*Anatopynia plumipes*
Ophidonais serpentina	*Limnephilus stigma*	*Camptochironomus tentans*
Marionina riparia	*Molanna angustata*	*Ceratopogonidae indet.*
Stylaria lacustris	Coleoptera	*Chaoborus flavicans*
Uncinais uncinata	*Haliplus confines*	*Chironomus commutatus*
Hirudinea	*Haliplus* gr. *Ruficollis*	*Chironomus plumosus*
Erpobdella octoculata	*Graphoderus cinereus*	*Cladopelma viridula*
Glossiphonia complanata	*Ilybius ater*	*Cladotanytarsus mancus*
Helobdella stagnalis	*Ilybius fenestratus*	*Cricotopus sylvestris*
Hemiclepsis marginata	*Porhydrus lineatus*	*Cryptochironomus* gr. *Defectus*
Crustacea	*Rhantus* sp.	*Dicrotendipes lobiger*
Argulus foliaceus	*Enochrus testaceus*	*Dicrotendipes pulsus*
Asellus aquaticus	*Hydrophilidae* indet.	*Dicrotendipes tritomus*
Gammarus lacustris	Heteroptera	*Dixidae indet.*
Arachnida	*Ilyocoris cimicoides*	*Einfeldia carbonaria*
Argyroneta aquatica	*Mesovelia furcata*	*Endochironomus albipennis*
Hydracarina indet.	*Nepa cinerea*	*Endochironomus tendens*
Ephemeroptera	*Notonecta* sp.	*Glyptotendipes glaucus*
Caenis horaria	*Cymatia coleoptrata*	*Glyptotendipes paripes*
Caenis robusta	*Sigara distincta*	*Microtendipes pedellus*
Cloeon dipterum	*Sigara falleni*	*Parachironomus arcuatus*
Odonata	*Sigara fossarum*	*Paratanytarsus confuses*
Aeshna cyanea	Bivalvia	*Polypedilum cultellatum*
Coenagrion lunulatum	*Amesoda scaldiana*	*Pentapedilum sordens*
Cordulia aenea	*Anodonta cygnea*	*Polypedilum tetracrenatum*
Enallagma cyathigerum	*Euglesa* (= *Pisidium*) *rivularis*	*Procladius choreus*
Epitheca bimaculata	*Sphaerium corneum*	*Psectrocladius simulans*
Erythromma najas	Gastropoda	*Tanypus kraatzi*
Sympetrum vulgatum	*Bathyomphalus contortus*	*Tanypus vilipennis*
Megaloptera	*Bithynia tentaculata*	*Tanytarsus lugens*
		Tanytarsus mendax (= *holochlorus*)

Hydrobiologia (2005) 547:197–203
I. Ott & T. Kõiv (eds), Lake Verevi, Estonia – A Highly Stratified Hypertrophic Lake
DOI 10.1007/s10750-005-4160-z

Diel migration and spatial distribution of fish in a small stratified lake

Ain Järvalt*, Teet Krause & Anu Palm
Institute of Zoology and Botany, Estonian Agricultural University, Võrtsjärv Limnological Station, 61101, Rannu, Tartu County, Estonia
*(*Author for correspondence: E-mail: ajarvalt@zbi.ee)*

Key words: *Perca fluviatilis*, *Rutilus rutilus*, diel migration, spatial distribution, stratified lake, multimesh gillnet

Abstract

The diel migration and spatial distribution of fish were explored using six sequential 4-h sample gillnettings in the pelagic and littoral zones of Lake Verevi (Estonia, 12.6 ha, max. depth 11 m, hard-water, deoxygenated hypolimnion) in August 2001 and July 2002. Considering abundance, two-thirds of the total fish moved to the littoral zone. The biomass of fish was distributed evenly between the littoral and pelagic zones, where the topmost epilimnion accounted for 80–85% leaving 10–15% for the lower epilimnion in the pelagic zone. Just above the thermocline only some large specimens of perch *Perca fluviatilis* (L.) and roach *Rutilus rutilus* (L.) (1–5%) during the daytime were captured. No fish movements were recorded under the thermocline. Rudd *Scardinius erythrophtalmus* (L.) inhabited only the littoral zone; all the other species were captured in both zones. Juvenile perch stayed in the littoral zone, whereas juvenile roach was caught in both zones and was active over a 24-h period. Piscivores, perch and pike *Esox lucius* L., were inactive in the dark. Perch inhabited mostly the littoral zone and the duration of its activity increased with age. In summer-stratified Lake Verevi, sharp change in the values of oxygen in the metalimnion along with species interaction affected the spatial distribution of fish, while diel migration was light-dependent.

Introduction

In lakes, the biota is mostly determined by water quality and lake morphometry, while fish composition and biomass are related to the content of dissolved solids *per* mean depth of the water column (Ryder et al., 1974; Moss, 1998). The catchability of passive fishing gears is directly related to moving activity of fish. The activity of fish depends on season, light conditions and temperature, besides mutual interaction of fish species (Davenport & Sayer, 1993; Rowe, 1994; Jurvelius & Sammalkorpi, 1995; Persson et al., 1996; Ekloev, 1997; Horppila, 1998; Dörner et al., 1999; 2001; Jepsen et al., 1999; Hölkner & Breckling, 2002). Vegetated sites had higher densities of fish, specially smaller fish and greater species richness than unvegetated sites (Randall et al., 1996; Jacobsen et al., 2002a). Inter-annual variation in fish community structure, in biomass–size distributions of benthic lake fish communities, in mutual interaction of species and in activity are well-known (Holmgren, 1999; Holmgren & Appelberg, 2000; Olin et al., 2002). The dominant fish species in Lake Verevi as in most eutrophic small Estonian lakes were roach and perch (Mäemets, 1977; Pihu 1993). Our goal was to study the diel migration and spatial distribution of fish in this small hard-water lake at the time of sharp summer stratification.

Material and methods

Verevi (South Estonia) is a small (12.6 ha), slightly exorheic, sheltered, and hence a stratified lake with a steep thermocline, and a small drainage basin of 1.1 km^2. The maximum depth is 11 m, and the average depth 3.6 m. At the time of the study, the Secchi disc transparency of water in Lake Verevi was 3 m, bottom was covered with submerged plants to a depth of 3.5 m, and values of pH ranged from 8.2 in the upper epilimnion to 7.5 at 4 m, and to 7.0 at 5 m.

According to the literature data, 12 fish species inhabited L. Verevi (Eesti järved, 1968; Mäemets, 1977). Roach and perch are still the most abundant in the lake, while pike, rudd, tench *Tinca tinca* (L.) and crusian carp *Carassius carassius* (L.) are of second-rate abundant. The common benthophagous species, bream *Abramis brama* (L.) and ruffe *Gymnocephalus cernuus* (L.) of L. Verevi between 1950 and 1980, have obviously disappeared by now.

The fish composition was explored on 2–3 August in 2001, and on 8–9 July in 2002. We used Danish type of multi-mesh nylon monofilament gillnets (of 14 randomly placed 3 × 1.5-m mesh panels). The gillnets were arranged to catch at depths of 0–1.5 , 2.5–4 , 4.5–6 , and 6.5–8 m in the pelagic zone. In every depth one gillnet was used. Two gillnets were placed at a depth of 1–2.5 m in the littoral zone. This zone was characterised by the *Typha–Phragmítes–Chara–Potamogeton–Nuphar* complex. All captured fish were sorted by mesh size and species, and measured by total length (TL, to the nearest 1 mm), and total weight (TW, to the nearest 0.1 g); sex and consumed prey (fish) were identified. The age of perch by *operculum* and of roach by scales was determined.

Since noon, the gillnets were checked in sequential 4 h over a 24-h period (altogether six times). In both years, the days of experiment were sunny. The specific feature of weather in the morning of the experiment day in 2001 was a weak thunderstorm. In both years, the mornings were foggy and direct sunlight irradiated the water column for 14 h in 2001 (the morning was foggy from 4 to 8 a.m.), and only for 11.5 h in 2002 (clouds dispersed at 10 a.m.) out of possible 16–17 h characteristic of the season. The water temperature of the topmost epilimnion was 22.4 °C in August 2001 and somewhat lower (21.6 °C) in July 2002. The temperature and oxygen gradients over the water column in the pelagial at gillnetted depths are presented in Figure 1.

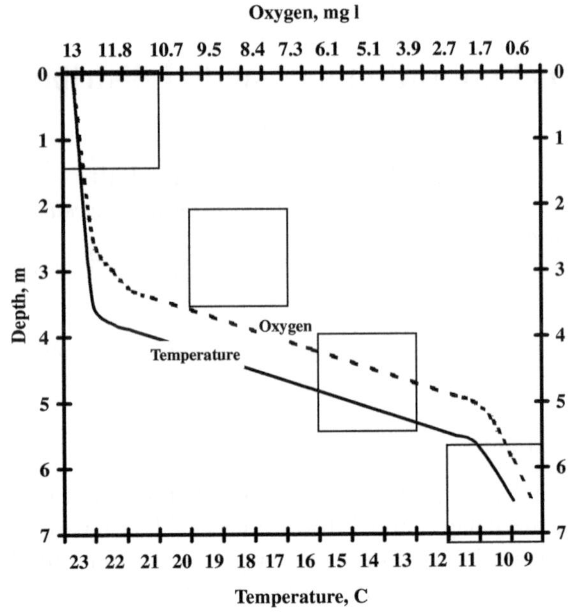

Figure 1. Temperature and oxygen gradients over a water column at the gillnetted (transparent squares) station.

Results

In August 2001, the total landing of 24-h gillnetting (13.8 kg) at two stations (littoral and pelagic) comprised 376 fish from five species: roach, perch, tench, rudd, and pike. The biggest captured fish were a 1188 g tench and a 1090 g pike (accounting 18% for total catch). In July 2002, 301 fish (9.9 kg) were captured at the same stations. In comparison with the catches of the previous year, rudd and juvenile tench were absent, and a juvenile pike and a 1407 g tench were captured.

Spatial segregation

Rudd inhabited only the littoral zone; all the other species were captured in both zones. Roach and perch outweighed the other species accounting for 80% of the total catch. The lengths of the captured perch and roach were distributed evenly in both years (Fig. 2), as did the catches between the net panels of different mesh size (Fig. 3). In the littoral zone, juvenile roach and perch were caught as shoals; rudd and juvenile pike were captured in places with opulent water-plants, while tench inhabited shallow

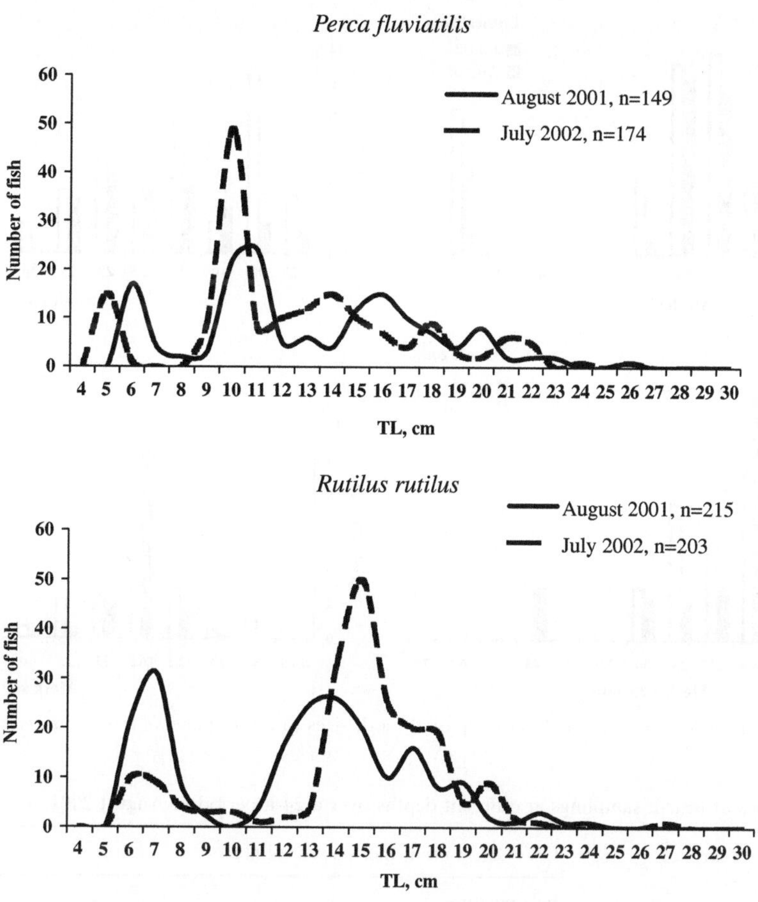

Figure 2. The length distributions of perch (*Perca fluviatilis*) and roach (*Rutilus rutilus*) in lake Verevi; comparison between two sequential summers.

waters with *Typha angustifolia* L. and *Phragmites australis* (Cav.) Trin. ex Steud. At the station of the pelagic zone, most fish inhabited the topmost 1.5 m of the epilimnion of the total 7 m water column (Table 1). In both years, roach and perch were abundant in this layer (Table 2). At a depth of 2.5–4 m, the abundance of fish decreased sharply (up to 10–12% of the total catch in the pelagic zone). Mostly roach, rarely perch and a 5 year-aged pike inhabited this water layer. As many as eight fishes were captured with five gillnets (out of 12) at a depth of 4–5.5 m. Perch (80–100 g, aged 4–5 years; 2001) as well as roach (20–30 g; 2002) were captured at this depth. Our experiments showed that fish avoided this depth in the dark and never occurred in the anoxic hypolimnion below the depth of 5.5 m.

Diel migration

Fish, mainly roach and perch, inhabited the topmost 1.5 m layer of the epilimnion throughout the 24-h study period in both years. In 2001, the TW of 4-h catch in the topmost layer ranged from 544 to 916 g. Under the conditions of intensive solar radiation in July 2002, the catch at 2–3.5 m outweighed the catch from the topmost layer. In two periods, one at sunrise and the other in the afternoon, 3–5-years-old roach and perch descended into the deeper water layers of the pelagial zone. In the littoral zone, juvenile fish was the most active and hence abundantly entrapped in the morning just after sunrise.

Gillnets caught most evenly roach that, unlike perch, was active in the pelagial even at night. Perch migrated most frequently at noon and at

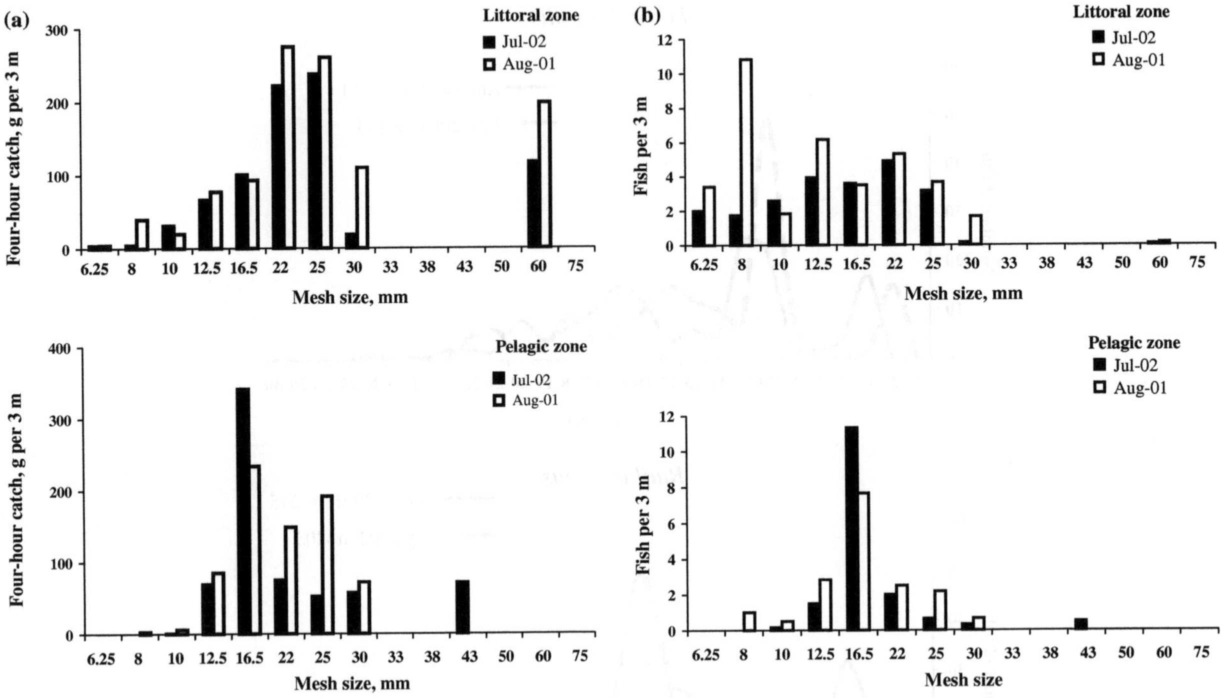

Figure 3. Comparison between the catches of explored mesh-sizes (3-m net-panels); a in biomass, b in abundance.

Table 1. The gillnet catches of four-h samplings at different depths over a 24-h period in August 2001 and July 2002

Time	Year	CATCH, g multi mesh-sized gillnet^{-1}			
		Pelagic zone			Littoral zone
		0–1.5 m	2–3.5 m	4–5.5 m	1–2.5 m
4–8 p.m.	2001	675	1290	0	845
	2002	911	157	51	1026
8–12 p.m.	2001	544	92	0	32
	2002	214	339	31	1246
0–4 a.m.	2001	701	0	0	795
	2002	691	112	0	631
4–8 a.m.	2001	856	39	34	860
	2002	571	233	0	580
8–12 a.m.	2001	675	0	0	1239
	2002	401	280	0	1882
0–4 p.m.	2001	916	779	201	3224
	2002	168	84	24	259
Total, g	2001	4367	2200	235	6995
	2002	2200	2035	106	5624

sunset. Whereas juvenile pike got entrapped in shallow waters either in the dark or in the morning, mature pike foraged on roach in the afternoon at a depth of 4 m. In the littoral zone, fish were the most frequent at noon and the scarcest long before sunset, when the nearby

Table 2. The abundances of perch (*Perca fluviatilis*) and roach (*Rutilus rutilus*) in lake Verevi at explored depths over a 24-h period in 2001 and 2002

Time	Pelagic zone				Littoral zone			
	2001		2002		2001		2002	
	Roach	Perch	Roach	Perch	Roach	Perch	Roach	Perch
0–4 p.m.	33	14	9	6	20	26	7	24
4–8 p.m.	14	9	30	4	6	25	17	45
8–12 p.m.	22	–	14	2	24	–	48	29
0–4 a.m	10	3	16	–	17	13	22	11
4–8 a.m.	18	5	7	10	27	16	16	26
8–12 a.m.	10	7	9	9	7	42	3	14
In 24 h	107	38	85	31	101	122	113	149
Both species	145		116		223		262	

forest overshadowed the station. Tench was active in the littoral zone in the afternoon, whereas juvenile tench was captured in the pelagial zone at sunset.

Discussion

Among the local fish species, roach is the most abundant planktivore in Estonian small lakes where it is followed by piscivorous perch (Pihu, 1993). While perch prefers and is more numerous in clear-water lakes (Diehl, 1988), cyprinids including roach, are known to thrive under eutrophic and turbid conditions (Helminen et al., 2000; Jeppesen et al., 2000; Jacobsen et al., 2002a, 2002b). In our study on Lake Verevi, roach and perch were co-dominants, the roach slightly outnumbering but clearly overweighing perch. A similar composition of fish is characteristic of mesotrophic lakes in southern Finland (Olin et al., 2002). In highly eutrophic lakes, the proportion of other cyprinids and percids, such as bream, white bream (*Blicca bjoerkna*) and ruff, increased (Michelsen et al., 1994; Olin et al., 2002). During the past years L. Verevi, especially its epilimnion, has changed from hypertrophic to mesotrophic with oligotrophic features (Ott et al., 2005) while bream and ruffe have disappeared by now.

Compared with other small Estonian lakes explored in summer, the catch per unit effort (CPUE) of multi-mesh gillnet calculated in kilograms per night (12 h) reached only the average of the littoral zone, and three times lower of the pelagial (Krause et al., 2001). The length of juvenile roach corresponded to that backcalculated for Estonian small lakes (Eesti järved, 1968) and for Finnish lakes (Horppila & Nyberg, 1999) and its length frequency distributions was similar to that in the eutrophic lake Frederiksborg Slotssø (Michelsen et al., 1994). Age composition showed a similar scarceness of 2- to 3-year-aged roach both in the pelagic and littoral zones, being obviously catchable as a refugee in reeds (Herzig et al., 2002), the biotope not explored in lake Verevi.

On a daily basis, the migration of roach depends mostly on temperature (Krause et al., 1998), and on the need to escape or forage (Ekloev & Perrson, 1995, 1996; Jachner, 2001). According to our investigation, juvenile roach (TW less than 10 g) was active in the dark and early in the morning in the pelagic zone. At this time, juvenile pike was foraging in the littoral zone. Roach, with TW over 30 g, was active over a 24-h period, and catchable conditioned by light. Roach inhabited the upper layer of the pelagial in the dark and descended in the daylight, contrary to the previously described pattern based on the movement of zooplankters (Rowe, 1994; Dörner et al., 1999; Lauridsen et al., 1999; Romare et al., 1999; Burks et al., 2002).

In Lake Verevi, the abundance and biomass of large copepods was the highest in the lower layer of the metalimnion (Kübar et al., 2005) avoided by fish. Although the highest biomass of planktivores was located in the topmost epilimnion, decreasing sharply with each next metre in depth, the biomass of *Daphnia cucullata* remained unchanged up to a depth of 5 m, where fish occurred extremely rarely. This might indicate strong predation on juvenile perch that consumes predominantly cyclopoid copepods in the absence of predators and macroinvertebrates in the presence of predators (Persson & Ekloev, 1995). Juvenile roach feeds on phytoplankton, then switches to rotifers and microcrustaceans, maintaining the capacity to feed on blue-green algae and higher plants (Hofer & Wiezer, 1987; Persson, 1987). Roach remains among the vegetation in the presence of perch, while both roach and perch remain in the vegetation in the presence of pike (Ekloev & Persson, 1996). Roach is reported to move vertically in turbid lakes – ascending in the daylight and descending at night, contrary to the mainly horizontal movement between the pelagic and littoral zones described in a clearwater lake (Jacobsen et al., 2002b).

Piscivores, perch and pike, were inactive in the dark. Mature perch remains among the littoral vegetation overnight, avoiding shallow waters in the daylight. One-year-old perch changes its habitat due to predation risk in the morning, at mid-day, and in the evening, migrating from the pelagial into the macrophytes in the morning (Imbrock et al., 1996). In the lakes lacking pike, perch used both littoral and pelagial habitats of the lake. The relative number of adult perch and pike increases with depth, although in pike-lakes, perch occurs along the littoral zone but never below the thermocline (Persson et al., 1996). In lakes having low density of pike, as in L. Verevi, the abundance of perch was twice higher in the littoral zone.

According to Perrow et al. (1996), tench was generally active at night and almost completely inactive during daylight, resting in shoals among the littoral emergent plants, mostly *Typha angustifolia*. Conversely, in L. Verevi tench was active only in daytime.

Sharp gradients in the oxygen content, food availability and temperature in the metalimnion, evidently, determine the vertical distribution of fish in L Verevi.

Acknowledgements

This work was supported by the target financed project No 0370208s98 of the Ministry of Education "The influence of thermal stratification on biological matter circulation in lakes" and by grants Estonian Science Foundation No 3579 and 4835. We are grateful to Mrs Ester Jaigma for linguistic revision of the text.

References

Burks, R. L., E. Jeppesen, D. M. Lodge & T. Lauridsen, 2002. Diel horizontal migration of zooplankton: costs and benefits of habitating littoral zones. Freshwater Biology 47: 343–365.

Davenport, J. & M. D. J. Sayer, 1993. Physiological determinants of distribution in fish. Journal of Fish Biology 43: 121–145.

Diehl, S., 1988. Foraging efficiency of three freshwater fishes: effects of structural complexity and light. Oikos 53: 207–214.

Dörner, H., A. Wagner & J. Benndorf, 1999. Predation by piscivorous fish on age-0 fish: spatial and temporal variability in a biomanipulated lake (Bautzer reservoir, Germany). Hydrobiologia 408/409: 39–46.

Dörner, H., H. Schultz, T. Mehner & J. Benndorf, 2001. Interaction between prey availability and feeding behaviour of age-2 and age-2 Perch (*Perca fluviatilis* L.) in a Biomanipulated Lake (Bautzen Reservoir, Germany). Limnologica 31: 11–16.

Ekloev, P. & L. Perrson, 1995. Species – species antipredator capacities and prey refuges: interactions between piscivorous perch (*Perca fluviatilis*) and juvenile roach (*Rutilus rutilus*). Behavioral Ecology and Sociobiology 37: 169–178.

Ekloev, P., 1997. Effects of habitat complexity and prey abundance on the spatial and temporal distributions of perch (*Perca fluviatilis*) and pike (*Esox lucius*). Canadian Journal of Fisheries and Aquatic Sciences 54: 1520–1531.

Helminen, H., J. Karjalainen, M. Kurkilahti, M. Rask & J. Sarvala, 2000. Eutrophication and fish biodiversity in Finnish lakes. Verhandlungen Internationale Vereinigung für theoretische und angewandte Limnologie 27: 194–199.

Herzig, A., J. Kubecka & G. Wolfram, 2002. Fish distribution in the open water and littoral of Neusiedler See (Austria): a matter of habitat structure and abiotic factors. In: International Conference on Limnology of Shallow Lakes, Balatonfüred, Hungary 25–30 May (2002), Abstracts, Veszprm University Press Vp, 309 pp.

Hofer, W. W. R. & W. Wieser, 1987. Diet spectra and resource partitioning in the larvae and juveniles of three species and six cohorts of cyprinids from a subalpine lake. Oecologia (Berlin) 71: 388–396.

Holmgren, K., 1999. Between-year variation in community structure and biomass-size distributions of benthic lake fish communities. Journal of Fish Biology 55: 535–552.

Holmgren, K. & A. Appelberg, 2000. Size structure of benthic freshwater fish communities in relation to environmental gradients. Journal of Fish Biology 57: 1312–1330.

Horppila, J., 1998. Effects of mass removal and variable recruitment on nutrient excretion by a planktivorous roach stock. Journal of Fish Biology 52: 951–961.

Horppila, J. & K. Nyberg, 1999. The validity of different methods in the backcalculation of the lengths of roach – a comparison between scales and cleithra. Journal of Fish Biology 54: 489–498.

Hölkner, F. & B. Breckling, 2002. Influence of activity in heterogeneous environment on dynamics of fish growth: an individual-based model of roach. Journal of Fish Biology 60: 1170–1189.

Imbrock, F., A. Appenzeller & R. Eckmann, 1996. Diel and seasonal distribution of perch in Lake Constance: a hydroacoustic study and *in situ* observations. Journal of Fish Biology 49: 1–13.

Jachner, A., 2001. Anti-predator behaviour of nave compared with experienced juvenile roach. Journal of Fish Biology 59: 1313–1322.

Jacobsen, L., S. Berg, M. Broberg, N. Jepsen & C. Skov, 2002a. Activity and food choice of piscivorous perch (*Perca fliviatilis*) in a eutrophic shallow lake, investigated by radio-telemetry. Freshwater Biology 47: 2370–2379.

Jacobsen, L., N. Jepsen & S. Berg, 2002b. Roach (Rutilus rutilus) behaviour by use of mini-radio transmitters: a comparative study of a clearwater and turbid lake. In: Shallow Lakes 2002, International Conference on Limnology of Shallow Lakes, Balatonfüred, Hungary 25–30 May 2002, Abstracts, Veszeprm University Press Vp, 309 pp.

Jeppesen, E., J. P. Jensen, M. Søndegaard, T. Lauridsen & F. Landkildehus, 2000. Trophic structure, species richness and biodiversity in Danish lakes: changes along a phosphorus gradient. Freshwater Biology 45: 201–218.

Jepsen, N., A. Koed & F. Økland, 1999. The movements of pikeperch in a shallow reservoir. Journal of Fish Biology 54: 1083–1093.

Jurvelius, J. & I. Sammalkorpi, 1995. Hyrdoacoustic monitoring of the distribution, density and the mass-removal of pelagic fish in a eutrophic lake. Hydrobiologia 316: 33–41.

Krause, J., G. Staaks & T. Mehner, 1998. Habitat choice in shoals of roach as a function of water temperature and feeding rate. Journal of Fish Biology 53: 377–386.

Krause, T., A. Palm & A. Järvalt, 2001. Eesti väikejärvede kalastik. Annual Report No 2-11-14/767, Min. Envir. Dep. Fish Res., Tartu, 41 pp. [Fish in small Estonian lakes. In Estonian].

Kübar, K., H. Agasild, T. Virro & I. Ott, 2005. Vertical distribution of zooplankton in a strongly stratified hypertrophic lake. Hydrobiologia 547: 151–162.

Lauridsen, T. L., E. Jeppesen, S. F. Mitchell, D. M. Lodge & R. L. Burks, 1999. Diel variation in horizontal distribution of *Daphnia* and *Ceriodaphnia* in oligotrophic and mesotrophic lakes with contrasting fish densities. Hydrobiologia 408/409: 241–250.

Michelsen, K., J. Pedersen, K. Christoffersen & F. Jensen, 1994. Ecological consequences of food portioning for the fish population structure in a eutrophic lake. Hydrobiologia 291: 35–45.

Moss, B., 1998. Ecology of fresh waters: man and medium, past to future. Blackwell Science, 531 pp.

Eesti järved. 1968. Tallinn, "Valgus". 532 pp. [Estonian lakes. In Estonian].

Mäemets, A., 1977. Eesti NSV järved ja nende kaitse. Valgus, Tallinn, 263 pp. (Lakes of the Estonian S.S.R. and their protection. In Estonian).

Olin, M., M. Rask, J. Ruuhijärvi, M. Kurkilahti, P. Ala-Opas & O. Ylönen, 2002. Fish community structure in mesotrophic and eutrophic lakes of southern Finland: the relative abundances of percids and cyprinids along a trophic gradient. Journal of Fish Biology 60: 593–612.

Ott, I., T. Kõiv, P. Nõges, A. Kisand, A. Järvalt & E. Kirt, 2005. General description of partly meromictic hypertrophic Lake Verevi, its ecological status, changes during the past eight decades and restoration problems. Hydrobiologia 547: 1–20.

Perrow, M. R., A. J. D. Jowitt & S. R. Johnson, 1996. Factors affecting the habitat selection of tench in a shallow eutrophic lake. Journal of Fish Biology 48/5: 859–870.

Persson, L., 1987. Effects of habitat and season on competitive interactions between roach (*Rutilus rutilus*) and perch (*Perca fluviatilis*). Oecologia (Berlin) 73: 170–177.

Persson, L. & P. Ekloev, 1995. Prey refuges affecting interactions between piscivorous perch and juvenile perch and roach. Ecology 76: 70–81.

Persson, L., J. Andersson, E. Wahlstroem & P. Ekloev, 1996. Size-specific interaction in lake systems: predator gape limitation and prey growth and mortality. Ecology 77: 900–911.

Pihu, E., 1993. Distribution of fish species in Estonian lakes. Proceedings of the Estonian Academy of Sciences Biology Ecology 3: 181–186.

Randall, R. G., C. K. Minns, V. W. Cairns & J. E. Moore, 1996. The relationship between an index of fish production and submerged macrophytes and other habitat features at three littoral areas in the Great Lakes. Canadian Journal of Fisheries and Aquatic Sciences 53/1: 35–44.

Romare, P., E. Bergman & L.-A. Hansson, 1999. The impact of larval and juvenile fish on zooplankton and algal dynamics. Limnology and Oceanography 44(7): 1655–1666.

Rowe, K., 1994. Vertical segregation and seasonal changes in fish depth distributions between lakes of contrasting trophic status. Journal of Fish Biology 45: 787–800.

Ryder, R. A., S. R. Kerr, K. H. Loftus & H. A. Regier, 1974. The morphoedaphic index, a fish yield estimator – review and evaluation. Journal of the Fisheries Research Board of Canada 31: 663–688.

MIX
Papier aus verantwortungsvollen Quellen
Paper from responsible sources
FSC® C105338

If you have any concerns about our products,
you can contact us on
ProductSafety@springernature.com

In case Publisher is established outside the EU,
the EU authorized representative is:
Springer Nature Customer Service Center GmbH
Europaplatz 3, 69115 Heidelberg, Germany

Printed by Libri Plureos GmbH
in Hamburg, Germany